Climate Change Management

For further volumes:
http://www.springer.com/series/8740

Walter Leal Filho
Editor

Experiences of Climate Change Adaptation in Africa

Springer

Prof. Walter Leal Filho
Hochschule für Angewandte Wissenschaften Hamburg
Forschungs- und Transferzentrum
Applications of Life Sciences
Lohbruegger Kirchstr. 65
21033 Hamburg
Germany
e-mail: walter.leal@haw-hamburg.de

ISSN 1610-2010
ISBN 978-3-642-22314-3
DOI 10.1007/978-3-642-22315-0
Springer Heidelberg Dordrecht London New York

e-ISSN 1610-2002
e-ISBN 978-3-642-22315-0

Cover design: eStudio Calamar, Berlin/Figueres

Printed on acid-free paper

Springer is part of Springer Science+Business Media (www.springer.com)

Foreword

Africa is a continent blessed with a substantial amount of environmental resources, and many African countries are quite rich with respect to biodiversity, minerals and favourable environmental conditions, which may allow them to harness solar energy, for example. Yet, African countries are also among the most vulnerable ones in respect to climate change. This is the case for three main reasons:

1. Climate change and the various phenomena associated with it may lead to disruptions in water cycle and subsequent increase in droughts, floods, storms and other extreme events, which may rather damage the fragile infrastructure;
2. Climate change may affect food production and pose an additional burden to societies already experiencing serious economic and social problems;
3. The impact of climate change-related events may especially affect the poor and most socially vulnerable people, thus negatively influencing their lives.

It is widely acknowledged that, in addition to global and regional efforts to cope with climate change by means of mitigation measures, adaptation initiatives can and perhaps should play a key role in enabling communities from across Africa to better handle the problems related to it. Due to the fact that experiences in climate change adaptation in Africa are poorly documented, this book provides an attempt to address the perceived need for better documentation and dissemination of African experiences on climate change adaptation. This is not to say that mitigation is not needed: it is. But bearing in mind that many measures are needed here and now, instead of in 15 or 20 years time, this book will prove a useful tool in showing what can be done, how, and, by means of concrete case studies, it makes a convincing case for how useful and effective climate change adaptation may be.

This book contains some of the papers presented at the conference on "Climate Change and Natural Resource Use in Eastern Africa: Impact, Adaptation and Mitigation", organised by the Ecological Society for Eastern Africa (ESEA). The event was held at the Multimedia University College of Kenya in Nairobi on 19–21 May 2010 and drew nearly 200 delegates from across eastern Africa. It also incorporated a great many other papers prepared by a number of African-based organisations, reflecting the diversity of activities taking place in Africa today.

The book is structured in such a way that government agencies and aid donors on the one hand, and practitioners and members of NGOs and associations on the other, are able to better understand the scope of climate change adaptation in Africa, with a combination of background research and examples of practical projects.

I want to thank all the authors for sharing their knowledge and their experiences, as well as Dr. Mihaela Sima for the editorial support provided. Thanks are also due to the ESEA, especially Professor Otienoh Oguge, for organising the conference that marked the start of this book project. The kind support provided by Caroline Lumosi from ESEA is also gratefully acknowledged.

It is hoped that this book, which is volume 4 in the "Climate Change Management Series" initiated as part of the "International Climate Change Information Programme" (ICCIP), will be useful and that it will inspire and motivate more projects and initiatives which may help African countries to address the many challenges that climate change poses to them.

Summer 2011 Prof. Walter Leal Filho

Contents

Chapter 1
Links Between Capacity and Action in Response to Global Climate Change: A Climate Response Shift at the Local Level

O. W. Olowa, O. A. Olowa and Walter Leal Filho

Abstract Although the development and implementation of a global greenhouse gas reduction regime has dominated policy debates even before the advent of the Kyoto Protocol (and remains a critical element of effective mitigation), communities in developing countries do not have direct control of critical sources of emissions or are not fully aware of the scale at which the potentially catastrophic impacts of climate change will play out. This paper has undertaken a revision of scientific literature on climate change issues and has introduced the concept of response capacity, its links to adaptive and mitigative capacity, as well as action or behaviour change in response to climate change. As this paper shows, communities face a unique set of challenges as they navigate through the uncertain future presented by climate change. Even so, communities bring to the task of climate change adaptation and mitigation a unique set of tools and proficiencies that are often absent at the national and international levels. The paper investigates the complex relationship between capacity and action in response to global climate change, which represents a significant gap in the climate change literature and influences the ability of climate policies to build effective mitigation and adaptation strategies. It is the one aim of this paper to present some of the means via which communities might more effectively employ the various forms of capacity they possess to rise to the challenge presented by climate change.

O. W. Olowa · O. A. Olowa
Department of Agricultural Education, Federal College of Education (Tech.), Akoka, Nigeria
e-mail: Olowa1@yahoo.com

O. W. Olowa · W. Leal Filho (✉)
Faculty of Life Sciences, Hamburg University of Applied Sciences, Lohbruegger Kirchstraße 65, 21033 Hamburg, Germany
e-mail: walter.leal@haw-hamburg.de

W. Leal Filho (ed.), *Experiences of Climate Change Adaptation in Africa*,
Climate Change Management, DOI: 10.1007/978-3-642-22315-0_1,

Keywords Capacity · Climate response · Shifts · Global level

Introduction

The implications of unsustainable patterns of development are nowhere more evident than in the challenges presented by global climate change. The potential impacts cross generational and geographical divides to permeate ecological and human systems alike, while the causes cut to the core of our economies and shake the foundations of politics at all levels. The scale and nature of the transformation that may be required to adequately respond to this challenge appear to have defied comprehension, but climate change has risen to dominate policy agendas and drive innovation at unprecedented rates (Adger et al. 2004).

Novel efforts are being made to link previously disparate categories of climate change responses, such as adaptation and mitigation, within a framework of sustainable development (Beg et al. 2002; Berkes et al. 2000a, b). This new approach will assist in the development of local-level climate response strategies that consider inevitable trade-offs and potentially attractive synergies between responses that may influence the ability of a group to follow a sustainable development path. Finally, increasing attention is being given to the underlying characteristics of a community or society that either help or hinder responses to climate change. Among social scientists, the concept of capacity (Betsill 2001; Brooks et al. 2005) and its influence on action in response to climate change is important to these discussions.

Although the adaptive and mitigative capacity literature does not claim that building capacity will necessarily lead to improved responses to the climate change risk, little work has been done to explicate the wide variation in responses to climate change among nations and groups with similar levels of capacity. Analysis of the complex relationship between capacity and action in response to global climate change represents a significant gap in the climate change literature, and influences the ability of climate policies to build effective mitigation and adaptation strategies.

This paper builds on a rich tradition of integrated approaches to planning and decision-making as they relate to climate change. Since the mid 1990s, the enhanced greenhouse effect and global climate change have become issues around which much public debate has centred. Initially, this debate was characterised by heated discussions regarding the precision and relevance of temperature and impact models, even periodically calling into question the most fundamental assumptions of climate scientists. More recently, dramatic shifts have occurred within climate change discourse itself, quite separately from the increased prominence of climate change as a political issue in jurisdictions around the world. First, perhaps partly because of the difficulty of regulating climate change at the global level, many scholars are beginning to consider the implications of climate change on ever smaller scales. Local and regional responses to climate change, such as the US Mayors' Climate Protection Agreement and the Lagos State Climate Change Initiative in Nigeria, have become more common in the time since the negotiation of the Kyoto Protocol to the United Nations Framework Convention on Climate Change (Adger et al. 2004; Bullard 1994; CIA 2004). It is this issue that this paper will address in detail, with the goal of further elaborating the nature of the complex relationship between capacity and action.

Capacity and the Climate Change Problem

A central component of recent discussions regarding the human dimensions of climate change is the concept of capacity. This section will introduce the ideas of adaptive and mitigative capacity. The evolution of these concepts within the climate community will be considered, and a method of defining adaptive and mitigative capacity more carefully will be proposed in order to assist in the explanation of the links between capacity and action in response to climate change. Finally, this section will introduce response capacity as a useful idea through which adaptation and mitigation can be integrated in the context of sustainable development.

Adaptive and Mitigative Capacity

As the science of climate change risk becomes more established, and greater consensus is reached among climate system experts around the world, attention has shifted to the issue of responding to climate change. Responses are typically grouped into two categories: climate change mitigation, which reduces emissions or increases capture of greenhouse gases in order to reduce the magnitude of the future risk, and adaptation, which consists of adjustments in structures, practices, or processes, to respond to changing climate conditions (Cohen et al. 1998). Recent literature in the field of climate change response argues that adaptation to, and mitigation of, climate change take place within the context of adaptive and mitigative capacity, respectively (Brooks et al. 2005; Dang et al. 2003; Dietz et al. 1998). This paper takes as its starting point the definitions proposed by the IPCC third assessment report, which defines adaptive capacity as the ability to adapt to the impacts of climate change. We adopt a more recent definition of mitigative capacity, which describes it as the ability to reduce greenhouse gas emissions that lead to global climate change (Folke et al. 2002).

The concept of capacity only emerged with the IPCC's Third Assessment Report, and was a significant development on the path towards a more nuanced explanation of responses to climate change. Prior to the third assessment report, most analyses of human responses to climate change were limited to estimations of specific climate change impacts and proposals for mitigation and adaptation responses, rather than investigations into the socio-political and institutional precursors to these responses. This focus was established in the first and second assessment reports of the intergovernmental panel on climate change (IPCC), a collaborative effort between the United Nations environment programme and the World Meteorological Organization. Published in 1991 and 1995, these reports captured the focus of the research on climate change that had taken place during the preceding decade, and dealt heavily with issues of modelling the potential impacts of anthropogenic climate change, greenhouse gas reduction through mitigation, and issues of the cost-effectiveness or efficiency of mitigation policies (Grothmann and Patt 2005). This reflected the natural science-driven and somewhat technocratic views of the climate change community at the time (Haddad 2005). The assumption was made that science could be used to fill the knowledge gaps that plagued studies on climate change, and that policymakers could simply rationally apply this knowledge to the development and implementation of effective response strategies (Hawley 1986; Holling 1978; Holling 1986). A recognition of the limits of this approach led to an attempt by the IPCC, in the third and fourth assessment reports, to pay significantly more attention to questions concerning the need for consideration of more human issues, such as the social, cultural, political or institutional constraints on responses to climate change (Grothmann and Patt 2005). An example of this new focus was the introduction of the concept of adaptive and mitigative capacity in the third assessment report (Berkes et al. 2000b; Grothmann and Patt 2005). Even after this introduction occurred, however, the research response within the climate change community remained limited. As such, the concept of capacity, and its implications for climate change mitigation and adaptation, is only now gaining prominence.

In arenas other than that of traditionally natural science-dominated climate change research, however, the question of capacity for behaviour change has been investigated extensively. Early framing of the climate change problem, embodied in the first two assessment reports of the IPCC, underestimated the contributions of social sciences such as cultural anthropology, sociology, and social psychology to ‘understanding the processes by which societies recognise new threats to their security or well-being, formulate responses, and act collectively upon them’ (Hawley 1986).

The traditional, natural science-based climate change discourse, somewhat reductionist in nature, attempted to mould the climate change problem to suit the requirements of scientific analysis, and mostly ignored the political, social and cultural dimensions of the problem (Haddad 2005). The human-centred social science discourse, on the other hand, sought to explain the drivers of climate change in a more politically sensitive and geographically appropriate manner, but was often considered analytically vague by more quantitatively oriented scholars (Haddad 2005). It is out of this more human-centred approach that the concepts of adaptive and mitigative capacity have arisen. The concept of adaptive capacity was first brought to the climate change community because of its use in the field of ecology (IPCC Climate Change 2001). The ability of a system to withstand and adapt to external stresses had long been a subject of study in the scientific community, but the resilience of human communities and economic systems was not initially part of this analysis. Over the last several decades, however, a large literature has developed that investigates adaptive human responses in the realm of ecosystem management (Brooks et al. 2005; Irwin and Wynne 1996, Irwin 1995, Jasanoff and Wynne 1998, Jones et al. 2007, Kahneman and Frederick 2002) These elements were eventually added to the consideration of adaptive capacity in the climate change community, and articulated by Yohe (Brooks et al. 2005). Yohe (Brooks et al. 2005) suggested that adaptive capacity was determined by the following group-level characteristics:

- The range of available technological options for adaptation
- The availability of resources and their distribution across the population
- The structure of critical institutions and the derivative allocation of decision-making authority
- The stock of human capital, including education and personal security
- The stock of social capital including the definition of property rights
- The system's access to risk-spreading processes
- The ability of decision-makers to manage information, the processes by which these decision-makers determine which information is credible, and the credibility of decision-makers themselves
- Public perception of attribution.

The work of Gary Yohe (Brooks et al. 2005) again assisted in the initial elucidation of the determinants of capacity and posited that essentially the same set of characteristics help to determine the mitigative response to climate change (Brooks et al. 2005). The wide acceptability of the hypotheses put forward by Yohe is demonstrated by their integration into the third assessment report of the intergovernmental panel on climate change, and by their reiteration and further development by numerous climate change experts (Betsill 2001; Kempton 1997). A more integrated approach to the analysis of adaptation and mitigation is currently gaining momentum, leading to preliminary policy recommendations (Bullard 1994; Folke et al. 2002).

Response Capacity

As outlined above, recent research supports Yohe's view that adaptive capacity and mitigative capacity are essentially driven by the same factors (Dietz et al. 1998), which operate at a very high level of abstraction and seem to apply only to very large groups. Furthermore, these determinants yield little insight into crucial aspects of climate responses; namely, which party is initiating the response action, and how is that action carried out? In other words, more information is required about the institution or agency and the resultant policies and programmes that are geared towards adaptive and mitigative responses to climate change if the factors that engender effective and ineffective climate change responses are to be articulated. For these reasons, it may be fruitful to consider the broad determinants of capacity, as outlined by Yohe and others, to be part of a more general, development-related pool of resources called response capacity. A slightly different framing of the term 'response capacity' has been proposed by Tompkins and Adger (2005), Dietz et al. 1998 who suggest that response capacity can be thought of as the human ability to manage both the generation of greenhouse gases and the consequences of their production (Dietz et al. 1998).

Put differently, response capacity, in this context, represents simply the confluence of adaptive and mitigative capacity. This concept may be broadened, however, in order to represent the broad pool of resources that can be utilised to address any risk or challenge faced by a human society. The value of this broadening is that it allows response capacity to be connected to the underlying socio-economic and technological development path of a given society or community and thus provide a new focus for attempts to understand how capacity can effectively be translated into action. Response capacity, according to this view, is time and context-specific, and culture and region-specific.

The concept of response capacity simultaneously allows for the greater specification of mitigative and adaptive capacity, and reveals the importance of deeper socio-cultural trajectories that form the context within which action may occur. The examples of the interrelationships between response capacity, adaptive capacity and mitigative capacity that are discussed in this section serve to illustrate this claim. The generalised pools of resources that constitute response capacity might be utilised to produce an institution or policy that is geared towards the mitigation of, or adaptation to, climate change, which represents the formation of adaptive and/or mitigative capacity out of the pool of response capacity resources. For instance, generalised institutional capacity, in the form of government budgetary capacity and jurisdiction, might be activated in the creation of an agency or institution that is geared towards carrying out emergency measures in response to severe climate events.

Similarly, a corporate social responsibility policy or an environmental division of a large corporation might be formulated out of pre-existing institutional capacity and human capital, which then goes on to design effective policies geared towards the reduction of greenhouse gas emissions. In this case, it is unlikely that the embodiment of response capacity in the form of adaptive capacity would do much to build or enhance mitigative capacity. Similarly, the presence of technological innovation that has grown out of a socio-technical system might result in technologies that are applicable only to the reduction of greenhouse gas emissions, but might not contribute at all to the adaptation to climate change impacts. This clearly implies that trade-offs may exist in the way that general response capacities are transformed into more specific mitigative or adaptive capacities. For example, to the extent that climate change policy responses in general have only a finite amount of resources available to them within a given governmental policy framework, adaptive and mitigative measures may compete for these resources even if they are not substitutes in terms of their effects.

Trade-offs may also exist in the translation of response capacity into adaptive and mitigative capacity. For instance, response capacity may be utilised to form a flood protection and dyke management agency, geared explicitly towards implementing adaptation measures. This is likely in an area that has accumulated technical expertise to deal with such issues, constructed an institutional framework to support these actions, and entrenched flood protection responses in a system of policies and mandates. Not only does this necessarily take resources away from mitigation-oriented measures, but the construction and management of dykes will likely consume fossil fuels that actually contribute to the climate change problem (and thus the need for mitigation). Again, it is clear that the underlying technological, institutional and socio-cultural trajectories fundamentally shape the way that response capacity is transformed into mitigative and adaptive capacities, as well as into action. Recent work on the part of climate change scholars has begun to explore a more comprehensive framework for the consideration of interactions between adaptation and mitigation, as well as their respective capacities (Beg et al. 2002; Kollmuss and Agyeman 2002; Lowe et al. 2006; Moss et al. 2001; Olowa 2003). It is clear that the ways in which response capacity is transformed into adaptive and mitigative capacity are manifold, creating layers of interaction leading to very different climate response outcomes.

The relationship between response capacity, mitigative and adaptive capacities and actual mitigation and adaptation is a complex one. First, the schematic shows that all factors that contribute to human responses to climate change are embedded in the underlying development path. In other words, socio-cultural, technological, and institutional trajectories fundamentally shape the quality and quantity of response capacity resources, which are then available for mitigative and adaptive activities. Next, adaptive and mitigative capacities are shown to arise from these response capacity resources in the form of institutions and policies that are geared towards one or both of these responses. Finally, adaptive and mitigative capacities are utilised to produce adaptation or mitigation in response to the climate change risk.

Translating Capacity into Action

The discussion above, which provides a proposed clarification of the concepts of response capacity and adaptive and mitigative capacity, lays the groundwork for addressing the central question, often left unasked, with regard to human responses to climate change: Does a group with larger stocks of capacity necessarily respond more effectively than a group with lower stocks of capacity? More generally, one must ask: What influences the relationship between capacity and action? This section will first consider criticisms of, and additions to, the traditional formulation of the determinants of adaptive and mitigative capacity, which provide the roots of the capacity/action question. Next, the translation of capacity into action will be considered, and one important factor will be introduced that may shape this relationship.

Although the determinants of capacity, as laid out by Gary Yohe, have come to be widely accepted in the IPCC and in global climate research circles, recent work has suggested that Yohe's list of determinants is incomplete. Haddad (Robinson et al. 2006), for instance, suggests that the traditional measures of adaptive capacity do not consider the normative or motivational context of adaptation. Specifically, Haddad examines the effect of national goals and aspirations on adaptation choices. Teleological legitimacy, procedural legitimacy and norm-based decision rules are three broad categories of goals that Haddad argues might lead nations to make different decisions in response to the climate change risk (Robinson et al. 2006). Although he deals specifically with adaptation and makes no claims about the effects of national aspirations on mitigative responses to climate change, one might argue that the motivational context behind mitigation is equally important. But from the point of view of the approach taken in this paper, these factors have more to do with the process of turning capacity into action than with expanding the list of determinants of capacity.

Grothmann and Patt (Satterfield et al. 2004) stress the need for an examination of responses to climate change, which, instead of considering resource constraints as the most significant determinant of adaptation, separates out the psychological steps that precede action in response to perception of the climate change risk. A large literature pertaining to human decision-making and action, traditionally outside of the climate change realm, suggests that both motivation and perceived abilities are important determinants of action (Slovic et al. 1992; Slovic 2002; Smit et al. 2001) Thus, both risk perception and perceived adaptive capacity, for instance, may enhance or inhibit adaptive responses to climate change (Satterfield et al. 2004).

These criticisms of, and additions to, Yohe's determinants of adaptive and mitigative capacity point to the need to revise the previously deterministic view of capacity and action, and to consider more carefully the intricacies of human behaviour. They suggest the need to relate adaptive and mitigative capacity to the concrete institutional and socio-technological contexts in which these capacities are embedded. To the extent that these capacities are part of a trajectory of decisions and behaviours that prioritise or even make conceivable only certain forms of action, proposals for policy responses that are incompatible with such a trajectory are much less likely to succeed or even be seriously considered. An example of such path dependence might be policy regimes that subscribe to a strongly market-oriented approach to policy formation. In such regimes, response capacity is much more easily mobilised for forms of adaptive or mitigative responses that reflect such priorities (e.g. market-based instruments) than for forms of response characterised by more traditional command and control policies. In this way, the development pathway may strongly condition the types of responses considered legitimate.

Although the adaptive and mitigative capacity literature does not claim that building capacity will necessarily lead to improved responses to the climate change risk, little work has been done to explain the widely noted variation in response to climate change among communities and nations with similar capacities. For instance, Canada and Sweden are remarkably similar with regard to a variety of economic, demographic and geographic indicators. Canada's GDP per capita is 29,000 USD, while Sweden's is 26,000 USD per capita; Canada and Sweden have similar northern hemisphere climates; and are currently passing through similar stages of their demographic transitions, marked by ageing populations and slow growth (Stern 2000). Canada and Sweden also possess similarly high literacy rates, similar distributions of GDP by sector, and so on. These two nations are exposed to the same internationally endorsed climate change science through the IPCC and have access to essentially the same mitigative and adaptive technologies via open markets and international trade. Such similarities indicate that Canada and Sweden possess very similar levels of response capacity. These two countries, however, have very different levels and types of climate change-related institutions and policies (adaptive and mitigative capacity), and very different success rates in reducing greenhouse gas emissions. Canada, despite having ratified the Kyoto Protocol, has experienced a 24.2% increase in emissions since 1990, while Sweden has managed to reduce its emissions by 2.3% (Swart et al. 2003). This variation in response and the potential influence of varying perceptions of risk reveal that capacity is a necessary, but not sufficient, condition for mitigative action (Folke et al. 2002). Clearly additional factors are influencing the complex and non-linear relationship between response capacity and behaviour change.

In order to tease out some characteristics of the relationship between capacity and action in climate change, one factor—risk perception—will be provided as an example of a context-specific, culturally variable factor that may play some role in variance in climate change responses among countries.

Risk Perception and Varying Responses to Climate Change

Studies of the perception of risk offer considerable insight into common patterns of behaviour that individuals and groups might follow in response to a risk such as climate change. Although most of the work in this area has been carried out to explain technological disasters (such as chemical spills or nuclear disasters) or, more strictly, the conditions for mitigative action (Folke et al. 2002), many other factors influence the complex and non-linear relationship between response capacity and behaviour change.

Natural disasters (such as an earthquake or a tsunami) provide good examples of policy responses to climate change. Research dealing with the perception and characterisation of risk can be grouped into two approaches: psychological or psychometric, and socio-cultural. Each of these will be addressed below. Although scientific experts have often considered the responses of policymakers and the lay public to risk to be irrational, key scholars in the sub-field of psychological or psychometric studies of risk perception argue otherwise (Tompkins and Adger 2005; UNFCCC 2005; Wilbanks and Kates 1999) Rather than responding to some 'true' level of risk that is inherent in changing climate, this literature posits that the lay public creates perceptions of risk that are based on different criteria and thus may differ from those of experts. These perceptions are still rational, however, and are based on two factors, each of which is made up of a combination of characteristics (Wilbanks et al. 2007).

In particular, these scholars posit that perceptions of risk are derived essentially from feelings of dread (resulting from a risk that is perceived to be severe, catastrophic or uncontrollable), and the unknown (often resulting from risks that are perceived to be unfamiliar, unobservable or new to science) (Wilbanks et al. 2007). These perceptions are individual, however, and are therefore strongly affected by the socio-economic standing of the individual perceiving the risk. It has been found, for instance, that economically and socially disadvantaged populations, such as visible minorities and women, are likely to perceive risks to be greater than their more empowered counterparts (Wilbanks and Sathaye 2007).

These disadvantaged groups possess much less power in their socio-political surroundings, and thus have less reason to believe that they can control or recover from a risk. It has often been noted that less-wealthy and minority communities are less likely to receive protection from harm (Winkler et al. 2007), and are more likely to be located in environmentally unstable or unsafe locations, such as cliffs or low-lying areas prone to flooding and inundation. Climate change may prove to be a very real risk to these groups, who, as the victims of environmental injustice, are least likely to be well-served by the political and economic services that are at the disposal of others. Thus, the groups that have the highest concerns about risks related to climate change may be least empowered to translate those concerns into policy, leading to a systematic under-representation of risk concerns. Risks related to climate change are especially susceptible to variation in perception, in part because of the scientific controversy discussed previously and in part because of the geographical variation in vulnerability, provision of scientific information, and economic stability. As such, a community (Group A) that is economically resilient, with few marginalised groups, high education and literacy levels, and minimal risk of extreme exposure to the impacts of climate change, might associate low levels of dread and 'unknowability' with climate change. This group might perceive the risk of climate change to be relatively minor, thereby carrying out few adaptation or mitigation options. Group B, on the other hand, might be characterised by groups that are economically or socially disadvantaged, privy to conflicting information about the potential risks of climate change, and physically vulnerable. This group, according to Slovic's psychometric factor space, would perceive the risks from climate change to be much greater. If this group possesses considerable response capacity, then it may be more likely to carry out adaptive or mitigative actions. One could easily imagine communities with a mixture of these characteristics as well. For instance, Group C might be highly vulnerable (for instance a low-lying small island state), leading to high estimations of the severity and fatality of climate change, but might also consider climate change to be a familiar and observable risk. Group D, alternatively, experiences low levels of dread in relation to the climate change risk, but regards climate change as a mysterious and unobservable force that is unknown and new to science. The levels of mitigation in these latter, more ambiguous, groups might depend more strongly on other factors such as political structures, political will and capacity.

The groups described above, however, clearly represent idealised versions of reality. Since most jurisdictions consist of mixtures of these groups, power and representation become the resources that most directly influence the responses of a group to climate change. For instance, the risk perception of the less-empowered majority may not be the risk perception that characterises the more affluent or empowered minority. As a result, the response of the group as a whole may be far from representative of the majority.

Social theories of risk also lend considerable insight into variation in perceptions of the climate change risk that might lead to different mitigative behaviours. Wynne (Wilbanks and Kates 1999), for instance, demonstrates that conflict between experts and the lay public may result from competition or clashes between their respective cultures. Wynne shows that because scientists have been socialised to evaluate phenomena empirically and claim objectivity, they are not receptive to the contributions of local 'experts', which may lack traditional academic credentials. Local lay people, in turn, view scientists as agents of those in power, and do not trust their methods. As a result, conflicts arise that appear to be about knowledge, when the catalyst for the conflict is actually threatened identities (Wilbanks and Kates 1999). As climate change researchers begin to learn the value of local knowledge, similar conflicts may arise between scientists who may be viewed as capitalising on the recent explosion of interest in the climate change issue, and local individuals who want to preserve (for instance) agriculture in vulnerable areas.

Although, as mentioned, weather patterns may not be indicative of broader climate patterns, local and traditional knowledge may provide valuable insights into the nature of varying perceptions of the climate change risk, and varying responses. This knowledge, however, and the concerns of those who possess it, is often devalued by expert-driven cultures and is thus under-represented in policy. Added to this rational or analytical way of interpreting risk is an intuitive or affective layer of response, which has been termed the 'experiential system' (Wynne 1992). This more rapid, mostly subconscious evaluation of risk is thought to operate in parallel with our more logical assessments of risk, and the two systems fundamentally guide and inform one another. This role of affect in thinking and information processing has gained visibility in the risk community(Wynne 1992; Yohe) and draws our attention to the ways in which rationality and emotion are inextricably linked. This complicated relationship may shed some light on varying responses to the risk of climate change (which few of us have experienced) and may help explain why, in cases such as this where consequences are new or unexpected, the 'affect heuristic' fails to enable us to be rational actors.

Risk perception, whether defined through psychological or socio-cultural models, is not incorporated into the current definition of response capacity, but can clearly influence behaviour. A high perception of the risks associated with global climate change, for instance, might provide the foundation of interest in climate change adaptation or mitigation and knowledge of the benefits of adaptation or mitigation that is needed to effect behaviour change. Similarly, a community that perceives a high level of risk might also utilise the social forces that encourage and reinforce adaptive or mitigative behaviour.

Conclusions and Future Directions

This paper has introduced the concept of response capacity and traced its links to adaptive and mitigative capacity and ultimately to action or behaviour change in response to climate change. This complex and dynamic set of relationships is deeply embedded within the underlying development path, pointing to the need to consider integrated adaptive and mitigative responses. Risk perception has been presented, for illustrative purposes, as only one of many socio-cultural characteristics that may influence the relationship between capacity and action, and thus shape responses to climate change. It is necessary to further elaborate upon other factors that may also fundamentally alter human responses to this risk, in order to stimulate sufficient greenhouse gas reduction and adaptation strategies. For this, one must look to the literature of institutional genesis and change, socio-technical systems, social movements and collective behaviour change theory (to name but a few) to shed light on the underlying development paths that influence both capacity and action.

Similarly, from the point of view of policy, and moving more readily from capacity to action, the concept of response capacity and the socio-economic and technological development pathways such capacity is embedded within, suggest the need to consider carefully the socio-technical context within which climate policy responses must be undertaken. Actions inconsistent with development path trajectories are likely to face greater hurdles in implementation and may indeed not be given serious consideration. Whether adaptive and mitigative measures and actions are likely to compete for resources or else reinforce each other will likewise depend in part on the nature of the development path within which they are expressed.

To the extent that climate change policies are increasingly framed in terms of sustainability goals, the arguments presented in this paper suggest that a crucial consideration must be the question of how it may be possible to make a transition from the currently dominant development paths to sustainable ones. This in turn not only shifts the focus of concern somewhat from climate policy to sustainability policy (Berkes et al. 2000a; Yohe and Strzepek 2007) but also suggests the importance of investigating and developing effective means by which perceived barriers can be overcome at the municipal, regional, national and international levels.

References

Adger NW, Brooks N, Kelly M, Bentham S, Eriksen S (2004) New indicators of vulnerability and adaptive capacity. Technical Report 7. Tyndall centre for climate change resource

Beg N, Morlot JC, Davidson O et al (2002) Linkages between climate change and sustainable development. Clim Policy 2:129

Berkes F, Colding J, Folke C (2000a) Rediscovery of traditional ecological knowledge as adaptive management. Ecol Appl 10:1251–1262

Berkes F, Folke C, Colding J (2000b) Linking social and ecological systems: management practices and social mechanisms for building resilience. Cambridge University Press, Cambridge

Betsill M (2001) Mitigating climate change in US cities: opportunities and obstacles. Local Environ 6:393–406

Brooks N, Adger WN, Kelly PM (2005) The determinants of vulnerability and adaptive capacity at the national level and the implications for adaptation. Global Environ Change 15:151–163

Bullard R (1994) Decision making. Environ 36:11–20, 39–44

CIA [homepage on the Internet]. CIA World Factbook; c (2004). Available from http://www.cia.gov/cia/publications/factbook/geos/

Cohen S, Demeritt D, Robinson J, Rothman D (1998) Climate change and sustainable development: towards dialogue. Global Environ Change 8:341–371

Dang HH, Michaelowa A, Tuan DD (2003) Synergy of adaptation and mitigation strategies in the context of sustainable development: the case of Vietnam. Climate Policy 3(1):81–96

Dietz T, Stern PC, Guagnano G (1998) Social structure and social psychological bases of environmental concern. Environ Behav 30:450–471

Folke C, Carpenter S, Elmqvist T, Gunderson L, Holling C, Walker B (2002) Resilience and sustainable development: building adaptive capacity in a world of transformations. Ambio 31:437–440

Grothmann T, Patt A (2005) Adaptive capacity and human cognition: the process of individual adaptation to climate change. Global Environ Change 15:199–213

Haddad B (2005) Ranking the adaptive capacity of nations to climate change when socio-political goals are explicit. Global Environ Change 15:165–176

Hawley AH (1986) Human ecology: a theoretical essay. University of Chicago Press, Chicago

Holling CS (1978) Adaptive environmental assessment and management. Wiley, New York

Holling CS (1986) The resilience of terrestrial ecosystems: local surprise and global change. In: Clark W, Munn R (eds) Sustainable development of the biosphere. Cambridge University Press, Cambridge, pp 292–317

IPCC. Climate Change (2001) Impacts, adaptation and vulnerability. Contribution of working group II to the third assessment report of the intergovernmental panel on climate change. Cambridge, United Kingdom and New York

Irwin A (1995) Citizen science. Routledge, New York

Irwin A, Wynne B (1996) Misunderstanding science? The public reconstruction of science. Cambridge University Press, Cambridge

Jasanoff S, Wynne B (1998). Science and decisionmaking. In: Rayner S, Malone E (eds) Human choice and climate change. 1. The societal framework. Batelle Press Ltd., Columbus, Ohio, pp 1–87

Jones RN, Dettmann P, Park G, Rogers M, White T (2007) The relationship between adaptation and mitigation in managing climate risks: a regional response from North Central Victoria, Australia. Mitig Adapt Strat Glob Change 12:685–712

Kahneman D, Frederick S (2002) Representativeness revisited: attribute substitution in intuitive judgement. In: Gilovich T, Griffin D, Kahneman D (eds) Heuristics and biases: the psychology of intuitive judgment. Cambridge University Press, New York, pp 49–81

Kempton W (1997) How the public views climate change. Environment 39:13–21

Kollmuss A, Agyeman M (2002) Mind the gap: Why do people act environmentally, and what are the barriers to pro-environmental behaviour? Environ Educ Res 3:239–260

Lowe T, Brown K, Dessai S, de França Doria M, Haynes K, Vincent K (2006) Does tomorrow ever come? Disaster narrative and public perceptions of climate change. Public Underst Sci 15:435–457

Moss R, Brenkert AL, Malone EL (2001) Vulnerability to climate change: a quantitative approach. Pacific Northwest National Laboratory

Olowa OW (2003) Global warming: implications for freshwater and marine fish supply in Nigeria. Afr J Vocat Educ 4(1):5–6

Robinson J, Bradley M, Busby P et al (2006) Climate change and sustainable development: realizing the opportunity. Ambio 35:2–8

Satterfield T, Mertz C, Slovic P (2004) Discrimination, vulnerability, and justice in the face of risk. Risk Anal 24:115–129

Slovic P (1992) Perceptions of risk: reflections on the psychometric paradigm. In: Krimsky S, Golding D (eds) Social theories of risk. Praeger, New York, pp 117–152

Slovic P, Finucane ML, Peters E, MacGregor DG (2002) The affect heuristic. In: Gilovich T, Griffin D, Kahneman D (eds) Heuristics and biases: the psychology of intuitive judgement. Cambridge University Press, New York, pp 397–420

Smit B, Pilifosova O, Burton I et al (2001) Adaptation to climate change in the context of sustainable development and equity. In: Mccarthy J, Canziani O, Leary N et al (eds) Climate change 2001: Impacts, adaptation and vulnerability. Cambridge University Press, Cambridge

Stern PC (2000) Towards a coherent theory of environmentally significant behaviour. J Soc Issues 56:407–424

Swart R, Robinson J, Cohen S (2003) Climate change and sustainable development: expanding the options. Climate policy. Special issue on climate change and sustainable development 3:19–40

Tompkins E, Adger WN (2005) Defining response capacity to enhance climate change policy. Environ Sci Pol 8:562–571

UNFCCC [homepage on the Internet]. Key greenhouse gas emissions data; c2005. Available from http://unfccc.int/ghg_data/items/3800.php

Wilbanks T, Kates RW (1999) Global change in local places: how scale matters. Clim Change 43:601–628

Wilbanks T, Sathaye J (2007) Integrating mitigation and adaptation as responses to climate change. Mitig Adapt Strat Glob Change 12:957–962

Wilbanks T, Leiby P, Perlack RD, Ensminger JT, Wright SB (2007) Toward an integrated analysis of mitigation and adaptation: some preliminary findings. Mitig Adapt Strat Glob Change 12:713–725

Winkler H, Baumert K, Blanchard O, Burch S, Robinson J (2007) What factors influence mitigative capacity? Energ Policy 35:692–703

Wynne B (1992) Misunderstood misunderstanding: Social identities and public uptake of science. Public Underst Sci 1:281–304

Yohe G, Strzepek K (2007) Adaptation and mitigation as complementary tools for reducing the risk of climate impacts. Mitig Adapt Strat Glob Change 12:727–739

Author Biographies

Dr O W Olowa work at the Department of Agricultural Education, at the Federal College of Education (Tech.) in Akoka, Nigeria.

Dr O A Olowa work at the Department of Agricultural Education, at the Federal College of Education (Tech.) in Akoka, Nigeria.

Prof. Walter Leal Filho is Head of the "Applications of Life Sciences" Research and Transfer Centre at the Hamburg University of Applied Sciences in Germany, and Chairman of the International Climate Change Information Programme (ICCIP).

Chapter 2
Key Themes of Local Adaptation to Climate Change: Results from Mapping Community-Based Initiatives in Africa

Franziska Mannke

Abstract In Africa as well as in other developing countries, small communities are often the most severely impacted, yet often the least equipped to cope with the effects of climate variability and climate change. Even though the question of what exactly constitutes a successful adaptation to climate change is still unanswered and calls for further research, there is some evidence that many rural and indigenous communities are actively putting into place initiatives to adapt to climate change. But even in today's Internet age it seems difficult to identify these communities properly as information from Africa is fragmented, outdated or simply not to be found. Yet knowing what kind of adaptation is taking place on the local level is a prerequisite for successful approaches to state-of-the-art adaptation which link bottom-up action with top-down strategies. Addressing this knowledge need, this chapter presents the outcome of a 2010 mapping study on local community-based adaptation to climate change in Africa, illustrating how communities are integrated into local adaptation actions that tackle both development and climate change objectives. To reflect the complex nature of adaptation, the identified initiatives were grouped into two categories—social and economic resilience—and a range of emerging adaptation themes were identified. Comprising 39 local and regional initiatives, the analysis provides some recent examples of concrete local adaptation that may possess a certain potential for replication, upscaling and mainstreaming. In this way, and combined with the continued

F. Mannke (✉)
Applications of Life Sciences Research and Transfer Centre,
Hamburg University of Applied Sciences, Lohbruegger Kirchstrasse 65,
21033 Hamburg, Germany
e-mail: franziska.mannke@haw-hamburg.de

F. Mannke
Centre of International Business and Sustainability,
London Metropolitan University, London, UK

W. Leal Filho (ed.), *Experiences of Climate Change Adaptation in Africa*,
Climate Change Management, DOI: 10.1007/978-3-642-22315-0_2,

demand for enhancing the scientific data base to reduce the uncertainty of current climate predictions for the African continent in particular, further research is needed to allow the design of tailored measures to respond to often unique regional adaptation needs.

Keywords Resilience · Community-based adaptation · Africa · Local-level action · Mapping

Introduction

Africa is very vulnerable to current climate sensitivity, and this vulnerability is exacerbated by existing developmental challenges, such as:

• Low GDP per capita	• Widespread, endemic poverty
• Weak institutions	• Low levels of education + high illiteracy
• Low levels of primary health care	• Ecosystems degradation
• Little consideration of women and gender balance in policy planning	• Limited access to capital, including markets, infrastructure + technology
• Poor management capacity	• (Armed) conflicts, complex disasters

This situation contributes to the weak overall adaptive capacity, increasing Africa's vulnerability to projected climate change, which is expected to exacerbate the current challenges even further (Boko et al. 2007). Moreover, research shows a lack of, or at least weak, mechanisms to raise the continent's coping capacity (Kinyangi et al. 2009).

In a world that is predicted to warm by to 4°C or even more, the impacts of climate change on agricultural production are likely to be severe in certain places, for example in Africa. Here, climate change may considerably reduce the options available for crop and livestock production. However, to meet increasing food demand, food production will need to be increased at the same time as the climate is changing (Thornton et al. 2011).

Small communities are often the most severely impacted, yet often the least equipped to cope with the effects of climate variability and climate change. As such, community-based adaptation (CBA) to climate change has evolved as a systematic, participatory, bottom-up approach. Community-based adaptation strengthens the resilience of communities and the ecosystems upon which they rely in light of climate impacts.

The term community is in itself multifunctional; depending on the context in which it is used, it may have various meanings. Thus, in the absence of a clear definition of 'community-based adaptation', the following working definition will be used in this chapter: "Community-based adaptation to climate change is an action *by or for* a community to alleviate or respond to the negative impacts of increasing climate dynamics in order to maintain human security and enhance levels of social and economic development". This basically provides an umbrella definition for all kinds of initiatives, on various implementation levels, including a wide variety of stakeholders, thus not limiting the research to community-driven initiatives.

Dedicated web-based adaptation knowledge repositories, e.g. africa-adapt.net, weadapt.org, adaptationlearning.net, eldis.org, reveal that a range of rural and indigenous communities are currently putting into place initiatives to adapt to climate change. However, even in today's Internet age it seems difficult to identify these communities properly as information is fragmented, outdated or simply not to be found at all. Yet knowing what kind of adaptation is taking place on the local level is a prerequisite for an integrative approach to adaptation which links bottom-up action with top-down strategies.

Addressing this knowledge need, this chapter presents the outcome of a 2010 mapping study on local community-based adaptation to climate change in Africa (Mannke 2010). It illustrates how communities nowadays are involved in local adaptation actions that tackle both development and climate change objectives. To reflect the complex nature of adaptation, the identified initiatives were assessed according to Boko et al. (2007) framework and grouped into two categories—social and economic resilience—so as to identify a range of emerging adaptation themes. Comprising 39 local and regional initiatives, the analysis provides some recent examples of concrete local adaptation that may possess a certain potential for replication, upscaling and mainstreaming. Moreover, the assessment lists a number of critical success factors and indicates which inherent characteristics contribute to creating a potential for replication, upscaling and mainstreaming.

Funded by the Arkleton Trust, a Web-based mapping project on community-based adaptation to climate change in Africa took place in 2010, illustrating how local and indigenous knowledge is embedded in these initiatives to tackle both broader development and climate change objectives. To reflect the complex nature of adaptation, the identified initiatives were grouped into two categories—social and economic resilience—and a range of emerging adaptation themes were identified. The local and regional level initiatives identified from the African continent cover a wide spectrum of adaptation measures. In this context, a database with 39 identified initiatives has been set up, complemented by a number of detailed case studies, to underline the importance and to assess critical success factors.

After introducing the methodology and the conceptual framework upon which this mapping study relies, this chapter will provide some evidence on the importance of an integral approach to adaptation and demonstrate how current CBA initiatives in Africa deal with climate change and development challenges by identifying characteristic features of social and economic resilience. Finally, current challenges and limitations for mainstreaming and upscaling these local adaptation measures in Africa will be explored and recommendations given for further research.

Methodology

In methodological terms, a structured Internet-based assessment of community-based climate change initiatives in Africa was undertaken in the frame of a 12-month mapping research fellowship.

The following limitations apply to this research: The first list of approximately 100 identified projects—approximately two per African country—which were in the first instance thought to fit needed to be adjusted in the end to provide an up-to-date, meaningful picture of recent initiatives.[1] The resulting database has thus been reduced and cleared of those countries where no initiatives could be identified; projects where climate change was not explicitly mentioned; those projects where information on status of implementation is lacking; those projects which were implemented before 2000; those projects which have been terminated in the meantime; and those projects for which no further information could be obtained to qualify for inclusion in the evaluation.

From the resulting final list of 39[2] CBA projects, each of the ones identified was contacted by the Arkleton fellow to obtain further insights, e.g. to verify distinctive aims, identify critical success factors, disclose the involvement of stakeholders and actual outcomes, and assess the replication potential of the project.

Despite the limited scope of the above-mentioned mapping, the study provides some concrete examples of how African communities tackle the complex challenge of sustainable development in the face of a changing environment. The findings may provide a starting point for further in-depth research on and understanding of the particular opportunities for local-level adaptation options in Africa.

Conceptual Framework

Community-based adaptation may foster the resilience of rural communities, but also the resilience of the ecosystems on which these communities rely for a living. In this way, the IPCC defines resilience in the glossary to its fourth assessment report (Baede et al. 2007) as "the ability of a social or ecological system to absorb disturbances while retaining the same basic structure and ways of functioning, the capacity for self-organisation, and the capacity to adapt to stress and change".

For evaluation purposes, Boko et al. (2007, p. 453) classification approach provides the conceptual framework, characterising the initiatives, i.e. adaptation actions, according to their social or economic resilience focus. The advantage of

---

[1] Minimum information included: title; geographical location; aims; duration; outputs; funding bodies; owners; and involved stakeholders. "Recent" in this context refers to initiatives implemented in, or after, the year 2000.

[2] The detailed list of the initiatives identified is provided in the full fellowship report, which can be found on the Arkleton Trust website at www.arkleton.co.uk

Boko's approach is that it allows the grouping of complex adaptations not only in response to climate change, but also integrates further developmental dimensions, i.e. taking into account that adaptation actions in Africa consider and tackle more than just climatic conditions to be successful.

The sub-themes within the *social* or *economic* resilience dimension relate to the following:
Social Resilience

- *Social networks and social capital*, e.g. perceptions of risk by rural communities or local saving schemes which determine behaviour towards adaptation
- *Institutions*, e.g. design, function and governance of institutions which enhance or constrain adaptive capacity

Economic Resilience

- *Equity*, e.g. ability to access external funding sources or donor aid on various levels
- *Diversification of livelihoods*, e.g. agricultural intensification or income diversification to strengthen resilience to shocks
- *Technology*, e.g. seasonal forecasts or improvements of rain-fed systems through water harvesting and conservation techniques, use of new crop varieties to enhance resilience to shocks
- *Infrastructure*, e.g. communication lines, road networks and other physical infrastructure to improve adaptive capacity

In addition, the projects were distinguished between their *reactive*, i.e. backward-oriented, and *proactive*, i.e. anticipatory, nature of adaptation as projects often showed an overlap in terms of addressing social and economic resilience, thus an alternate distinction could be made according to their reactive or proactive nature.

Research Results

The results of the evaluation of initiatives shows a clear connection in terms of building social as well as economic resilience, reflecting the much-needed integrated approach of community-based adaptation projects to respond to the increasing challenges of climate change and development.

From the 39 in-depth initiatives analysed, slightly more than half (20 projects) were classified as social resilience-building projects whereas 19 projects were assigned to the category of economic resilience-building initiatives. Concerning the distinction between proactive and reactive nature, the majority of initiatives (35) showed a predominantly proactive nature. However, many projects utilised historical time series data on climate change and climate variability and respective impacts, indicating that these projects are grounded in sound scientific and complementary information (Mannke 2010, pp. 19–25).

When relating the characteristic features of the 20 social resilience-building projects to key themes from the conceptual framework, the majority of themes clearly intend to evoke individual behavioural changes towards adaptation as well as to raise institutional capacity and providing improved information for decision-makers. However, the focus of initiatives connected with technology or radio infrastructure pointed towards the fact that building economic resilience went hand in hand with building social resilience and vice versa.

Referring to the predominant sub-theme within the framework of analysis, the majority of the 19 economic resilience projects addressed the livelihood dimension. Technology appeared to be a key feature in all identified initiatives, whereas the capacity building of communities and their members resembled a cross-cutting issue related to all project activities.

In terms of concrete topics, the identified initiatives range from sharing joint knowledge through oral histories within communities, rural communications using radio and the Internet, promoting technological solutions such as dams to trap silt and water, sustainable shelter to face desertification, or improving the decision-making capacity of smallholder farmers in drought-prone districts.

In the following section, the key themes that emerged from the evaluation will be elaborated further.

Key Themes of Social Resilience

Four key themes of social resilience building emerged from the analysis of the local-level initiatives (Mannke 2010, p. 26):

- Awareness-raising on climate change
- Improving institutional capacities
- Livelihood, ecosystem and vulnerability assessments
- Pilot adaptation actions (of a technical nature)

The perception of climate change impacts is critical for effective adaptation approaches. Accordingly, the most prominent theme is related to assessing and *raising awareness* on climate change and its potential impacts.

Also, many initiatives are aimed at *improving institutional capacities*, either through the development of new strategies which incorporate climate change or the improvement of institutional structures. The water sector appears to be the most relevant in this respect. The importance of individual capacity building in terms of strengthening the technical expertise of further stakeholders in particular was also observed.

Concerning establishing a common baseline as a foundation for the development of appropriate adaptation strategies and measures, projects covered *assessments of livelihoods, ecosystems and respective vulnerabilities*; in some cases, the integration of indigenous knowledge was mentioned.

Another issue often referred to is *pilot adaptation actions* such as testing agricultural practices, or water supply and quality-related measures.

Key Themes of Economic Resilience

In the following section, five key themes of economic resilience building emerged from the analysis of the local-level initiatives (Mannke 2010, pp. 26–27):

- Introducing new technologies
- Access to energy, jobs and risk reduction options
- Building knowledge and skills
- Small-scale infrastructure and maintenance
- Awareness-raising

The introduction of *new technologies* resulting for example in more sustainable production methods, improved soil and water management practices or increased diversification suggested that technological adaptations are key when it comes to risk reduction and food security. In this respect, many initiatives also address improvements in forecasts by means of better data handling or even supplementing data with indigenous observations.

Risk reduction is also the centre of attention for those actions that improve livelihood, e.g. ensuring *accessibility to and affordability of energy services*, or even promoting the use of renewable energy. Also, fostering *sustainable job sectors*, e.g. climate-proof construction, and the provision of *micro-climate risk management options* may result in improving rural livelihoods.

Capacity building entails not only the provision of information and funding, but also building stakeholder-relevant *knowledge and sustainable skills*, for example, skills that are needed for the maintenance of new technology, skills that are needed for implementing recently learned solutions, etc.

The issue of *infrastructure* appears to be marginally important as the initiatives are in general limited to setting up pilot infrastructure, albeit on a rather small scale, or covering *maintenance* work to make existing infrastructure more climate-proof.

Finally, all the projects identified in this category show a *built-in awareness-raising* component, e.g. the communication of weather and climate information, awareness and capacity-building, the dissemination of lessons learned as well as the provision of climate-relevant data, tools and strategies for decision-makers.

Critical Success Factors

In the context of the research, the identified initiatives were approached with the objective of disclosing critical success factors. Due to a response rate of less than 36%, the number of individual replies needed to be viewed with caution, yet many overlaps in responses suggested that the following list covers some key aspects for the success of projects (Mannke 2010, pp. 27–28):

- *Engagement of target beneficiaries*: letting farmers experiment with new crops and technology, and engaging them in further risk diversification actions
- "*What you see is what you get*": awareness creation, provision of convincing information and strategic mobilisation facilitated by having community members 'witness' adaptation actions
- *Strategic partnerships*: functioning and continuous community support through local-level participation and local stakeholder involvement; involve scientific institutions and local authorities closely in project implementation; involvement of sub-country leadership in mobilisation of communities; using community women's groups to sensitise communities
- *Practicality*: translating theoretical knowledge into practical tools that secure livelihoods as well as ecosystems; apply methodology which matches best available science with resource and capacity constraints and lack of data
- *Proper contextualisation*: designing project objectives and actions in line with climate-related as well as socio-economic issues and seizing opportunities in the local environment in which the project is to take place
- *Staffing*: paying attention to setting up a professional, well-organised coordination team to ensure reliable project management and implementation
- *Dissemination lever*: utilising the media as a multiplier of project information and results, e.g. collaboration with local radio stations
- *Ownership*: integrating activities into local-level resource management planning, e.g. embedding local expertise such as extension staff, technical experts, etc. in project implementation
- *Solidarity*: the fact that communities are suffering due to impacts of climate change may facilitate the will to work together

Concerning the sustainability of the projects, interestingly, only one project explicitly mentioned the financial objective of securing external funding in order to enable project operations to continue beyond the actual project duration.

From the comprehensive responses, three illustrative examples are showcased below. More detailed examples can be found in the full Arkleton mapping report (Mannke 2010).

Box 1: Some Examples of Community Involvement, Critical Success Factors and Transferability of African CBA Projects

Building Resilient Marine Protected Areas—Madagascar (by WWF, January 2008–to date)

- *Key objectives*: (a) climate change response integrated into MPA design and management + development of sound monitoring programmes; (b) streghthening key stakeholder capacity—marine/coastal managers;

(c) identification of feasible financing mechanisms; (d) developing marine biodiversity and MPA knowledge base for use as an educational tool.
- *Community involvement*: community involvement has been through the application of WWF's Climate Witness approach, a participatory assessment tool that increases awareness of climate change issues and develops locally specific climate change adaptation strategies with local communities. The tool has been applied in three villages in the vicinity of the marine protected area.
- *Critical success factors*: this is a successful pilot project that aims to translate theoretical knowledge on climate change impacts into practical tools that protect not only ecosystems, but the local communities that depend on their goods and services.
- *Transferability*: the lessons learnt from the project will be used to inform similar activities in other protected areas throughout Madagascar.

Bravos do Zambeze—Mozambique (by Community Media For Development/UN Delivering as One Joint Programme on Disaster Risk Reduction, May 2009–February 2010)

- *Key objectives*: producing a high-quality, 26-episode radio drama with training for community radio journalists to convey information about disaster risk reduction and build local capacity for reporting disasters and climate change adaptation
- *Community involvement*: communities affected by flooding in the Zambezi river region were visited by researchers and case studies were drawn up which fed into the design of the radio drama. Focus groups, integrating feedback from government representatives, local journalists, and NGOs and further knowledge sources ensured the accuracy of the overall storyline.
- *Critical success factors*: local participation and local stakeholder involvement made the project a success. Through these networks and contacts, valuable feedback and input was gathered to ensure that the drama was as locally relevant as possible. In addition, formative research provided a good foundation for decision—making in the context of the project.
- *Transferability*: a major benefit is that radio dramas are extremely transferable and easily adapted to other communities, contexts and languages.

Climate Change Awareness Creation and Adaptation for Improved Livelihoods Among Rural Communities—Uganda (by AUPWAE Association of Uganda Professional Women in Agriculture and Environment/AFRICA-ADAPT, January–June 2010)

- *Key objectives*: (a) improving food security of smallholder farmers through creating awareness about causes and effects of climate change within farming communities and schools; (b) motivating stakeholders to mitigate and adapt to changing climate conditions.
- *Community involvement*: in the context of the project, farming communities which are organised in groups were sensitised. About 80% of farming work is done by women, who benefited from the awareness raising, information and knowledge, and multiplied the new information within their own (informal) peer network. Entertainment, e.g. drama and songs were used to pass on the message, audience views were obtained through participatory methods, policymakers were involved in discussions. Moreover, schoolchildren were sensitised as 'information dissemination agents' to reach parents and local FM radio stations disseminated information to a larger audience.
- *Critical success factors*: the project implementers regard the involvement of sub-country leadership in the mobilisation of the communities, using the community women's group members to sensitise the overall community and the cooperation with local radio stations, as critical success factors.
- *Transferability*: women's groups were trained to give standard performances and are encouraged to perform these in other sub-countries for a small fee to sustain themselves. District leaders involved in the discussions are expected to disseminate the information in other areas. Schools are encouraged to develop drama and perform at other schools.

Source: Mannke (2010)

Discussion

For many of the African states, adaptation to climate change has become an urgent priority, even though it appears to be marginally important compared to the other serious developmental challenges that Africa is facing in general. In this respect, a recent World Bank report highlighted 60 countries most threatened by future climate change impacts (see Table 1). Grouped according to five main 'climate threats', a total of 24 African countries are among the states listed—some of them threatened by more than one impact (McKinnon et al. 2009). Out of these 24

Table 1 Five climate threats, and the 12 countries most at risk

Drought	Flood	Storm	Coastal 1 m	Agriculture
Malawi[a]	Bangladesh[a]	Philippines	All low-lying island states	**Sudan**[a]
Ethopia[a]	China	Bangladesh[a]	Vietnam[a]	**Senegal**[a]
Zimbabwe[a]	India[a]	**Madagascar**[a]	**Egypt**	**Zimbabwe**[a]
India[a]	Cambodia[a]	Vietnam[a]	**Tunisia**	**Mali**[a]
Mozambique[a]	**Mozambique**[a]	Moldova	Indonesia	**Zambia**[a]
Niger[a]	Laos[a]	Mongolia[a]	**Mauritania**[a]	**Morocco**
Mauritania[a]	Pakistan[a]	Haiti[a]	China	**Niger**[a]
Eritrea[a]	Sri Lanka	Samoa	Mexico	India[a]
Sudan[a]	Thailand	Tonga	Myanmar[a]	**Malawi**[a]
Chad[a]	Vietnam[a]	China	Bangladesh[a]	**Algeria**
Kenya[a]	**Benin**[a]	Honduras	**Senegal**[a]	**Ethiopia**[a]
Iran	**Rwanda**[a]	Fiji	**Libya**	Pakistan[a]

(Boko et al. 2007; McKinnon et al. 2009, p. 19)
Bold text indicates African countries
[a] indicates low-income countries

African states, 16 are low-income countries, with a gross national income per capita of less than USD 995.

Underlining the potential of climate change to impact all aspects of sustainable development, the United Nations Framework Convention on Climate Change (UNFCCC 2010) warns that climate change may push developing countries back into the poverty trap and revoke recent progress in reaching the Millennium Development Goals (MDG). Therefore, adaptation to climate change should be in line with sustainable development priorities and incorporated on all levels—from national to local—and across sectors to avoid shortfalls and maladaptation.

Both academia and practitioners increasingly recognise that *top-down* adaptation processes need to be complemented with *bottom-up* information in the way local communities cope with climate change as adaptation actions converge on a local level: Faced with local shifts in weather patterns which impacts the environment, the local population is required to find locally appropriate solutions to adapt to these challenges (McQueen 2008). Besides well-established top-down approaches, community-based adaptation has therefore come quickly to the forefront of the international community's adaptation agenda as a systematic and well-planned bottom-up approach (Gutiérrez and Mead 2007).

According to Ramasamy et al. (2009), *community-based adaptation* at a local level aims at enhancing the adaptive capacity of communities to reduce their vulnerability to adverse climate impacts from current and future climate hazards. Concerning the previously mentioned local dimension, the Web-based analysis of 39 local-level adaptation projects and initiatives suggests that—in these and similar projects—current initiatives embrace both the economic resilience and the social resilience dimensions to enhance local adaptive capacity.

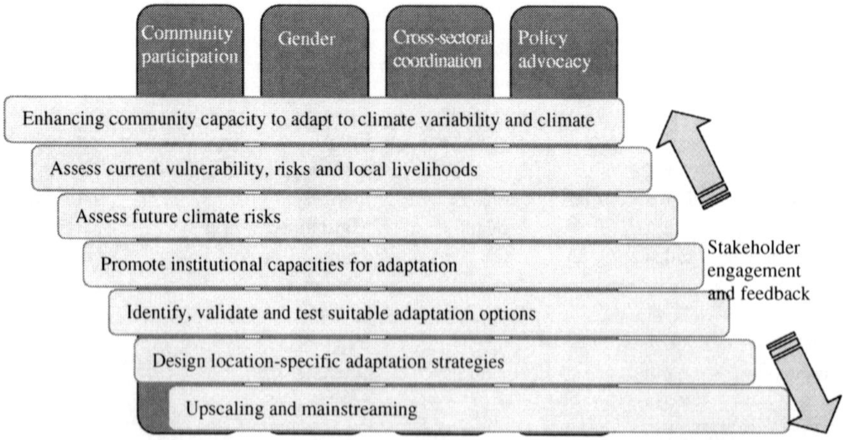

Fig. 1 Main steps, key elements and cross-cutting issues of CBA planning adapted from (Ramasamy et al. 2009)

Moreover, the *multiple challenges* tackled by the projects indicate that, besides clearly defined climate-related activities, developmental objectives are not neglected but are an integral part of the project from the design stage all the way through to the final implementation with the ultimate goal of raising the resiliance of local communities to shocks and disasters.

In terms of *community involvement*, it can be observed that participatory approaches seem to have become best practice. However, the level of involvement of stakeholders differs considerably, from being actively involved in project design or execution to being involved as rather passive stakeholders or final beneficiaries.

Relating to the *transferability* of project results, respondents indicated that their projects could be replicated easily in other parts of the state in question, or even in other countries, indicating an upscaling potential of these kinds of approaches. For example, a radio drama approach such as the "Bravos de Zambeze" project may easily be adapted to other regions and translated into other local languages. Also, proven local methodologies such as the one which was developed by the WWF in its project "Mangrove conservation in Madagascar" can be rolled out in similar environments and subsequently disseminated to further parts of Africa.

All project coordinators pointed out the key role of *stakeholders* in terms of working with and within the initiatives, becoming multipliers of project results for other regions and stakeholders, supporting the project with resources of all kinds, and promoting the uptake of similar activities in further regions, stressing the crucial importance of not only incorporating stakeholders fully into the design, planning and implementation of such local initiatives, but also keeping them motivated and involved in the process.

In terms of *mainstreaming* and exchanging experiences, many projects use Internet and communication technologies (ICT) to network, share lessons learned or, at the very least, for the dissemination of general project information via Web-based knowledge repositories.

Finally, the mapping exercise demonstrated that the projects identified all display many if not all of the main steps and essential elements which should be considered when planning and implementing community-based adaptation actions, depending on their scale (see Fig. 1). In general, those initiatives with longer project durations and a larger funding base incorporated all steps—from enhancing community capacity to upscaling and mainstreaming. They also engage a broad stakeholder base and tackle all four cross-cutting issues, whereas other shorter-term, smaller initiatives with a project duration of less than one year may concentrate on only one or two cross-cutting issues such as community participation and gender.

Some current limitations to successful adaptation

Knowing *how* the climate will change in the future, *which* projected impacts can be associated with it and *when* as well as *where* to expect these impacts to happen can greatly facilitate adaptation planning. However, serious limitations for this still exist, such as a lack of awareness of the local population, a lack of institutional capacity, and a lack of general funding. Besides socio-economic and political constraints, this relates first and foremost to the enhancement of the scientific data basis for decision-making to reduce uncertainty and allow the design of tailored measures to often unique regional adaptation needs which take into consideration not only climate impacts, but multiple interacting factors (Boko et al. 2007; UNFCCC 2007; Speranza 2010).

Also, there seems to be increasing evidence that—because they are constrained in the demographic component—recent climate models might have underestimated the impacts of population on climate change (Liang and Hardee 2009). In addition, climate information needs to be put in context to avoid treating climate change in isolation of other developmental issues. In this respect, Ziervogel and Zermoglio (2010) found the following three core barriers for such contextualisation, which point to future key research needs including beyond the African context:

1. General tendency to isolate climate change impacts from the broader context in which developments are taking place
2. Vulnerability context often not fully understood
3. Climate change adaptation efforts often fail to contextualise climate change risks within the set of other climate information used in decision making (including historical data, real-time data as well as local knowledge)

Moreover, in his frequently cited paper, Agder (2003, p. 192) states that: "much adaptation in the developing world will rely on past experience of dealing with climate-related risks". Farmers, fishermen, coastal dwellers and residents of large cities are expected to adapt autonomously to a large extent, i.e. utilising their own resources and social capital. Yet, before any adaptation can take place and to avoid high transitional costs of adaptation, it is important that stakeholders are aware of climatic changes in order to be able to decide whether or not to adopt appropriate measures. In this respect, exploring how individuals and institutions anticipate or respond to reduce risks, and how policy influences those actions, are stated as key objectives for capturing these adaptation responses (Agder 2003; Maddison 2007).

Conclusion

The detailed experience gathered as part of the mapping study allows a set of conclusions to be drawn:

Firstly, more than any other continent, Africa is characterised by its low adaptive capacity, which limits its ability to cope with climate variability and change. Moreover, this pivotal vulnerability is aggravated by existing developmental challenges.

Secondly, knowing how the climate will change in the future, which projected impacts can be associated with it and when as well as where to expect these impacts to happen can greatly facilitate adaptation planning. One key challenge lies in the method of dealing with the current uncertainties in climate predictions as well as how to reach the very poor and very marginal areas to enable them to adapt to, cope with and develop in a changing environment. In this respect, the scientific data base used as foundation for decision-making needs to be further improved to reduce uncertainty and to design appropriate regional adaptive measures.

Thirdly, the evaluation of the initiatives finds that rural communities are generally integral stakeholders of any adaptation initiatives taking place at a local level. Local stakeholders are actively engaged in the actions taking place, in particular when it comes to awareness-raising activities. All adaptation projects identified show a capacity-building component that may be key to strengthening the resilience of the community involved in order to be able to cope with climate change impacts as well as overall development challenges.

Fourthly, all the initiatives identified are united by the fact that they are not restricted only to adaptation to climate change but tackle general development issues as well, leading to improved social as well as economic resilience to cope with disasters, including those related to climate change and variability, thus raising the adaptive capacity on the local level.

Fifthly, the main steps and essential elements to be considered when planning and implementing community-based adaptation actions (enhancing community capacity; assessments of risks, vulnerabilities and livelihoods; identification of future climate risks; institutional capacity building; matching adaptation options to local needs; design of location-specific adaptation strategies, and upscaling and mainstreaming) were observed in the community-based initiatives identified. Therefore, it appears to be well recognised that community-based adaptation initiatives are a sustainable approach to fostering the resilience of communities but also the resilience of the ecosystems on which communities rely for a living.

Finally, even though communities adapt autonomously to a changing environment to a certain extent, it is necessary that actors are aware of and understand climate change and the associated implications so as to allow them to choose the most appropriate options for adaptation. Even though uncertainties about the extent of climate change impacts remain and a lack of human capacity and financial resources for action may exist, the projects identified can provide some insights into how local stakeholders in Africa anticipate and respond to climate and associated developmental risks today. Combined with the continued demand for enhancing the scientific data base in order to reduce the uncertainty in current climate predictions, this calls for further bottom-up research to foster the design of tailored adaptation measures so that unique regional adaptation needs may be met and answers provided to the remaining question of what constitutes successful adaptation in the light of a constantly changing environment.

Acknowledgments The author wants to thank the Arkleton Trust, which funded this research as part of its 2010 Fellowship Programme. The full fellowship report on community-based adaptation in Africa is available on the Trust's website (www.arkletontrust.co.uk). I would also like to thank my supervisors from the Research and Transfer Centre at the Hamburg University of Applied Sciences (Germany) and the Centre of International Business and Sustainability at London Metropolitan University (UK) for the support and guidance provided as well as for providing the resources needed to conduct and publish this research.

References

Agder NW (2003) Adaptation to climate change in the developing world. Prog Dev Stud 3(3):179–195

Baede APM, vd Linden P et al (2007) Annexe II, international panel of climate change, pp 76–89

Boko M, Niang I et al (2007) Africa. Climate change 2007: impacts, adaptation and vulnerability. In: Parry ML, Canziani OF, Palutikof JP, vd Linden PJ, Hanson CE (eds) Contribution of working Group II to the fourth assessment report of the intergovernmental panel on climate change. Cambridge University Press, Cambridge, pp 433–467

Gutiérrez M, Mead L (2007) Community-based adaptation to climate change bulletin. A summary of the second international workshop on community-based adaptation to climate change, International Institute for Sustainable Development (IISD), vol 135

Kinyangi J, Herrero M et al (2009) Vulnerability to climate change and climate variability in the Greater Horn of Africa. Mapping impacts and adaptive capacity, International Livestock Research Institute

Liang L, Hardee K (2009) How do recent population trends matter to climate change?. Population Action International, Washington, D.C.

Maddison D (2007) The Perception of and adaptation to climate change in Africa. Policy research working paper. D. R. Group, Washington

Mannke F (2010) An overview of community-based adaptation to climate change in Africa. Issue paper. The Arkleton Trust, Berks, pp 1–144

McKinnon K, Hickey V et al (2009) Convenient solutions to an inconvenient truth: ecosystem-based approaches to climate change, International Bank for Reconstruction and Development/ The World Bank

McQueen D (2008) Springing back: climate resilience at Africa's grassroots. Sustainable development opinion, International Institute for Environment and Development (IIED), May 2008

Ramasamy S, Wild M et al (2009) Planning for community-based adaptation to climate change in agriculture. Session 1: Community-based adaptation. WEBGEO Retrieved 13/04/2010, from http://www.webgeo.de/fao-webgeo/

Speranza CI (2010) Resilient adaptation to climate change in African agriculture studies. G. D. Institute, Bonn

Thornton PK, Jones PG et al (2011) Agriculture and food systems in sub-Saharan Africa in a 4°C + world. Philos Trans R Soc A(369):117–136

UNFCCC (2007) Climate change: impacts, vulnerabilities and adaptation in developing countries. Bonn, Germany, UNFCCC

UNFCCC (2010) Fact sheet: the need for adaptation. Retrieved 11/09/2010, from http://unfccc.int/press/fact_sheets/items/4985.php

Ziervogel G, Zermoglio F (2010) Informing adaptation. Tiempo, International Institute for Environment and Development (IIED), Stockholm Environment Institute (SEI) 75:12–16

Author Biography

Franziska Mannke (MBA, BSc, BA int.) coordinates international development projects in the ACP region at the Research and Transfer Centre "Applications of Life Sciences" of the Hamburg University of Applied Sciences. Her current research activities focus on adaptation to climate change and sustainable agriculture. Moreover, she is coordinating the Centre's worldwide online climate conference series CLIMATE and organises online climate teach-in days.

Chapter 3
Potential Climate Change Impacts on Direct Economic Values From Wildlife in the Kilombero Ramsar Site, Tanzania

Siima Bakengesa, Pantaleo Munishi and Stale Navrud

Abstract Tanzania is one of the world's leading nations in terms of wildlife conservation, with rich and diverse wildlife resources. Game controlled areas in Tanzania are used for wildlife conservation and most of them were set aside when human populations were low and global climate was stable. Under the climate change scenario realised for Tanzania for the next few decades, a 10% increase in annual inflow is predicted at the Kilombero Ramsar site. This may have varied impacts on the wildlife populations with consequences for the potential direct economic values from wildlife hunting. The current study assessed how rainfall may influence wildlife populations and their contribution to the national economy. Data was collected from discussions with game officials, literature searches, field observations and data was recorded for weather and hunting licences. We established a rainfall pattern based on trends observed over 40 years (1968–2008), and its correlation with wildlife outtake by both tourist and local hunters. The mean annual rainfall was 1,600 mm, with a probability of 0.90 of receiving ($100 \leq 300$) of the mean annual rainfall especially for March and April point rainfall. Increased inflow of water is likely to be exacerbated by inflow from surrounding catchments. There were a total of 258 local and 78 tourist hunters respectively in the period from 2001 to 2008. There was a positive correlation between the number of animals hunted per species and point annual rainfall for buffalos, reedbuck, hippos,

S. Bakengesa (✉)
Tanzania Forestry Research Institute, PO Box 1854, Morogoro, Tanzania
e-mail: siima_b@yahoo.com

P. Munishi
Faculty of Forestry and Nature Conservation, Sokoine University of Agriculture, PO Box 3009, Morogoro, Tanzania

S. Navrud
Department of Economics and Resource Management, University of Life Sciences Norway, PO Box 5003, As, Norway

W. Leal Filho (ed.), *Experiences of Climate Change Adaptation in Africa*, Climate Change Management, DOI: 10.1007/978-3-642-22315-0_3,

puku, warthog, crocodiles and hartebeest. Conversely, the availability of game birds declined with increased point rainfall. This would mean that revenues from buffalo, reedbuck, hippos, puku, warthog, crocodiles and hartebeest are likely to increase or remain the same with increasing point annual rainfall. On the other hand, hunting revenues from game birds is likely to decrease with point annual rainfall. The predicted hydrological change in the Kilombero River is likely to affect wildlife populations and the contribution of hunting industry to national earnings. Thus climate adaptation measures need to be instituted in order to accommodate climate-induced economic loses.

Keywords Climate change · Direct wildlife economic value · Ramsar site · Ecological change

Introduction

The history of wildlife conservation in the country dates back to the late 1890s when laws controlling hunting were enacted during German rule which focused on off-take, hunting methods and trade. Management also continued under British rule in the 1920s through the establishment of game controlled areas in 1946, which opened the way for the tourist hunting industry (URT 1998). The present wildlife framework in protected areas (PAs) covers 28% of 945,000 km^2 of the land surface. Currently, the framework of PAs comprises 12 national parks (4%), 31 game reserves (15%), the Ngorogoro conservation area (1%) and 38 game controlled areas (8%). The revised wildlife policy (2007) has led to the establishment of a new category of PAs known as wetlands and wetland reserves (Tarimo 2009). The wildlife utilisation industry is currently operated through this network of PAs devoted to wildlife conservation. Game viewing, tourist hunting, resident hunting and ranching and farming are practised.

Consumptive utilisation of wildlife is the country's chosen policy, with the exception of 5 years in the period from 1973 to 1978 when hunting was banned. When the ban was lifted, the Tanzania Wildlife Company (TAWICO) was the authority to allocate quotas and hunting blocks until 1984 when economic liberalisation was introduced and the power to allocate hunting blocks and determine quotas was vested to the Director of Wildlife, who is Secretary to the Hunting Block Allocation Advisory Board under the Minister for Natural Resources and Tourism (Wildlife Act No. 5, 2009). Hunting is practised by both traditional and tourist hunters in game reserves, special open areas and in game controlled areas which are subdivided into hunting blocks where professional hunters and their clients may hunt trophy animals.

National earnings from game hunting were USD 48 million for the period between 2004 and 2007 (Mande 2009). Some 1,654 game hunting tourists in 2004 earned the Government USD 9 million, and by 2007, annual earnings increased to

USD 15 million and the number of game hunting tourists increased to 3,233. Also, the number of hunting firms operating in the country increased from 21 in 1988 to 54 in 2008, while hunting blocks increased from 128 in 1988 to 158 in 2008. A total of 158 hunting blocks in 42 districts were allocated to 54 licensed hunting companies. Fees and other charges are as stipulated in the Tanzania Tourist Hunting Regulations 2003 and the new 2008 schedule of trophy fees. Since 1994, there has been a consensus agreement between the Director of Wildlife and the Tanzania Hunting Operators' Association (TAHOA) on procedures for allocating hunting blocks. This was further consolidated in the 2009 Wildlife Act No. 5, resulting in increased hunting activities countrywide.

Tanzania's great reservoir of wildlife and biological diversity is facing management challenges as a result of ecosystem fragmentation, overutilisation of resources and conflicts between agriculture and wildlife. Persistent drought due to increases in temperature and unreliable rainfall patterns in the country is expected to affect the behaviour of most of the migratory wild species, in particular the wildebeest and some bird species (NAPA 2007). Wildlife is supported by different niches and conditions, which depend on a combination of climatic factors (Holmes 1995). These include the duration and intensity of sunlight, temperature range, rainfall humidity and winds. As for Tanzanian flora, rainfall is the most important factor, taking into account both total annual rainfall and respective monthly distributions. Habitat fragmentation has been linked to human populations and their anthropogenic activities, resource overexploitation, climate change and invasive species. Habitat destruction can also be linked to pollution, fragmentation and degradation, making it difficult for plants, animals and other organisms to survive (Scholes and Biggs 2004). The study by WRI (2000) indicated that more than 50% of wildlife habitats in old tropical countries suffer from habitat alterations. In Tanzania, PAs covers about 28% of the land surface but they suffer from encroachment and other activities. Experience from the US reveals that the effect of global climate change is seen through alpine plants to be growing at higher altitudes on mountains and migrating birds spending longer time in their summer breeding grounds (Walther et al. 2002).

Alongside other attributes, wetland ecosystems are rich in wildlife biodiversity. At the global level, efforts to conserve them were initiated by the Ramsar Convention of 1971, which called for the wise use of all wetlands through local, regional and national actions and international cooperation. The government of Tanzania ratified the Ramsar Convention in 2000 and the Wildlife Division of the Ministry of Natural Resources and Tourism (MNRT) has a facilitating role in implementing the Ramsar Convention on the wise use of wetlands (Tarimo 2009). The country is endowed with exceptional wetland resources, ranging from lake systems, river floodplains and deltaic mangrove formations that cover about 10% of 945,000 km^2 of the land surface. Currently, four sites have been designated and cover about 5 million ha, of which Kilombero Ramsar Site covers about 1 million ha. The wetland is the largest inland freshwater wetland at low altitude (200–400 m.a.s.l.) and has a wide variety of wetland habitat types and high concentrations of mammals, which include puku (75% of the world's wetland-dependent populations), buffalo, hippo, elephant, three endemic bird species, fish (including two endemic species *Citharinus congicus* and *Alestes stuhlmanni*) (Ramsar 2008) (http://ramsar.org/sitelist.pdf). Sustainable management of these resources depends, among other factors, on understanding their economic values—consisting of both use and non-use values—which are ignored in most valuation studies (de Groot et al. 2006; Schuyt 2005).

The report by the IPCC (2001) indicates that there will be an increase in global temperature by an additional 1.40–5.80°C by 2100 and rainfall will continue to increase but vary by region, with some regions showing decreasing rainfall. IPCC (2001) further reports increases in extreme weather events such as hurricanes, flooding and drought associated with warming. Climate change poses major risks to ecosystems including wetlands and their services, which may result in their decline, with multiplier effects on wildlife populations. Climate predictions in Tanzania (TMA 2005) show that the mean daily temperature will rise by 3–5°C throughout the country and the mean annual temperature by 2–4°C. There will also be an increase in rainfall in some parts while other parts will experience decreased rainfall. Predictions further show that areas with a bimodal rainfall pattern will experience increased rainfall of 5–45% and those with unimodal rainfall pattern will experience decreased rainfall of 5–15%. The Initial National Communication (INC) shows that rainfall patterns and soil moisture will vary due to changes in mean temperature, hence affecting the runoff of major rivers. The report indicates that changes in precipitation, both increases and decreases, are likely. This will have implications for the frequency and intensity of weather events such as storms and floods. For instance, the increase in temperature between 1.8 and 3.6°C in the catchment areas of the River Pangani in the north and north east of the country, and the resultant decrease in rainfall, will lead to a decrease of 6–9% of the annual flow of the river. The Rufiji River, which houses the Mtera and Kidatu hydropower stations, is expected to experience an increase in river flow by 5–11% due to low temperature fluctuations of between 3.3 and 4.6°C and will hence see an increase in rainfall and floods.

This study was carried out to establish patterns of rainfall at the Kilombero Ramsar site, and provides findings on its influence on wildlife populations and possible impacts on the revenues derived from wildlife hunting. This will perhaps provide a picture of climate-related economic trend variations with changing climate patterns, particularly with regard to rainfall and overall national economy.

Materials and Methods

Study Area

The Kilombero Valley inland floodplain at the Ramsar site in the Morogoro region of Tanzania covers 796, 735 ha, is approximately 260 km long and 52 km wide, with a catchment area of about 40,000 km^2. The central point coordinates are 8°40′S and 36°10′E (Fig. 1).

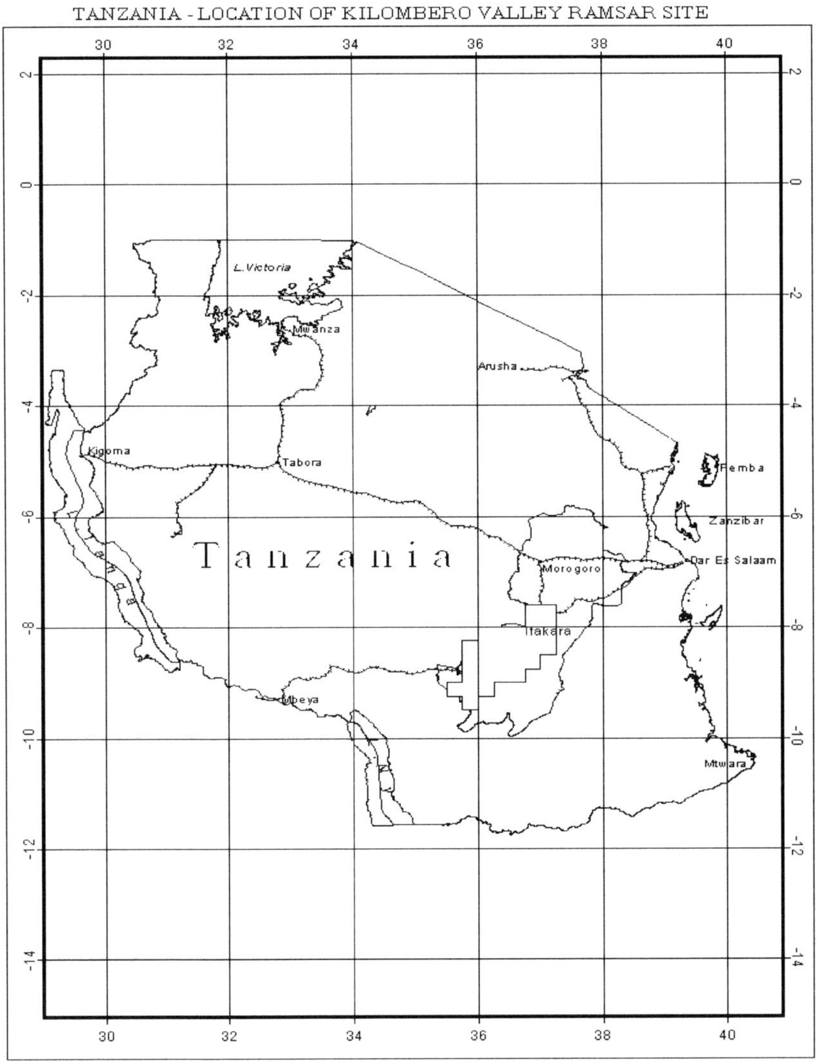

Fig. 1 Map showing Kilombero Valley Ramsar site, Morogoro region, Tanzania

The floodplain lies between 210 and 400 m.a.s.l. The valley is divided by the Kilombero River and falls within the two administrative districts of Kilombero and Ulanga. North Kilombero Game Controlled Area is situated in Kilombero district. The district is situated in vast floodplains between Kilombero River in the south-east and the Udzungwa mountains in the north-west. A considerable proportion of the district falls within the designated Ramsar site. Generally, the district has high temperatures (hot weather conditions) and has bimodal rainfall patterns. Short rains begin towards the end of November and end in January or February. Long rains usually start in March and end in May or June. The average temperature in the district ranges from 26 to 32°C. The average rainfall ranges from 1,200 to 1,600 mm. Kilombero experiences seasonal flooding which causes some parts of the district to be inaccessible during the long rainy season.

The Kilombero Game Controlled Area is divided into two blocks (Fig. 2). The northern block is in Kilombero district while the southern block is in Ulanga district. The study was conducted in the northern block, which is subdivided into the two hunting blocks of Kilombero North Mlimba and Kilombero North Mngeta. Plans are underway to further subdivide existing hunting blocks and establish a new one at the confluence of the Mnyera and Ruhudji rivers (Kilombero District Profile 2009). Human population (2005) is estimated at 400,000 with an annual population increase of about 3%.

Data Collection and Analysis

The study involved a range of data collection methods. Data was collected from discussions with game officials and was analysed using qualitative methods, literature searches, field observations and the use of recorded data for weather and hunting licences. Climatic weather data was obtained from TMA for local weather station. Data on wildlife outtake was obtained from Kilombero Game Office archives.

We established a rainfall pattern based on a trend observed over 40 years for mean monthly and mean annual time series data (1968–2008), and its corresponding deviations. The probability of having the computed mean annual rainfall was calculated based on the following equations (Alder and Roessler 1972):

$$\mathbf{z} = (\mathbf{x} - \mathbf{m})/\mathbf{s} \tag{1}$$

where:

- z area under a normal frequency curve which corresponds to a certain probability
- x mean annual rainfall
- m selected amount of rainfall deviations which the probability calculation assesses (in our case, the deviation of 100 and 300)
- s the standard deviation

The probability of proportional years receiving a certain amount was computed using binomial frequency

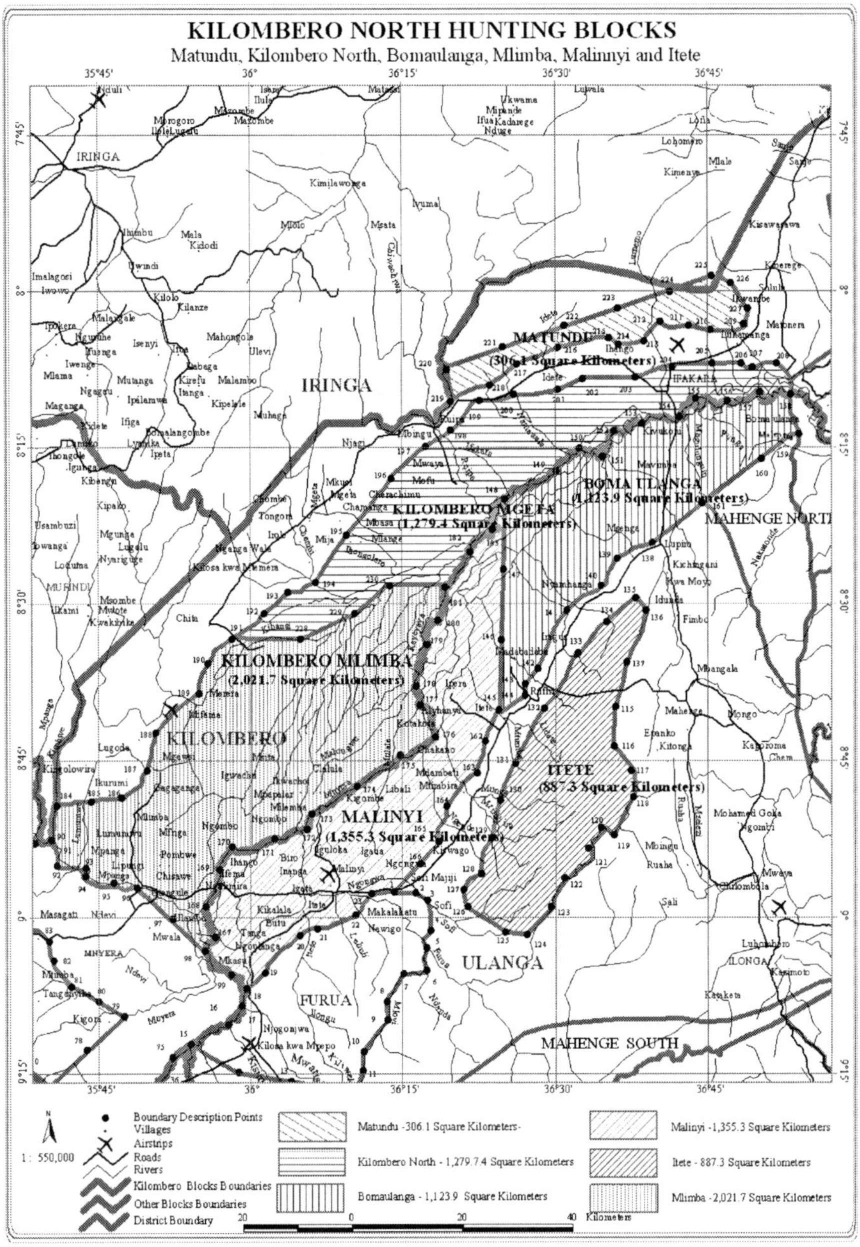

Fig. 2 Map showing North Kilombero hunting blocks, Morogoro region, Tanzania

$$(\mathbf{p} + \mathbf{q})^{\mathbf{n}} \quad (2)$$

where:

p the general probability of the selected amount of rainfall as assessed from the normal distribution

q the probability of not receiving the mean annual rainfall (1−p)

n number of years

The relationship between the annual wildlife outtake and annual rainfall (mm) was analysed using regression analysis to detect parameters and trends.

Computations of direct wildlife economic values were based on hunting-related revenues which include hunting licence fees, game fees, block fees, permit fees, Professional Hunter (PH) examination fees, PH licence fees and trophy handling fees as provided by the Tanzania Hunting Schedule, 2003.

Results

We present rainfall pattern, its prediction, wildlife outtake by both local and tourist hunters, the trend between total annual rainfall and wildlife outtakes and direct economic value derived from wildlife hunting as well as possible impacts on wildlife populations and earnings as a result of climate change.

Rainfall Pattern and Probability

A bimodal rainfall pattern was observed with average annual rainfall of 1,617 mm. The highest total annual rainfall recorded over the period was 2,388 mm in 1989 while the lowest was 942 mm recorded in 2003. On average, April was the wettest month with average monthly rainfall of 413 mm. On the other hand, the driest month was September with average monthly rainfall of 4 mm (Table 1).

The probability of receiving a calculated mean rainfall was 0.90. This suggests a stable rainfall distribution for Ifakara and correspondingly stable weather predictions. An extension of such prediction using bimodal distribution is correspondingly academic as p is almost one unit, thus making the q value close to zero.

Wildlife Hunting Licenses and Outtakes in the North Kilombero Hunting Block

Hunting licences and outtakes both for resident and tourist hunters are presented in order to indicate the economic contribution of these two groups of hunters.

Table 1 Average monthly and mean annual rainfall for North Kilombero in Kilombero Ramsar site, Morogoro, Tanzania (1968–2008)

Months	Average monthly rainfall (mm)	Standard deviation
January	210	106
February	211	117
March	363	145
April	413	160
May	121	82
June	17	16
July	7	14
August	14	18
September	4	6
October	7	7
November	61	50
December	189	129
	Mean 1,617	393

A total of 258 hunting licences were issued to local hunters and a total of 873 outtakes Respective outtakes and species are presented in Table 2. Big game such as buffalo (*Syncerus caffer*), hartebeest (*Alcelaphus buselaphus*), eland (*Taurotragus oryx*), reedbuck (*Redunca* spp) and game birds such as guinea fowls (*Numida*), Egyptian and spurwing geese (*Alopochen* and *Plectropterus*), francolins and pigeons were hunted.

A total of 78 hunting licences were issued to tourist hunters with a total outtake of 289 (Table 3). Leading game animals were buffalos, puku, warthog, hartebeest and crocodiles.

The relationship between total annual rainfall and wildlife outtakes was evaluated by combining leading wildlife outtakes for both local and tourist hunters (Table 4).

The number of animals hunted per species correlated positively with annual rainfall for buffalos (Fig. 3), reedbuck (Fig. 4), hartebeest (Fig. 5), hippos (Fig. 6), crocodile (Fig. 7), puku (Fig. 8), warthog (Fig. 9). The respective outtakes were magnified to facilitate interpretations. Magnification factors are presented in the respective figures. Conversely, the availability of geese (Fig. 10) and guinea fowl (Fig. 11) declined with point annual rainfall above average rainfall.

Direct Economic Value of Wildlife

Calculations of the direct economic values are as presented in Table 5. Fees are as stipulated in the Government hunting schedule of 2003. Different tables for local and tourist hunters are presented. This is because local and tourist hunters pay different fees. In terms of revenues, a total of USD 13,000 was earned from local hunters. Buffalo hunting earned a total of about USD 9,000 in big game and in

Table 2 Wildlife outtakes by local hunters in the North Kilombero hunting block

Year	Hunting licences	Buffalo	Hartebeest	Eland	Reedbuck	Warthog	Bushbuck	Guinea fowl	Geese	Francolins	Pigeons	Dikdiki	Duiker	Impala	Total
2001	45	39	15	1	18	7	4	16	12	1	0				113
2002	21	20	2	0	2	0	2	2						1	29
2003	33	30	14	0	5	0	0	6	24						79
2004	28	17	7	1	10	1	3	24	20	14	3				100
2005	47	47	10	2	11	3	1	36	32	12					154
2006	38	22	14	2	11	0	1	40	50	5					145
2007	22	16	7	6	7	4	2	45	30	35	40		9	2	203
2008	24	17	7	0	7	2	0	10	5	0	0	4	0	1	53
Total	258	208	76	12	71	17	13	179	173	67	43	4	9	4	873

Table 3 Wildlife outtakes by tourist hunters in North Kilombero Hunting block

Year	P.H	Buffalos	Lions	Sable antelope	Eland	Hartebeest	Hippopotamus	Leopard	Puku	Reedbuck	Waterbuck	Baboon	Bushbuck	Bush pig	Crocodile	Duiker	Hyena spotted	Warthog	Total
2006	23	43	2	5	2	7	5	0	19	5	4	2	0	0	6	1	1	8	110
2007	26	31	2	4	1	7	2	3	8	3	4	0	1	0	4	0	1	6	77
2008	29	49	3	3	1	4	5	0	10	1	5	0	3	2	7	1	2	6	102
Total	78	123	7	12	4	18	12	3	37	9	13	2	4	2	17	2	4	20	289

Table 4 Influence of total annual rainfall on wildlife outtake

Year	Total annual rainfall (mm)	Buffalo	Guinea fowl	Geese	Hartebeest	Reedbuck	Puku	Crocodile	Warthog	Hippo
2001	1,427	39	16	12	15	18				
2002	1,728	20	2	0	2	2				
2003	942	30	6	24	14	5				
2004	1,449	17	24	20	7	10				
2005	1,077	47	36	32	10	11				
2006	1,621	71	40	50	18	12	10	7	6	5
2007	1,655	47	45	30	14	10	8	4	6	2
2008	1,643	60	10	5	14	12	19	6	8	5

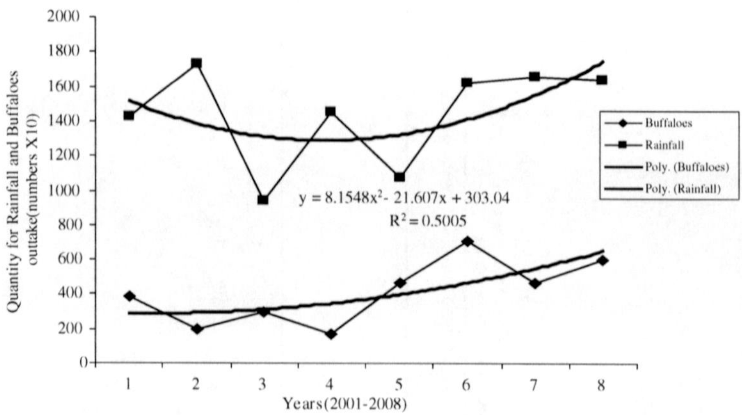

Fig. 3 Relationship between total annual rainfall and buffalo outtake in North Kilombero game controlled area, Morogoro, Tanzania

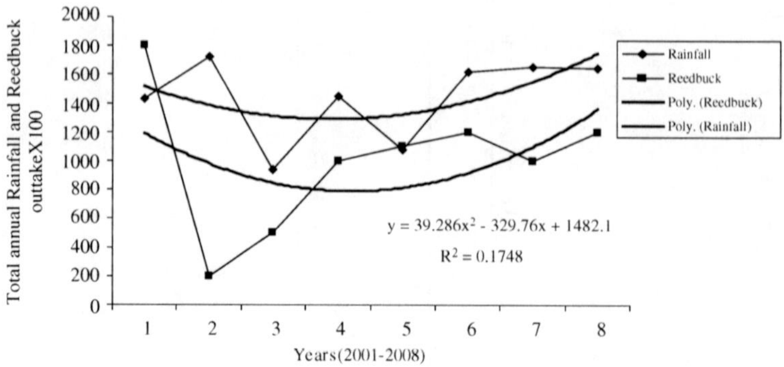

Fig. 4 Relationship between total annual rainfall and reedbuck outtake in North Kilombero game controlled area, Morogoro, Tanzania

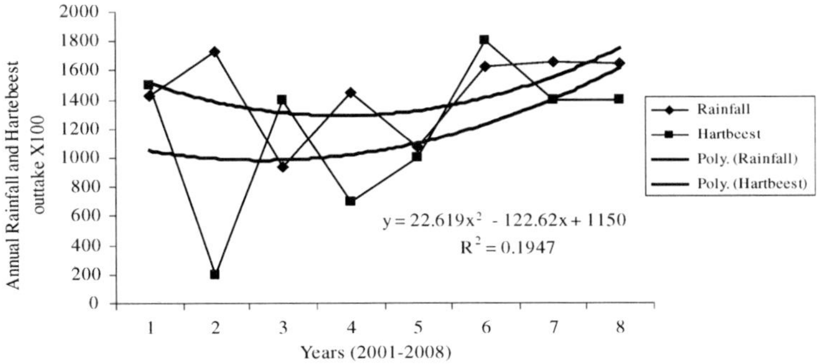

Fig. 5 Relationship between total annual rainfall and hartebeest outtake in North Kilombero game controlled area, Morogoro, Tanzania

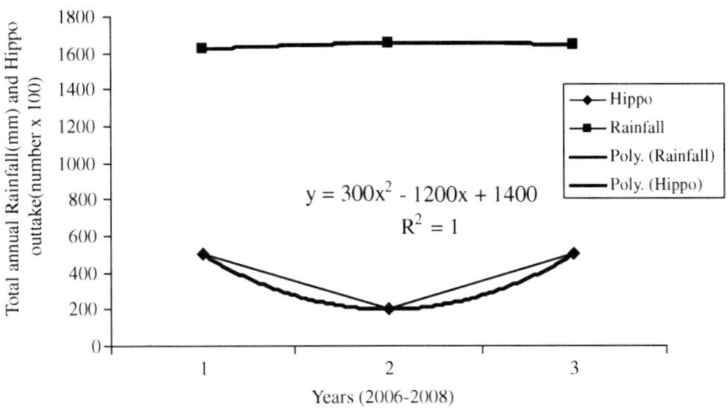

Fig. 6 Relationship between total annual rainfall and hippo outtake in North Kilombero game controlled area, Morogoro, Tanzania

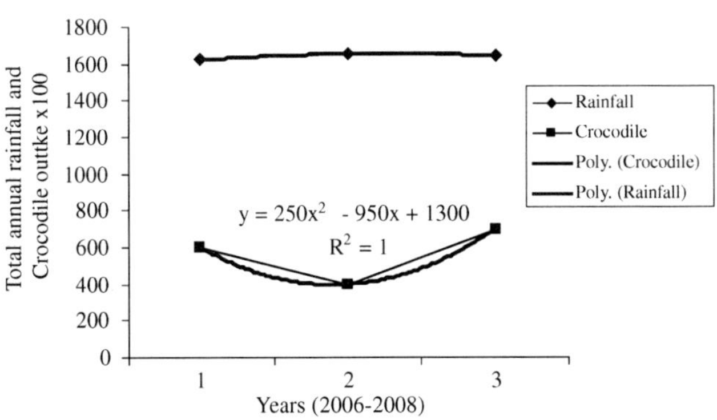

Fig. 7 Relationship between total annual rainfall and crocodile outtake in North Kilombero game controlled area, Morogoro, Tanzania

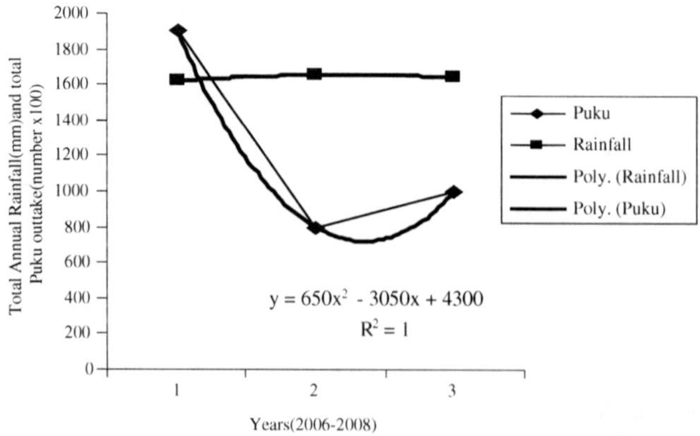

Fig. 8 Relationship between total annual rainfall and puku outtake in North Kilombero game controlled area, Morogoro, Tanzania

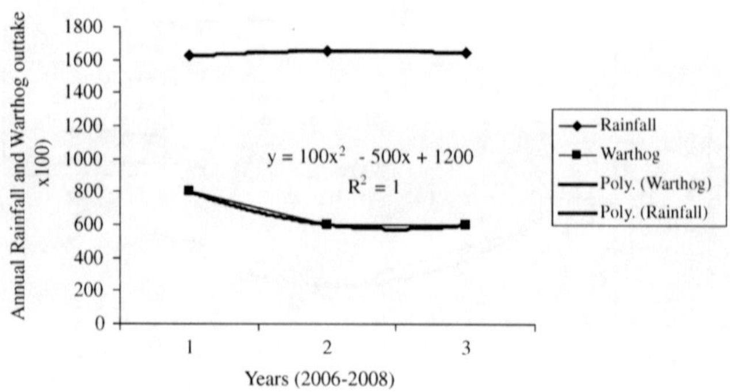

Fig. 9 Relationship between total annual rainfall and warthog outtake in North Kilombero game controlled area, Morogoro, Tanzania

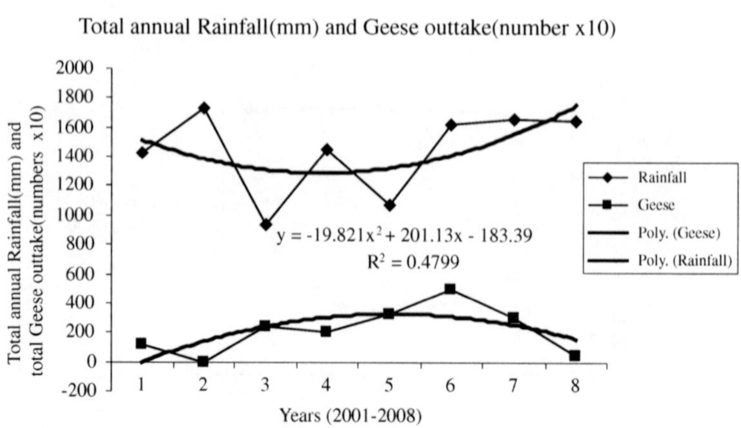

Fig. 10 Relationship between total annual rainfall and geese outtake in North Kilombero game controlled area, Morogoro, Tanzania

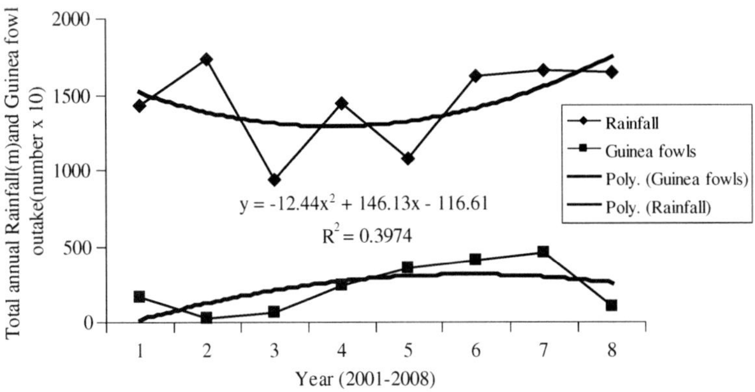

Fig. 11 Relationship between total annual rainfall and guinea fowl outtake in North Kilombero game controlled area, Morogoro, Tanzania

terms of game birds, goose hunting earned a total of about USD 363. Respective income from other wildlife is as indicated in Table 5.

Wildlife economic value in the study period fetched USD 245,600. With the highest income earned from buffalo (USD 92,250), the respective incomes from lion, crocodile, hippopotami and puku hunting are as indicated per hunted category in Table 6.

Discussion

Rainfall probability indicates a more stable pattern at the Kilombero Ramsar site. This stable rainfall pattern is in close agreement with findings of Nshubemuki et al. 1978 (1931–1973), except for the increased amount of rainfall in the month of April. A variety of atmospheric pressure and winds in Tanzania are mostly represented in the months of January, April, July and October (Jackson 1971, 1972) as cited by Nshubemuki et al. 1978. It is evident that a much wetter scenario is building up at the Kilombero Ramsar site, especially if one considers the surrounding catchments causing 10% increased inflows as predicted (TMA 2005). According to Mitsch and Gasselink (1993), the contents of the water stored within the wetland constitutes rainfall falling directly on the wetland (P) and groundwater (Qin) minus evaporation (E) from wetland surface area water minus evaporation from wetland surface area and outflow from wetlands (Qout). The equation of the change in water stored within the wetland is summarised as $\Delta S = P + Q_{in} - E - Q_{out}$. The stored water constitutes the ecology and consequential wetland biodiversity. In another study by Kashaingili et al. (2005) in Usangu, it was indicated that a flow of 0.5 m^3/s is required to maintain the habitat and ecology of the Ruaha National Park, taking into account other anthropogenic activities taking place in the wetland.

Table 5 Direct economic value of wildlife as earned from residential hunters in the North Kilombero hunting block, Morogoro, Tanzania

Year	Hunting licences	Buffalo	Hartebeest	Eland	Reedbuck	Warthog	Bushbuck	Guinea fowl	Geese	Francolins	Pigeons	Dikdiki	Duiker	Impala	Total outtake annual
2001	45	39	15	1	18	7	4	16	12	1	–				113
2002	21	20	2	–	2	–	2	2	–	–				1	29
2003	33	30	14	–	5	–	–	6	24	–					79
2004	28	17	7	1	10	1	3	24	20	14	3				100
2005	47	47	10	2	11	3	1	36	32	12					154
2006	38	22	14	2	11	–	1	40	50	5					145
2007	22	16	7	6	7	4	2	45	30	35	40		9	2	203
2008	24	17	7	0	7	2	–	10	5	–	–	4	–	1	53
Total outtake	258	208	76	12	71	17	13	179	173	67	43	4	9	4	873
Fee USD/ animal		42	21	70		10.5	8.4	6.3	2.1	1	1	3.1	4.2	14.1	
Total income (USD)		8,736	1,596	840		178.5	109.2	1,128	363.3	67	43	12.4	37.8	56.4	13,167.6

Table 6 Direct economic value of wildlife as earned from tourist hunters in the North Kilombero hunting block

P·H	Buffalos	Lions	Sable antelope	Eland	Hartebeest	Hippopotamus	Leopard	Puku	Reedbuck	Waterbuck	Baboon	Bushbuck	Bush pig	Crocodile	Duiker	Hyena spotted	Warthog	Total
23	43	2	5	2	7	5	0	19	5	4	2	0	0	6	1	1	8	110
26	31	2	4	1	7	2	3	8	3	4	0	1	0	4	0	1	6	77
29	49	3	3	1	4	5	0	10	1	5	0	3	2	7	1	2	6	102
78	123	7	12	4	18	12	3	37	9	13	2	4	2	17	2	4	20	289
Fees in USD 450	750	2,500	1,500	1,050	465	1,050	2,500	275	350	550	110	425	240	1,050	220	230	400	
Total 35,100	92,250	17,500	18,000	4,200	8,370	12,600	7,500	10,175	3,150	7,150	220	1,700	480	17,850	440	920	8,000	245,605

Thus inflows tend to be counterproductive. In this regard, bad climatic conditions in 1997 are of note, in which the hunting company Wengert Windrose of Arusha could not meet the required 40% utilisation rate in three out of five hunting blocks (Rugemeleza 1999). Predicted increased inflow in the wetland may result in destructive flooding. Others, including Jones et al. (1997), have observed climate change as a factor altering the distribution of wildlife and habitat coupled with expanding settlements and agriculture in the Kilombero district.

With respect to wildlife outtake and potential contribution to the Government economy, keeping all other factors constant (which is not the case in the actual situation), there may be variations that may reduce wildlife populations and the overall wildlife hunting industry. Total earnings of about USD 260,000 were recorded, with average annual earnings of USD 1,600 from local hunters and average annual earnings of USD 82,000 from tourist hunters. The predicted trend for the number of animals hunted per species is positively correlated with point annual rainfall for buffalos, reedbuck, hartebeest, hippos, puku, warthog and crocodiles. While the availability of geese and guinea fowl are predicted to decline with point annual rainfall above average rainfall. This would mean that revenues from buffalo, reedbuck, hippos, puku, warthog, crocodile and hartebeest are likely to increase or remain the same with increasing point rainfall. On the other hand, hunting revenues from geese and guinea fowl are likely to decrease with increased point rainfall. In the other scenario, with the predicted increase in river inflow of 10% within the Kilombero River under the scenarios of climate change in Tanzania, earnings from wildlife in Tanzania are likely to be affected through change in habitat ranges and niches of wildlife, ranging from open grasslands (wildbeest), to tall grass and open woodland (hartebeest), and wet niches (waterbucks and buffaloes). It is instructive that flooding will selectively favour certain species at the expense of others (Primark 2006; Holmes 1995).

Conclusions

Different wildlife populations adapt differently to environmental changes. In the light of hydrological change in the Kilombero River (10%), increase of inflow from surrounding catchments may affect wildlife populations. The Kilombero Ramsar site covers about 20% of 5 million ha of wetlands declared as Ramsar sites in Tanzania. It is the largest inland fresh water wetland at low altitude (200–400 m.a.s.l) and has a wide variety of wetland habitat types and high concentrations of mammals. Conservation and active utilisation are the basic tenets based on presence of high concentrations of mammals, which include puku (75% of the world wetland-dependent populations), buffalo, hippo, elephant, three endemic bird species, fish (including two endemic species *Citharinus congicus* and *Alestes stuhlmanni*) (Ramsar 2008). Active utilisation is shown by parcelling the area into hunting blocks. They represent protected areas in the country where tourist hunters make available some USD 82,000 annually in Government revenues and resident hunters some USD 1,600. Thus, its contribution to the national economy is significant.

Rainfall has a stable pattern. However, given the predicted 10% hydrological change, coupled with the increase of inflow from surrounding catchments and the results of this study, it is evident that increased rainfall and inflow will affect habitats and dependent wildlife as we have observed in the example of the availability of buffaloes and geese with changing rainfall patterns (Figs. 3, 10). Lessons learnt from Kilombero Ramsar site can be applied to other protected areas in Tanzania and elsewhere. Thus, climate adaptation measures supported by the internal and international community to maintain ecosystem balances and address sectors' vulnerability need to be instituted in Kilombero as well as other protected areas so that the wildlife sector can continue to contribute to the national economy through wildlife hunting.

Acknowledgments We are very grateful to the government of Tanzania and Royal Kingdom of Norway for financial support via a NUFU project. We are grateful to TAFORI, SUA and the Norwegian University of Life Sciences (UMB) for supporting their staff during project implementation in Kilombero as well as District Natural Resources, the Game department and Ramsar Project staff for assistance in enabling and participating in field data collection.

References

Alder LH, Roessler E (1972) Introduction to probability and statistics. W.H. Freeman and Company, New York, p 373

de Groot RS, Stuip MAM, Finlayson CM, Davidson N (2006) Valuing wet-lands: guidance for valuing the benefits derived from wetland ecosystem services, Ramsar Technical Report No. 3/CBD Technical Series No. 27. Ramsar Convention Secretariat, Gland, Switzerland and Secretariat of the Convention on Biological Diversity, Montreal, Canada

Holmes J (1995) Natural forest hand book for Tanzania: forest ecology and management vol I, chapter 5. Sokoine University of Agriculture, Faculty of Forestry, Morogroro, Tanzania, pp 395–416

Intergovernmental Panel on Climate Change (2001) Climate change 2001 synthesis report. Cambridge University Press, Cambridge, MA

Jackson IJ (1971b) Atmospheric pressure and winds, rainfall in Tanzania in maps. University of London Press; pp 34–39. Cited in Nshubemuki L, Somi FGR, Olotu C. A forester's view on average monthly and annual rainfall and number of rain days over Tanzania—regional comparisons. Tanzania Silviculture Technical Note. No. 41; 1978

Jackson IJ Mean daily rainfall intensity and number of rain days over Tanzania. Geografiska Anna-ler 54A (1971a); 1972. 4, pp 469–375. Cited in Nshubemuki L, Somi F G R and Olotu C. A forester's view on average monthly and annual rainfall and number of rain days over Tanzania—regional comparisons. Tanzania Silviculture Technical Note. No. 41; 1978

Jones T, Epps C, Mbano B, Coppolillo P, Mutayoba B, Rovero F (1997) Maintaining ecological connectivity between the protected areas of south-central Tanzania: evidence and challenges. Proceedings of the 6th TAWIRI scientific conference. 3–6 December 2007. Arusha International Conference Centre, Arusha, Tanzania, pp 541–554

Kashaigili J, Mahoo H, McCartney M, Bruce A, Lankford B, Mbilinyi P, Mwanuzi F (2005) Integrated hydrological potential of wetlands for environmental management. The case of the Usangu wetlands in the great Ruaha catchment. Proceedings of the 6th TAWIRI scientific conference. 3–6 December 2007. Arusha International Conference Centre, Arusha, Tanzania, pp 87–99

Kilombero district profile (2009) Kilombero District Excutive Director Office, p 83

Mande M (2009) Tanzania earns $48 m in four years from game hunting. The Online East African Magazine; Posted 16.01.2009

National Adaptation Programme—NAPA (2007) United Republic of Tanzania, Vice President Office, Division of Environment; Jan 2007. p 61

Nshubemuki L, Somi FGR, Olotu C (2008) A forester's view on average monthly and annual rainfall and number of rain days over Tanzania—regional comparisons. Tanzania Silviculture Technical Note. No. 41; 1978. Ramsar

Primark RB (2006) Essentials of conservation biology, 4th edn. Sinauer Associates Inc, Sunderland, MA 01375 USA

Ramsar (2008) Information sheet on Kilombero Ramsar site. (http://ramsar.org/sitelist.pdf)

Rugemeleza N'shala (1999) Granting hunting blocks in Tanzania—The need for reform. Lawyers' Environmental Action Team; 1999/2001. p 18

Scholes RJ, Biggs R (2004) Ecosystem services in southern Africa: regional assessment: the regional scale component of the millennium ecosystem assessment. CSIR, Pretoria, South Africa

Schuyt DK (2005) Economic consequences of wetland degradation for local populations in Africa. J Ecol Econom 2(53):177–190

Tanzania Meteorological Agency (TMA) (2005) Tanzania meteorogical annual report. Government Printers, Dar es Salaam

Tarimo E (2009) The new wildlife policy and wildlife conservation act and their implications for future research in wildlife and wetland conservation in Tanzania. The paper presented at the 7th TAWIRI scientific conference; 2–4 December 2009. (Proceedings in press)

URT (United Republic of Tanzania) (1998) The Wildlife policy of Tanzania. Ministry of natural resources and tourism. Government Printers, Dar es Salaam, p 36

Walther GR, Post E, Convey P, Menzel A, Parmesan C, Bee-bee T (2002) Ecological responses to recent climate change. Nature 416:389–395

World Resource Institute WRI (2000) World Resource 2000–2001. WRI, Washington, DC. www.Africanconservation.org/dcforum/DCForumID21/118html

Author Biographies

Siima Bakengesa is the senior researcher with Tanzania Forestry Research Institute in the Directorate of forest production research. She received her BSc and MSc in Forestry from Sokoine University of Agriculture (SUA), Tanzania. She has carried out research on ecosystem management in Tanzania. She is a recipient of a doctoral scholarship through the NUFU project "Integrating Livelihoods and Multiple Biodiversity Values in Wetlands Management in Tanzania" and is currently pursuing her Ph.D. in ecological economics via a sandwich programme organised by SUA and the University of Life Sciences (UMB) in Norway. She is interested in natural resource management, climate change and adaptation with a focus on the valuation and accounting of natural resources.

Prof. Pantaleo Munishi is currently a Professor in Natural Resources Ecology and Environmental Management at the department of Forest Biology in the Faculty of Forestry and Nature Conservation, Sokoine University of Agriculture, Morogoro, Tanzania. He undertakes teaching, research and outreach in diverse areas of natural resources and environmental management including forest ecology and vegetation science, wetlands conservation, climate change and livelihoods and coping/adaptation strategies. Professor Munishi has worked intensively in aspects linking climate change, natural resources, livelihoods and poverty to try to understand how climate change will likely influence this relationship.

Prof. Stale Navrud is a Professor in Natural Resource Economics at University of Life Sciences As, Norway. He has done extensive work on the valuation of natural resources in both developed and developing countries. His main areas of interest are biodiversity, air and water quality, marine pollution/ oil spills, landscape aesthetics, recreation, health effects and cultural heritage. He also carries out damage assessment and cost/benefit analyses and economic valuation of externalities of energy, wastewater treatment, and transportation, including noise.

Chapter 4
Improving the Participation of Agro-Pastoralists in Climate Change Adaptation and Disaster Risk Reduction Policy Formulation: A Case Study from Keiyo District, Kenya

Charles Kipkorir Songok, Emmanuel Chessum Kipkorir, Edward Musungu Mugalavai, Andrew Chepkok Kwonyike and Caroline Ng'weno

Abstract Growing scientific evidence suggests that climate change will accelerate weather extremes and increase human vulnerability to disasters. Exposure of agro-pastoralists to erratic rainfall and cyclical droughts leads to frequent crop failures and livestock losses, with additional shocks from economic perturbations and erosion of household assets depriving this vulnerable group of "insurance" against adverse risks, driving them from a state of vulnerability to destitution. Despite the growing global agitation to mainstream community-based strategies for climate change adaptation (CCA) and disaster risk reduction (DRR) in Kenya, existing institutional and policy formulation processes rarely incorporate the views and experiences of agro-pastoralists. This leads to policies that are neither responsive nor accountable to vulnerable groups. This paper identifies existing

C. K. Songok (✉)
Food Security and Nutrition Analysis Unit, Somalia, Nairobi, Kenya
e-mail: ngok4@yahoo.co.uk

E. C. Kipkorir
Department of Environmental Monitoring, Planning and Management, School of Environmental Studies, Moi University, Eldoret, Kenya
e-mail: emmanuel_kipkorir@yahoo.co.uk

E. M. Mugalavai
School of Environmental Earth Sciences, Maseno University, Kisumu, Kenya
e-mail: mugalavai_musungu@yahoo.com

A. C. Kwonyike
National Environment Management Authority, Nairobi, Kenya
e-mail: akwonyike@nema.go.ke

C. Ng'weno
Ol Pajetta Conservancy, Nanyuki, Kenya
e-mail: annwaret_2005@yahoo.com

W. Leal Filho (ed.), *Experiences of Climate Change Adaptation in Africa*, Climate Change Management, DOI: 10.1007/978-3-642-22315-0_4,

community strategies for CCA and DRR in Keiyo district and assesses their level of participation in policymaking, as well as key challenges and opportunities for mainstreaming agro-pastoralist involvement in policy formulation. The data was obtained through participatory approaches involving focus group discussions (FGDs), questionnaires and structured interviews. It is evident from the findings of the study that while agro-pastoralists have developed localised strategies for CCA and DRR, these are not fully recognised by existing government policy processes. Therefore, efforts should be made to mainstream the participation of agro-pastoralists in CCA and DRR policy formulation processes.

Keywords Agro-pastoralists · Livelihoods · Climate change adaptation · Disaster risk reduction · Policy formulation

Introduction

Growing evidence indicates that climate change will accelerate weather extremes and increase human vulnerability to catastrophic hazards (Heltberg et al. 2008; IPCC 2007). In particular, vulnerable communities in semi-arid lands (SALs) whose livelihood opportunities depend on rainfall will experience severe climatic shifts (Mitchell et al. 2008; HPG 2009). This, according to Mogaka et al. (2006) and Cline (2007) will destroy livelihood sources and household assets and impair existing coping/adaptive capacity, further aggravating food insecurity and retarding socio-economic development (Multi-Agency Report 2003; Beg et al. 2002; Stern 2006).

In recent years, climate change adaptation (CCA) and disaster risk reduction (DRR) have become the key focus of scientific and policymaking and are now a major area of discussion in multilateral climate change debates (Eriksen et al. 2006; IISD 2007). Adaptation has been implicitly and explicitly linked with development-focused action, particularly as the International Panel on Climate Change (IPCC) has underscored that developing countries are disproportionately vulnerable to climate change and lack adaptive capacity (IPCC 2007; Stern 2006).

While there is a growing global need to mainstream community-based strategies for CCA and DRR in Kenya, the main challenge is lack of a policy on climate change. Even where efforts have been made to address climate change, existing policy, legal and regulatory arrangements are segmented along various sectoral laws, which are not well coordinated (Walsh 2007; GoK 2003). In addition, processes for policy debate and formulation rarely incorporate the views, perspectives, perceptions and experiences of agro-pastoralists in policy formulation (Heltberg et al. 2008; Huq et al. 2004; Nelson et al. 2002).

Due to the limited or lack of participation by agro-pastoralists, resulting policies are neither responsive nor accountable to the needs of this vulnerable group (Hellmuth et al. 2007). Such policies as noted by O' Brien et al. (2008) cannot contribute, sustainably, to vulnerability or risk reduction or even bolster community resilience, without sinking deeper into poverty. This paper therefore identifies existing CCA and DRR strategies used by agro-pastoralists in Keiyo district, examines the level of agro-pastoralist engagement and assesses the challenges and opportunities for mainstreaming the participation of agro-pastoralists in CCA and DRR policy formulation.

Materials and Methods

Description of the Study Area

Keiyo district is a narrow strip of land that extends from latitude 0°10′ to 0°52′ N and longitude 35°25′ to 35°45′ E and covers an area of 1,450 km^2. The district is bordered by Markwet district to the north, Uasin Gishu to the west, Baringo North and Central to the east and Koibatek district to the south-east (Fig. 1).

The population in the district is estimated at 361,496 persons and is not evenly distributed, with the highland areas bordering Uasin Gishu district having higher population density due to its high potential for agriculture. The altitude ranges from 1,000 m in Kerio valley (Kerio River) to over 3,000 m.a.s.l at the peak of the Elgeyo/Marakwet escarpment. This elevation influences rainfall distribution. With a mean annual temperature of 23°C in Kerio valley, the eastern part of the district receives lower and more unreliable rainfall with occasional droughts. In the higher altitude areas, where mean annual temperatures are 13°C, average annual rainfall reaches 1,700 mm, with occasional frost conditions in some localised areas.

The district is divided into three agro-climatic zones (ACZs) that run almost parallel to each other in a north–south direction, namely: the sub-humid highland plateau (2,500–3,000 m), the thickly forested escarpment (1,300–2,500 m) and the lowland or valley floor to the east (semi-arid zone of altitude 1,000–1,300 m). The soils in the southwest are moist, deep and well drained, with calcareous topsoil. In the southern boundary, where rocky and stony soils dominate, agricultural production is difficult. In the highlands and around Tambach, steep slopes increase erosion risks, especially where vegetation is sparse, although natural soil fertility is generally good and soils are well drained and moderately deep. In the south-western and western highland areas, well drained, extremely deep dark soils are excellent for cultivation.

The main livelihood activity is agro-pastoralism. Most of the highlands are predominantly used for food crop farming, with the lower parts and valley floor used as extensive grazing areas of indigenous breeds of animals. The highlands and mid-highlands are suitable for mixed farming, where food crops such as maize, wheat, millet and sorghum are grown and dairy cattle reared. In the valley, cultivation of millet and groundnuts alongside keeping traditional zebu cattle, goats and sheep is practised with bee-keeping undertaken for food, drink and for medicinal purposes.

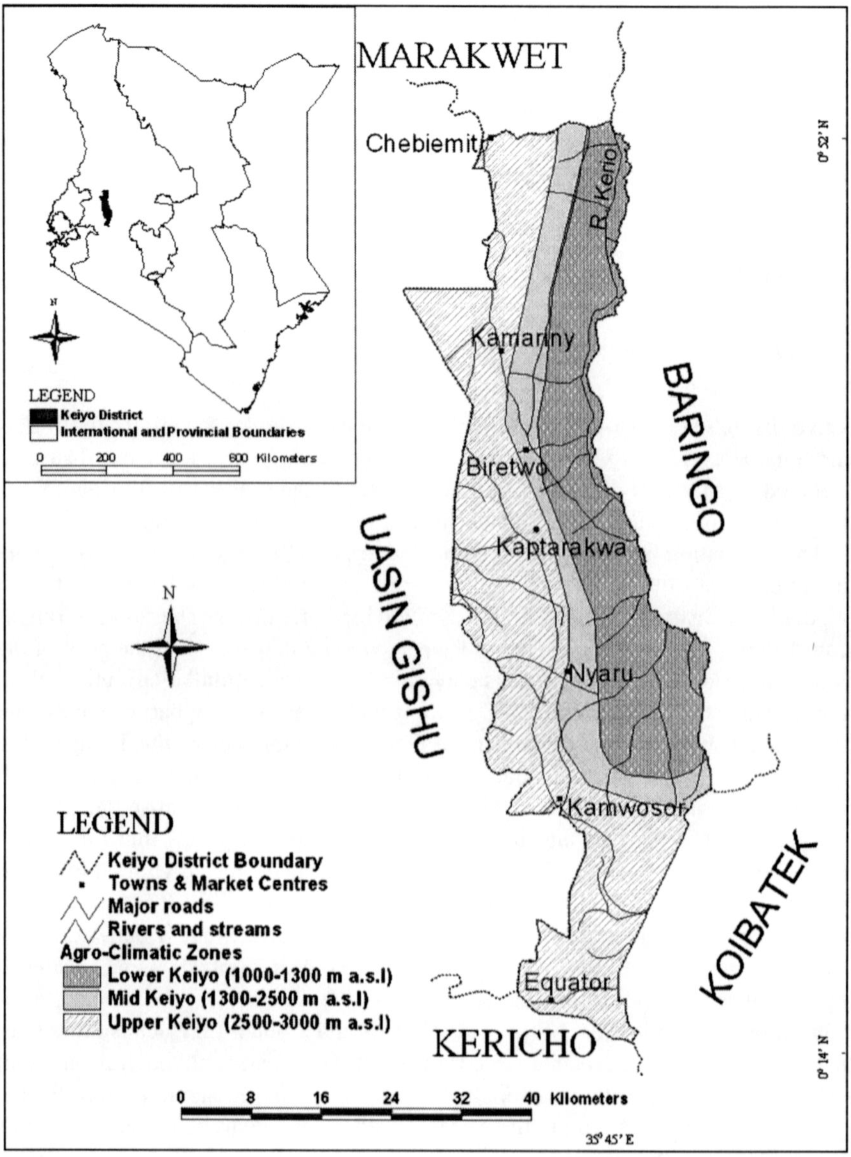

Fig. 1 Location of Keiyo district in Kenya

Table 1 Distribution of households sampled per agro-climatic zone

District	Agro-climatic zone	No. of households sampled
Keiyo	Lower Keiyo (semi-arid)	40
	Mid Keiyo (transitional)	57
	Upper Keiyo (humid)	68
Total		155

Research Approach

A combination of participatory approaches and literature review on policy formulation processes in Kenya was used in the study. Three focus group discussions (FGDs), corresponding to the three agro-climatic zones (ACZs) in the district, were used to collect primary and secondary data. Questionnaire surveys and structured interviews were also conducted with randomly selected agro-pastoralists and policy actors in the district. Representatives of NGOs and religious groups were also involved to identify challenges and opportunities for agro-pastoralists in existing institutional and policy formulation. A lot of information was collected by administering questionnaires distributed over the district mainly through purposive sampling based on age brackets.

The district was stratified into three agro-climatic zones (lower, mid and upper Keiyo) and from each strata, household distribution maps, based on projected population trends for 2010 (KNBS 2009), were used to randomly select 155 households (Table 1). A questionnaire was then administered to the selected household heads in order to assess their level of involvement in policy formulation, local strategies for adapting to climate change and reducing risks.

Results and Discussion

Local Strategies for Climate Change Adaptation and Disaster Risk Reduction

Although little has been documented on indigenous practices for CCA and DRR among agro-pastoralists in Keiyo district, the community has developed localised strategies for adapting to climate change. A significant finding that emerged from FGDs is the re-introduction of traditional cereal crops like finger millet (*Eleusine coracana*) and sorghum (*Sorghum bicolor*), and vegetable varieties such as spider plant (*Cleome gynandra*), black nightshade (*Solanum scabrum*), amaranthus (*Amaranthus blitum*), pigweed (*Amaranthus spp.*), pumpkin leaves (*Cucurbita moschata*) and jute mallow (*Corchorus olitorius*). These crops are preferred because of their ability to tolerate drought (Table 2).

Table 2 Common food and fruit crops produced

Name of crop	Type	Botanical name	Local name
Cereals			
Finger millet	Indigenous	*Eleusine coracana*	Kipsongik
Sorghum	Indigenous	*Sorghum bicolor*	Mosongik
Maize	Modern	*Zea mays*	Bandek
Legumes			
Cowpeas	Modern	*Vigna unguiculata*	
Beans	Modern		
Peas	Modern		
Green grams	Modern		
Groundnuts	Indigenous		
Vegetables			
Amaranth	Indigenous	*Amaranthus blitum*	
Pig weed	Indigenous	*Amaranthus spp.*	
Pumpkin leaves	Indigenous	*Cucurbita moschata*	Chebololet
Jute mallow	Indigenous	*Corchorus olitorius*	
Spider plant	Indigenous	*Cleome gynandra*	Isakiat
Black nightshade	Indigenous	*Solanum scabrum*	Isochot
Roots and tubers			
Sweet potatoes	Indigenous		Rabwonik

In addition, the community has recently adopted the cultivation of early-maturing crops such as cowpeas (*Vigna unguiculata*) and improved varieties of maize (*Zea mays*) developed by the Kenya Seed Company Limited and Kenya Agricultural Research Institute (KARI). Using the extension services provided by the crops department in the Ministry of Agriculture, the community has also shifted from cereal seed broadcasting, a traditional sowing technique, to planting crops in rows and the application of fertiliser.

Policy actors in the district indicated that food insecurity has been a major problem among agro-pastoralists. In the lower Keiyo for instance, food scarcity is experienced for over 8 months in a year, and worsens during periods of prolonged droughts. As a consequence of this, households have devised coping strategies such as reducing the number of meals per capita per day, excluding adults from meals, rationing food intakes, reliance on food assistance, migration in search of pasture and water for livestock and in extreme situations, sale of livestock to purchase food commodities (Table 3).

During seasons of poor harvest, millet is stored in traditional granaries with the stalks to delay consumption and saving for extreme times of acute food shortage. Recently, bee-keeping has been practised to produce honey, which is used to supplement vegetables and preserve meat. Communities residing along perennial streams and rivers practice small-scale irrigation farming of vegetables and cereal crops. This allows for diversification of early-maturing and long-duration crop varieties. Most of these irrigated farms have been established by self-help youth and women's groups as a way of broadening household income and livelihood

Table 3 CCA and food insecurity coping strategies

Coping strategy	Response (N)	(%)
1. Change cropping patterns	12	8
2. Reduce consumption	23	15
3. Purchase/borrow food and reduce consumption	27	17
4. Intensify secondary occupation[a]	18	12
5. Support from kin and friends	4	2
6. Borrow grain and money	9	6
7. Sale of animals	7	5
8. Temporary migration	2	1
9. Rationing consumption[b]	12	8
10. Eat wild fruit	6	4
11. Sale of wood, charcoal, grass and handcrafts[c]	8	5
12. Depend on relief	7	4
13. Sale of productive assets	9	6
14. Begging	3	2
15. Sale of household assets[d]	4	3
16. Distress migration and family separation[e]	3	2
17. Others[f]	2	1
Total (N)	155	100

[a] Includes petty trading, family labour hire and basket making from wild plants, brewing local alcoholic drinks by women for sale as a major household task

[b] Rationing consumption, small distributions of available food among family members as food crisis deepens

[c] Handcraft work includes making household utensils, spears, clubs, bows and arrows (by men) and pottery, collection of wood and grass (by women) to provide income, which is then used to purchase food

[d] Key non-productive sales of dwelling house, household furniture, etc., adopted as a pre-migration survival strategy and disposal of other assets

[e] Destitution and family dislocation faced by the very poor

[f] Others—collection of wild food, water, marrying off daughters at tender age

opportunities. These farms are labour intensive and thus provide wage employment for communities living in the lower parts of the district, with the earnings used to meet household food needs. These strategies assist in mitigating unexpected shocks and risks from drought.

Agro-Pastoralist Involvement in CCA and DRR Policy Processes

In Kenya, most polices are formulated with limited or no involvement of communities. Even as climate change and the need to better prepare for disasters is a major development concern, responses from key informant interviews confirm that the country does not have a policy on either CCA or DRR. In this regard, a common sentiment expressed by respondents during the FGDs and echoed by policy actors was that policymaking processes in Kenya are generally more inclined towards protecting political and institutional interests, with the needs of vulnerable agro-pastoralists less prioritised.

Although previous efforts have been made to involve community representatives at the district level, with policy processes such as the poverty reduction strategy, the economic recovery strategy for wealth creation and employment, the constituency development fund and the agriculture revitalisation strategy, they are over-politicised and consultations with agro-pastoralists are limited. This top-down approach often results in selective appointing of "community representatives" and fails to accommodate the actual needs of agro-pastoralists.

In addition, although agro-pastoralists have a representative at the divisional level who participates in decision-making, such individuals normally do not reside among the community and might not represent their views accurately. As such, the government's preference to use a top-down approach emerged as a key reason that explains the inherently problematic inability of existing policy processes to account for the actual needs of agro-pastoralists.

Reviewed literature shows that the livelihood systems of agro-pastoralists are constantly exposed to socio-climatic risks. This is compounded by negative perceptions and stereotypes of agro-pastoralists held by policy actors in government. They may be seen as having low economic potential and a livelihood system that is 'backward' and environmentally destructive. This reinforces the marginalisation of agro-pastoralists from policymaking, resulting in unsustainable and inappropriate policy interventions.

While the involvement of agro-pastoralists in policymaking is a fundamental pillar of democracy and a prerequisite for ensuring public accountability, efforts to consult the community are piecemeal. Respondents in the study observed that since changes in climate affect their lives and livelihoods, incorporating their experiences and understanding of local realities should form the cornerstone for establishing community-based strategies for CCA and DRR. In spite of this, current policy mechanisms are being replicated based on past models and approaches that are characterised by lack of involvement of vulnerable communities. This has systematically undermined the perspectives and failed to harness already established practices of agro-pastoralists, further limiting their involvement in policy processes.

Challenges to Agro-Pastoralist Involvement in Policy Processes

Respondents from NGOs in the district revealed that despite the benefits of participatory policy formulation, in practice, agro-pastoralists continue to suffer prejudicial treatment reinforced by discriminatory legal and institutional arrangements. In most cases, the constraint on participatory policy formulation seems to be the weak capacity of both civil society and the government to undertake the task, in addition to the lack of obvious avenues for citizens to give feedback to policymakers as well as a lack of clarity on how participation should be conducted. So, when the government fails to harness local realities and the experiences of agro-pastoralists, the community is left to depend on civil society as avenues for communicating to the policymakers, hence running the risk of being misrepresented in the process.

At the national level, climate change and risk reduction are recognised as cross-sectoral concerns. However, two key challenges conspire to set back the participation of agro-pastoralists in policy formulation. First, existing policy processes, which exist in sector ministries such as agriculture, environment and special programmes, are non-participatory. This limits the capacity of the overall national climate change policy formulation process. Second, where there is no tradition of participation in Kenya, it is important to tie participatory policy formulation into other relevant processes such as decentralisation in order to generate accountability for the process.

It is always assumed in Kenya that public participation in the formulation of policies will enhance local ownership. However, in many instances, public participation is limited to consultation and the public is defined as government, business and capital city-based NGOs. The poor people and civic associations working in marginalised areas are left out and women are rarely part of the consultation. Moreover, there is little to no feedback from the government about the trade-offs and on what basis they have been made. In fact, experience in Kenya shows that the closer the policy formulation process is to finalisation, the more secretive and exclusive the process becomes. This leads to the concern of the dominant role of government agencies in the process.

In Keiyo district, the influence of political motivations, characterised by safeguarding self-interests, has eroded the participation of agro-pastoralists in policy formulation. While politicians advocate community participation, they use it rhetorically to strengthen their power bases. This is due to the perceived political insecurity from an empowered community. Practically, the lack of political will by the central government to implement participatory governance has become a major impediment to the effective participation of agro-pastoralists in policy formulation.

The breakdown of traditional structures that enforced compliance with norms and values regarding resource use is another challenge. While these structures acted as repositories for traditional knowledge, an increasing emphasis on formal governance structures, riddled with poor governance, corruption and self-interests, continues to weaken the capacity of vulnerable communities to lobby and advocate effectively with policymakers. Agro-pastoralists are instead exposed to stereotyping by government officials, development agencies, NGOs and civil society organisations (CSOs).

Reviewed literature on agro-pastoralism indicates that since agro-pastoralists do not understand the complexities in which policy actors operate, they lack political clout and the resources to represent themselves, thus they remain vulnerable to other people's interpretation of what is best for them. Their relatively weak 'voice' in the context of rigid policy processes limits their contribution. As a result, policies made under these conditions are non-responsive to their needs, are not transparent and do not create lasting effects towards CCA and DRR efforts. Other constraints include: increasing poverty and marginalisation, lack of community cohesion, resource scarcity and environmental degradation, intra-partisan rivalries among leadership, general policy apathy towards participation and a lack of coherence between research and policymaking.

Opportunities for Enhancing Agro-Pastoralist Involvement in Policy Processes

Despite the above constraints, opportunities for improving agro-pastoralists' involvement in CCA and DRR policy processes exist. The use of NGOs, CSOs and national lobby forums such as the Pastoral Policy Framework for Africa, Pastoral Week and the Pastoralists' Parliamentary Group and the Constituency Development Fund have been adopted by pastoralist communities as successful entry points for bottom-up interaction, advocacy and lobbying with policymakers.

Currently, a number of international organisations (UN agencies and NGOs) are independently conducting their own early-warning systems (EWS) on natural occurrences such as famine, drought and floods. Information generated by these agencies is largely inconsistent, inaccessible and inappropriate to agro-pastoralists. However, the establishment of the Kenya Food Security Steering Group (KFSSG) by the government of Kenya in collaboration with World Food Programme (WFP), alongside initiatives such as the Strategy for Revitalising Agriculture and Vision 2030 provide opportunities for the establishment of community-based early-warning systems. This could provide useful information sources for agro-pastoralists to learn, be informed and influence policy decisions related to CCA and DRR.

Conclusion and Recommendations

The findings show that weak, centralised and non-participatory structures of governance constrain the effective involvement of agro-pastoralists in formulating policies on CCA and DRR. Additionally, agro-pastoralists do not understand the complex nature of the policy formulation process within government. While agro-pastoralists in Keiyo district have, for a long time, developed local strategies to cope with erratic environmental shocks and reduce climate-induced risks, existing governance structures have failed to harness and integrate these efforts in policy formulation and decision-making. Therefore restructuring governance structures to respond effectively to the increasing challenges of climate change will not only empower agro-pastoralists and elevate their role in policy formulation, it will also provide an impetus to accelerate rural development and establish sustainable livelihood solutions for disaster risk reduction. It is therefore recommended that:

- There is a need for practical action research to synergise local strategies and indigenous knowledge with scientific research in order to complement existing advocacy work by non-governmental and civil society organisations. This is because evidence-based advocacy is important in combating negative stereotypes of agro-pastoral livelihood systems and developing positive counter-narratives. In addition, this should enhance the internalisation of agro-pastoralist participation in national policy arrangements and strategically position their 'voices' in public policy debate, thus enabling them make informed contributions in the policy formulation process.
- Public awareness should be created alongside the enhancement of agricultural extension services in order to persuade, inform and improve the knowledge base of agro-pastoralists. Local level capacity building should include building strategic coalitions, networks, functional alliances and coordination structures that promote a common climate change adaptation agenda and positive direction for agro-pastoralists to better lobby and advocate with policymakers. This should take advantage of the opportunities of information technology to get information to villages and investing in the youth and women through education, awareness creation and training as critical avenues for lobbying and advocacy.
- Complementary policies and strategies should be formulated in consultation with agro-pastoralists with the objective of promoting livelihood diversification and sustainable exit strategies. Public education can play a key role in increasing people's ability to diversify, and in the need for access to credits, loans, institutions and social networks. Awareness creation initiatives should take cognizance of the need to reduce the pressure on declining arable land and facilitate the engagement of agro-pastoralists in alternative and sustainable income-generating activities. Initiatives such as innovative rainwater harvesting techniques (earth dams, rock catchment) for use in agro-pastoral production of early-maturing food, fruit and vegetable crops will enhance household food security during climate-stress years as well as improve livelihood opportunities and incomes.
- At the national level, there is a need for institutional capacity building based on existing capacity in agro-pastoral communities. Historically, government-led development investment in agro-pastoral areas in semi-arid zones has been based on economic rates of return rather than on human development and welfare. As a result, civil society organisations assist in addressing the needs of poor agro-pastoralists where government has failed to harness existing local strategies and prioritise development investment. Since community-based strategies for CCA and DRR cannot operate in a vacuum, efforts should be made to remove key structural barriers, such as insecure land tenure systems and increase commitment for policy action from governments and civil society. Additionally, existing policy formulation processes should be reviewed with the

aim of ensuring that agro-pastoralists are not only actively engaged in policy planning and implementation but that their rights, experiences and needs are duly integrated into current and future policies.

- Mainstreaming of community-based approaches to CCA and DRR at the national and local levels and, in particular, within current government efforts to formulate a national policy on climate change. Considering successes within some sectors, such as agriculture, health and water, mainstreaming CCA and DRR at the national level provides an effective means of incorporating knowledge and risks into development planning while highlighting national adaptation programmes of action as the first systematic effort to mainstream adaptation into national activities. Due to the traditional sector-based approach, care must be taken to curb the lack of coordination between government and non-governmental agencies working on climate change and exclusion of socio-economic aspects in addressing climate change and disaster risk reduction, hence the need for stronger linkages between agro-pastoralists and government interests through dialogue, negotiations and connecting solutions to current problems with long-term objectives for increased resilience. At the local level (the entry points of risk reduction), difficulties regarding vulnerability and building resilience to future climate change and climate variability should be addressed through capacity building programmes that use CBOs, CSOs and NGOs as intermediaries in building community trust.
- Community-based monitoring and surveillance systems should be developed and integrated with early-warning systems (EWS) established by international and national agencies such as the World Food Programme (WFP), the Famine Early Warning Systems Network (FEWSNET), the Kenya Food Security Steering Group (KFSSG) and FAO to monitor famine, food security, drought and floods. Information generated through such collaborative efforts should enrich the climate change information base of vulnerable communities, influence informed decision-making capacity at the community level and increase the ability of agro-pastoralists to learn and influence CCA and DRR policy decisions.

References

Beg N, Corfee Morlot J, Davidson O, Afrane-Okesse Y, Tyani L, Denton F, Sokona Y, Thomas JP, La Rovere EL, Parikh JK, Parikh K, Rahman AA (2002) Linkages between climate change and sustainable development. Clim Policy 2(2):129–144

Cline WR (2007) Global warming and agriculture: impact estimates by country. Centre for Global Development and Peterson Institute for International Economics, Washington, DC p 250. ISBN paper 978-0-88132-403-7

Eriksen S, Ulsrud K, Lind J, Muok B (2006) The urgent need to increase adaptive capacities: evidence from the kenyan drylands. African Centre for Technology Studies, Conflicts and Adaptation Policy Brief 2, Nairobi

Government of Kenya (GoK) (2003) Kenya economic recovery strategy for wealth and employment creation 2003–2007. Ministry of Planning and National Development, Nairobi, p 92

Hellmuth ME, Moorhead A, Thomson MC, Williams J (eds) (2007) Climate risk management in Africa: learning from practice. International Research Institute for Climate and Society (IRI), Columbia University, New York, p 116

Heltberg R, Jorgenson S, Seigal P (2008) Climate change: challenges for social protection in Africa. World Bank Working Paper Series. SSRN: http://ssrn.com/abstract=1174774

Huq S, Rahman A, Konate M, Sokona Y, Reid H (2004) Mainstreaming adaptation to climate change in least developed countries (LDCs). Clim Policy 4(1):25–43

International Institute for Sustainable Development (IISD) (2007) A summary of the second international workshop on community-based adaptation to climate change. Community Based Adapt Clim Chang Bull 135(1)

IPCC (2007) Summary for policymakers. In: Parry ML, Canziani OF, Palutikof JP, van der Linden PJ, Hanson CE (eds) Climate change 2007: impacts, adaptation and vulnerability, contribution of working group ii to the fourth assessment report of the intergovernmental panel on climate change. Cambridge University Press, Cambridge, p 7–22

Kenya National Bureau of Statistics (KNBS) (2009) Economic survey 2009. Nairobi. ISBN:9966-767-16-9

Mitchell T, Sabates-Wheeler R, Devereux S, Tanner T, Davies M Leavy J (2008) Rural disaster risk—poverty interfaces. Report prepared for global assessment report on disaster reduction, University of Sussex, Institute of Development Studies

Mogaka H, Samuel G, Davis R, Rafik H (2006) Climate variability and water resource degradation in Kenya: improving water resources development and management. World Bank Working Paper No. 69. The World Bank, USA, pp 1–105

Multi-Agency Report (2003) Poverty and climate change: reducing the vulnerability of the poor through adaptation. VARG multi-development agency paper prepared by AfDB, ADB, DFID, EC DG Development, BMZ, DGIS, OECD, UNDP, UNEP and the World Bank, Washington, DC, USA, xii and pp 43. http://www.undp.org/energy/povcc.htm

Nelson V, Meadows K, Cannon T, Morton J, Martin A (2002) Uncertain predictions, invisible impacts, and the need to mainstream gender in climate change adaptation. Gend Dev 10:51–59

O' Brien K, Sygna L, Leichenko R, Adger WN, Barnett J, Mitchell T, Schipper L, Tanner T, Vogel C, Mortreux C (2008) Disaster risk reduction, climate change adaptation and human security. A report prepared for the Royal Norwegian Ministry of Foreign Affairs by the Global Environmental Change and Human Security (GECHS) Project, GECHS Report No. 3

Humanitarian Policy Group (HPG) (2009) Pastoralism and climate change: enabling adaptive capacity. Synthesis Paper, April 2009, Overseas Development Institute, London, pp 1–4. http://www.odi.org.uk/hpg

Stern N (2006) The economics of climate change: the Stern review. Cambridge University Press, Cambridge

Walsh MT (2007) Pastoralism and policy processes in Tanzania: case study and recommendations. A report to the Tanzania national resource forum, Arusha and contribution to the collaborative study: filling in the knowledge gaps to better understand policy options for pastoralism and rangeland management

Author Biographies

Charles Kipkorir Songok is a Junior Livelihoods Analyst at the Food Security and Nutrition Analysis Unit (FSNAU) Somalia and holds a Master of Philosophy degree in Environmental Information Systems from Moi University, Kenya (2009). His research focuses on GIS and remote sensing applications in land use/land cover mapping and watershed modelling; environmental risk, vulnerability and capacity assessment; climate change adaptation and disaster risk management. He is a member of the Ecological Society of East Africa (ESEA), Young Professional Involvement in Agricultural Research and Development (YPARD) and the Society for Conservation GIS (SCGIS) and has participated in research and consultancy projects in the fields of environmental management, climate change adaptation, food security risk and vulnerability assessment and watershed modelling. He has two scientific research project reports and five announcements at international conferences. Charles Kipkorir Songok is the corresponding author and can be contacted at: ngok4@yahoo.co.uk.

Emmanuel Chessum Kipkorir is a Senior Lecturer in the Department of Environmental Monitoring, Planning and Management at the School of Environmental Studies, Moi University, Kenya, with a Ph.D. in Applied Biological Sciences (2002) and an M.Sc. in Water Resources Engineering (1997) from KU-Leuven University (Belgium). His areas of expertise include water resource development, planning and management; water, health and environment; irrigation engineering; agro-meteorology, and climate change adaptation. He has been involved in a number of research and consultancy projects in the fields of water resource management and environmental monitoring, planning and management. He has already 14 scientific publications in international journals with reviewers, 15 announcements at international conferences and many articles published in journals and books.

Edward Musungu Mugalavai is an Assistant Lecturer at the School of Environment and Earth Sciences, Maseno University, Kenya, with a Master of Philosophy degree in Environmental Earth Sciences from Moi University (Kenya). He is a Ph.D. Student in Environmental Earth Sciences (Moi University, School of Environmental Studies). His areas of research interest are in agro-meteorology, hydrometeorology, oceanography, climatology, climate change assessment, water resources management and applied geography. He has been involved in major research projects in the fields of agro-climatology and climate change adaptation. In the above-mentioned fields, he has already four scientific publications in international journals with reviewers and eight announcements at international conferences.

Andrew Chepkok Kwonyike is the Principal Programmes Officer at the National Environment Management Authority (NEMA), Kenya with a Ph.D. (2009) and Master of Philosophy (2004) in Environmental Monitoring, Planning and Management, Moi University, Kenya. His research interests include: environmental planning and management, climate change adaptation, biodiversity conservation and human ecology. He has consulted in the fields of environmental management and development planning with UN agencies and other international bodies. He has four announcements in international conferences and has co-authored one book.

Caroline Ng'weno is an Assistant Research Scientist at Ol Pejeta Conservancy (Kenya), Department of Ecological Monitoring Department graduate with a Master of Philosophy degree in Environmental Studies (Biological Science) from Moi University, Kenya. She has been selected as a Darwin scholar in the Monitoring and Communicating Biodiversity course (August 2010 at Preston Montford Field centre, Shrewsbury, UK). Her areas of research interests are conservation biology, invasive plants, biogeography, sustainable development, and animal ecology and behaviour. She has one scientific publication in the African Journal of Ecology and two announcements at international conferences.

Chapter 5
Integration of Indigenous Knowledge Systems into Climate Change Adaptation and Enhancing Food Security in Nandi and Keiyo Districts, Kenya

Charles Kipkorir Songok, Emmanuel Chessum Kipkorir and Edward Musungu Mugalavai

Abstract The increasing changes in the frequency, intensity and persistence of rainfall and temperature extremes are key determinants of food security vulnerability in Kenya. Shorter growing seasons and prolonged intra-seasonal dry spells often trigger much larger and more frequent harvest collapses than subsistence households can cope with, leading to declining crop production and increased food insecurity risks. To survive these climate-induced shocks, communities in Nandi and Keiyo districts have developed indigenous practices that enable them reduce vulnerability to food insecurity and adapt to changing climatic conditions. Despite the critical role the indigenous knowledge system (IKS) plays in sustaining household food security, it is rarely considered or integrated into the design of scientific strategies for climate change adaptation. This paper highlights indigenous strategies used by the Nandi and Keiyo communities and explores opportunities for integrating IKS with scientific strategies for climate change adaptation in order to sustain efforts for climate change adaptation and food security risk reduction. Participatory rural appraisal comprising focus group discussions, key informant interviews and household questionnaire surveys were used in the study. The results indicate that while the impacts of climate change directly affect food security, most households integrate both indigenous and modern practices in

C. K. Songok (✉)
Food Security and Nutrition Analysis Unit, Somalia, Nairobi, Kenya
e-mail: ngok4@yahoo.co.uk

E. C. Kipkorir
Department of Environmental Monitoring, Planning and Management, School of Environmental Studies, Moi University, Eldoret, Kenya
e-mail: emmanuel_kipkorir@yahoo.co.uk

E. M. Mugalavai
School of Environmental Earth Sciences, Maseno University, Kisumu, Kenya
e-mail: mugalavai_musungu@yahoo.com

W. Leal Filho (ed.), *Experiences of Climate Change Adaptation in Africa*, Climate Change Management, DOI: 10.1007/978-3-642-22315-0_5,

adapting to climate change and coping with the resulting risks associated with food insecurity. As such, there is a need to reinforce the integration of indigenous and scientific techniques in order to enhance efforts for climate change adaptation, food security risk management and perpetuate intra/intergenerational transfer of IKS.

Keywords Integration · Indigenous knowledge systems · Scientific techniques · Food security

Introduction

Food security is increasingly becoming a critical global development concern. In sub-Saharan Africa, potential variations in the frequency, intensity and persistence of climatic extremes, such as erratic rainfall and drought, are key determinants of future impacts and vulnerability to food insecurity (IPCC 2007; Todisco and Vergni 2008), with the resulting impacts expected to seriously affect subsistence farming systems (Leary et al. 2008).

According to Food and Agriculture Organization (FAO) (2006) and Adger (2006), climate change will result in more erratic and irregular rainfall regimes, shorter growing seasons, prolonged intra-seasonal and inter-seasonal dry spells. Consequently, this is expected to decrease crop production, severely disrupt or destroy livelihood opportunities, increase local food prices, increase household vulnerability to food insecurity (Adger et al. 2003) and, where a combination of slow climatic changes and increasing frequency of sudden shocks occur, could trigger much larger and more frequent harvest collapses than communities can cope with Kipkorir et al. (2002).

In Nandi and Keiyo districts, risks to food insecurity are not only the result of prevailing agro-climatic conditions but a combination of various environmental, social and economic factors such as ecological health, food access, availability and affordability, and changing climate. Even as food productivity models show that poor nations will become more vulnerable to low crop yields (IPCC 2007; UNEP 2008; Stockholm environment Institute (SEI) 2009), for most food-insecure households in the two districts, a variety of indigenous food production systems contribute significantly in mitigating against the climate change-induced effects of food and economic insecurity.

Even as the importance of the indigenous knowledge system (IKS) in climate forecasting, crop production and food security issues are replete in scientific literature (Agrawal 1995; Briggs 2005; Fischer et al. 2002; Mogaka et al. 2006; Smit et al. 2001; UNEP 2008; Warren 1991) and are gaining prominence in the global climate change adaptation (CCA) debate, this technology is not readily acceptable and integrated with modern scientific techniques. However, to most rain-fed subsistence production systems in Nandi and Keiyo districts, IKS constitutes an invaluable, diversified, cost-effective and dynamic localised resource that enables them to survive and produce under risks, without exposing themselves to more risks or shifting towards maladaptation.

While numerous research efforts have been conducted into the importance of IKS and the need for its integration into development initiatives, there have been minimal attempts on the scope of its climatic knowledge (UNEP 2008). IKS is still intact among the elderly in Nandi and Keiyo communities, however, this knowledge is not well documented and it stands in danger of being lost as its custodians are passing away. Of critical concern is the perception among the younger generation of IKS as primitive, outdated and inefficient, with weak and/or limited institutional and policy mechanisms for its integration into current development programmes.

Recognising this 'uncertainty status' and the valuable insights on how indigenous communities have interacted with the changing environment over generations using IKS, this study attempts to harness and illustrate the importance of integrating IKS and modern techniques into CCA and enhancing household food security. This paper reinforces the need to harness, preserve, disseminate and integrate IKS with scientific techniques in order to improve approaches for observing and predicting climate change, as well as influence decision and policymaking processes on CCA and food security at the household level. This paper addresses three key objectives: the IKS practices and technologies related to climate change adaptation, crop production and food security; the role, reliability and level of application of IKS in household food security, and opportunities for integrating IKS with modern strategies in enhancing CCA and food security risk reduction.

Description of the Study Area

Keiyo District

Keiyo district is a narrow strip of land, extending from latitude 0° 10′ to 0° 52′ N and longitude 35° 25′ to 35° 45′ (Fig. 1) and covers an area of approximately 1,440 km^2. The altitude ranges from 1,000 m in Kerio valley to over 2,500 m in the Elgeyo/ Marakwet escarpment. Climatic conditions range from harsh to moderately cool, with mean annual temperatures of 23°C and 13°C at the escarpment and Kerio valley, respectively. In the eastern part, bordering Kerio valley, shorter and more unreliable rainfall of <1,000 mm is experienced, accompanied by occasional droughts. The highlands (Elgeyo/Marakwet escarpment) on the western part of the district receive between 1,000 and 1,400 mm annually with certain frost conditions.

Keiyo district falls in three agro-ecological zones (ACZs): highland plateau (over 2,500 m.a.s.l), Elgeyo/Marakwet escarpment (1,300–2,500 m) and the lowland (1,000–1,300 m.a.s.l). The lowland (Kerio valley) is characterised by harsh semi-arid conditions, frequent droughts and the main livelihood activity is extensive grazing of indigenous breeds of cattle, sheep and goats.

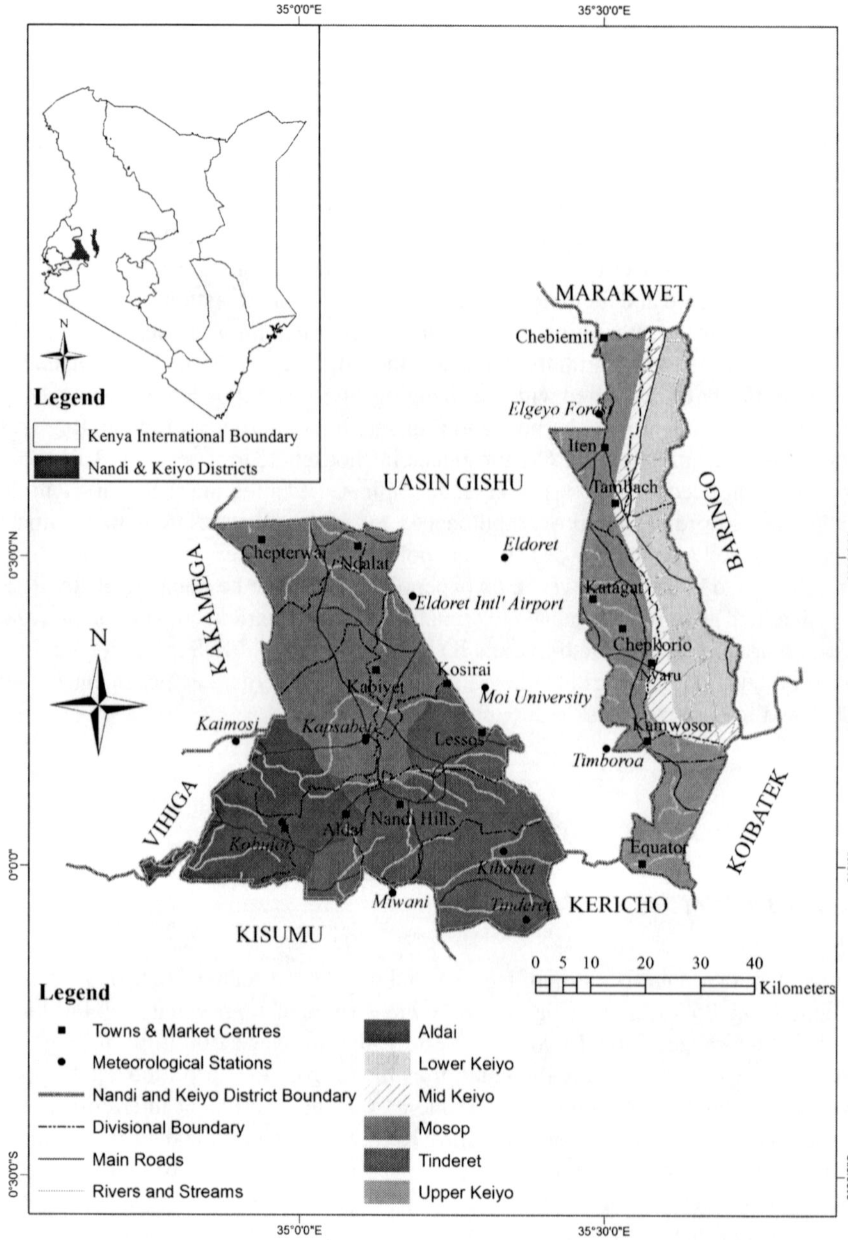

Fig. 1 Location of the study area

Nandi District

Nandi district lies between latitudes 0° 14′ to 0° 35′ N and longitudes 35° 00′ to 35° 25′ E (Fig. 1). Occupying area of approximately 2,880 km^2, the district falls in the high potential region with a fairly high relief of between 1,500 and 2,300 m.a.s.l. The climate is cool and moderately wet; mean annual rainfall ranges between 1,200 and 2,000 mm and mean monthly temperatures of 18–22°C.

The soils in most of the district are deep, well-drained and very suitable for crop cultivation. The district has the potential to produce diverse agricultural crops (cash, food and fruit) and dairy farming. The district is divided into seven agro-ecological zones, namely: lower humid highland (LH_1), lower sub-humid highland (LH_2), lower semi-humid highland (LH_3), upper highland (UH), upper humid highland (UM_1), upper sub-humid midland (UM_2), upper midland (UM_3). Agriculture plays a major role in the development of commerce, trade and services in the district, with the industrial sector mainly consisting of agro-based industries for tea and milk processing.

Methodology

The study used participatory rural appraisal (PRA) comprising organised focus group discussions (FGDs) with community members, household questionnaire surveys and in-depth interviews with a broad range of key informants, drawn from government agencies, representatives of NGOs, CBOs, self-help groups and religious institutions working in the two districts. The two districts were stratified into six agro-climatic zones (ACZs), namely Lower Keiyo, Mid-Keiyo and Upper Keiyo (Keiyo district) and Mosop, Tinderet and Aldai (Nandi district) (Fig. 2). From each stratum, GIS-based randomised sampling was done to obtain the sample size for each stratum (Table 1). Assisted by ten research assistants, a comprehensive household food security vulnerability and capacity assessment questionnaire was administered to household heads.

Convenient sampling was used to select participants, comprising traditional leaders, local chiefs/sub-chiefs, religious leaders, local professionals, agricultural extension officers, and representatives of women's and youth groups, for the FGDs. Elderly men and women of between 65 and 90 years old and who are considered custodians of IKS, were identified through the area chiefs and asked to identify their peers in a snowballing sequence. From each ACZ, six elderly men and women were sampled to form part of the FGDs. In addition, chronically poor (female headed households (HHs), widows, HHs with chronically sick, orphans or disabled members, households without a breadwinner), HHs without assets and HHs with very old/young members were also selected.

FGDs were used to assess the community's level of application of IKS in crop production and food security, prediction of rainfall onset, timing of sowing of crops, crop sequences, cropping patterns, food preservation, community

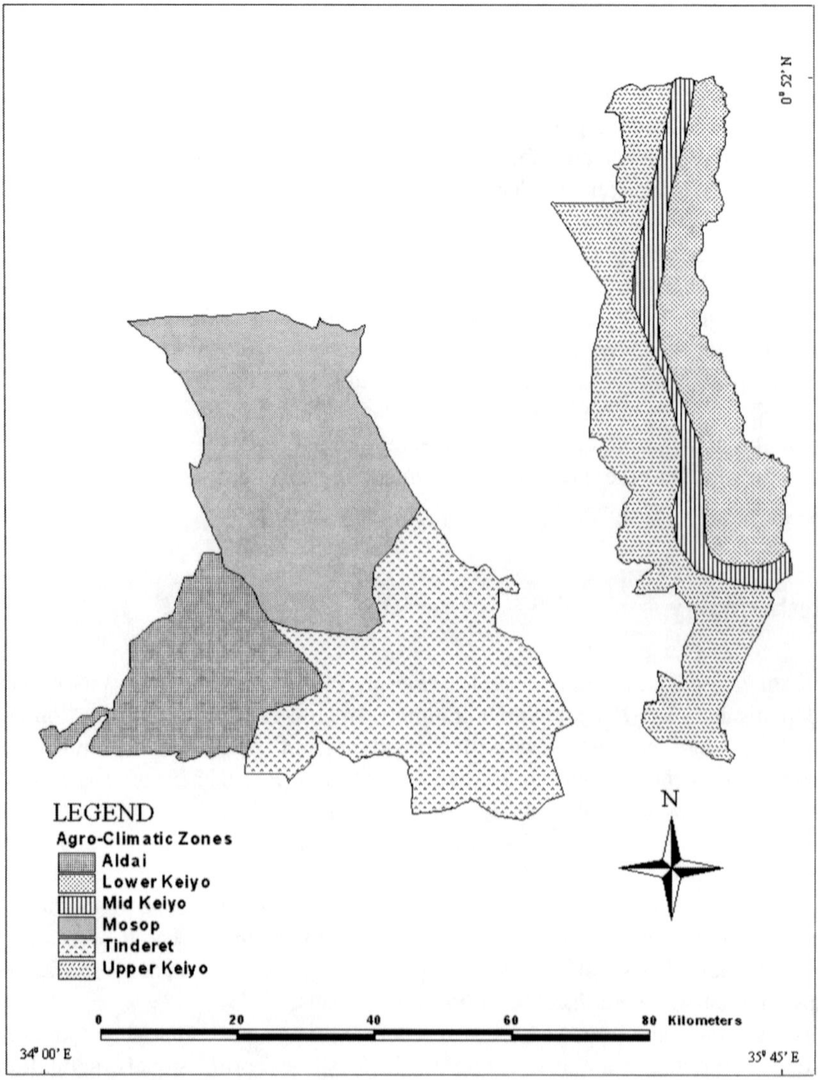

Fig. 2 Agro-climatic zones in Keiyo and Nandi districts

knowledge and gender differentials in knowledge of seed selection and preservation. Each FGD conducted three discussant groups, comprising men and women and defined by age as: 25–40, 40–60 and above 60, were formed. This provided insights into the intergenerational variability in the use of IKS. The choice of women in the FGDs was informed by the understanding that women are most knowledgeable about the distribution, utilisation and management of HH food resources.

Table 1 Distribution of sampled households according to agro-climatic zone

District	Agro-climatic zone	No. of households sampled
Keiyo	Lower Keiyo (semi-arid)	40
	Mid Keiyo (transitional)	87
	Upper Keiyo (humid)	90
Nandi	Mosop (sub-humid)	85
	Tinderet (humid)	68
	Aldai (humid)	36
Total		406

Results and Discussion

Household Demographics and Livelihood Activities

Households in Nandi and Keiyo derive their livelihoods from a wide range of activities (Table 2). A significant proportion of the HHs in Lower Keiyo engage, primarily, in the production and sale of animals/products (85.0%) and agricultural wage labour (77.5%), while incomes from informal mining (27.5%) and charcoal burning (57.5%) are only used to augment total household income. Due to the prevailing semi-arid conditions, the area cannot support rain-fed food crop production compared to Upper Keiyo (80%), Mosop (75.4%), Tinderet (52.5%) and Aldai (80.6%). In Mosop and Tinderet, production and sale of food crops (cereals, legumes and vegetables), cash crops (tea and coffee) and animals/products (beef and dairy) is practised, while in Aldai, the sale of food crops and cash crops is the main household income source.

From the findings, disparity in literacy levels across the six ACZs is more evident. Tinderet, Aldai, Lower and Mid Keiyo recorded less than 10% of the household heads (HHs) having acquired post-secondary education (certificate, diploma, higher diploma, postgraduate diploma, university degree), with the proportion in Upper Keiyo and Mosop ranging from 10 to 27.8%. Most of the HHs in Lower Keiyo (80%), Mid Keiyo (81.62%), Upper Keiyo (80.0%), Mosop (73.2%) and Tinderet (78.6%) are male-headed, while about 45% of HHs in Aldai are female-headed. This is because most women interviewed were widowed or the husbands had migrated to urban areas in search of employment opportunities. Considering that the majority of the household members are below 20 years of age, this age distribution indicates that most of the household members could be classified as children and/or youth (school-going or dropped out of school) and are expected to play crucial roles as dependants in the household, due to their limited contribution towards household food production. Furthermore, with a combined average family size of between six and ten members and most women having limited control over the productive use of land, the dependence of, in most cases one individual in the household, puts more strain on the food resources, sometimes affecting school attendance, which leads to higher drop-out rates and even increases poverty levels.

Furthermore, overreliance on income sources that are predisposed to weather risks and seasonal climatic variations, especially in Lower Keiyo and Aldai, is a key

Table 2 Household demographics and livelihood sources in Keiyo and Nandi districts

Variable	Category	Lower Keiyo (%)	Mid-Keiyo (%)	Upper Keiyo (%)	Mosop (%)	Tinderet (%)	Aldai (%)
Family type	Extended	22.5	8.0	34.4	37.6	19.1	13.9
	Nuclear	77.5	92.0	65.6	62.4	80.9	86.1
	No formal education	7.5	9.2	2.2	2.4	4.2	–
Education level of Household head	Primary	65.0	44.8	31.1	54.6	51.5	63.9
	Secondary	22.5	39.1	45.6	29.4	37.7	8.3
	Post-secondary	2.5	5.7	14.4	10.6	4.4	27.8
	Diploma	2.5	–	4.4	–	1.1	–
	University	–	1.1	–	3.0	1.1	–
Gender of household head	Male	80.0	81.6	80.0	73.2	78.6	55.6
	Female	20.0	18.4	20.0	26.8	21.4	44.4
Family size	<5	38.0	56.3	25.4	35.3	57.4	40.2
	6–10	42.5	37.2	62.3	59.3	39.7	53.1
	10–15	17.0	4.4	12.3	6.4	2.9	3.7
	>16	2.5	3.1	1.0	–	–	4.0
Age distribution	0–10	90.0	71.3	75.6	64.6	80.1	61.1
	11–20	72.5	72.4	82.2	68.9	85.8	72.3
	21–30	55.0	62.1	72.2	78.4	72.7	66.7
	31–40	37.5	45.9	55.5	77.9	60.2	44.4
	>40	12.5	32.2	61.1	74.1	70.8	61.2
Education levels of household members	Primary	88.0	88.0	79	89.0	96.0	94.0
	Secondary	25.0	54	74.0	72.0	59.0	53.0
	Post-secondary	3.0	7.0	28	18.0	6.0	6.0
	Diploma	2.5	12.6	27.8	25.9	7.4	5.56
	University	–	2.3	16. 7	8.24	1.5	5.56

(continued)

Table 2 (continued)

Variable	Category	Lower Keiyo (%)	Mid-Keiyo (%)	Upper Keiyo (%)	Mosop (%)	Tinderet (%)	Aldai (%)
Livelihood sources							
Production and sale of food crops		10.0	35.6	80.0	75.4	52.5	80.6
Production and sale of cash crops		7.5	3.4	15.5	15.7	41.1	55.5
Production and sale of animals/products		85.0	11.4	43.3	55.7	44.8	25.0
Trade		47.5	36.7	36.6	44.3	9.6	13.9
Agricultural wage labour		77.5	18.3	33.4	37.0	39.2	11.2
Self-employment		22.5	14.9	36.6	37.7	23.5	8.4
Salaried employment		5.0	13.8	22.2	31.4	7.6	8.3
Pension		–	1.1	1.1	4.9	2.1	2.8
Informal mining		27.5	8.0	1.1	4.4	2.3	3.8
Charcoal burning		57.5	4.6	18.9	6.0	–	2.8
Other sources		37.5	2.3	2.2	–	4.2	2.8
		N = 40	N = 87	N = 90	N = 85	N = 68	N = 36

Table 3 Shocks to household food security status ranked based on level of importance

Common shocks affecting HH food security	Rank
1. Bad climate resulting in poor crop yields and harvest	1
2. High food prices	1
3. Natural disasters such as floods, droughts, pest infestations	1
4. Sickness of household head or increase in health expenditure	1
5. Death of household member or head	1
6. Loss of employment or reduced salary of dependable household head or member	1
7. High fuel and transport costs	2
8. High house rentals frequently reviewed, pegged in forex	2
9. Debt to reimburse	2
10. Irregular and unsafe drinking water	2
11. Electricity cuts	3
12. Insecurity and thefts	3
13. Frequent school fees (tuition and boarding) reviews	3

1—Most Important 2—Important 3—Less important

indicator of vulnerability to food insecurity. In contrast, diversified livelihood sources in Upper Keiyo, Mosop and Tinderet, comprising the production and sale of food crops (cereals, legumes and vegetables), cash crops (tea and coffee), animal products (beef and dairy) and trade, guarantees a relatively stable household income. This provides alternative strategies for food security risk management in times of food scarcity.

According to Anderson and Cook (1999), household livelihood security encompasses food security, freedom from all dimensions of household poverty and ability to survive crisis and shocks, such as sickness, accidents, natural disasters, financial crisis and household food insecurity. The level of education of the HH is a key factor that determines the ability of the HH to acquire new skills and knowledge needed to boost food production activities, improve HH incomes and invariably enable the households to acquire an adequate food supply. Thus the low educational characteristics of the HHs, observed in Lower Keiyo and Aldai, influences the vulnerability of the HH to food insecurity by lowering the ability of HHs to acquire the skills and knowledge about food production. Therefore, for a household to enjoy livelihood security, which invariably would ensure household food security, it is necessary that they have reserves or other HH assets that can be used to meet contingencies. Hence, the more diverse the livelihood activities, even when exposed to climate risks, the more averse the HH is to food and income insecurity.

Common Shocks Affecting Household Food Security Status

Multiple response analysis of shocks (Table 3) experienced and which have important implications for household food security status revealed that bad weather and natural hazards (floods, drought and pest infestations) often predispose rain-fed food crop production systems to the weather risks resulting in poor food crop yields, harvest collapses or crop losses. This affects HH food sufficiency and adequacy, thus increasing food insecurity. Other shocks such as high market food prices, increased sickness and health expenditure, death of the household head, loss of employment and frequent reviews of tuition and boarding fees compound the problem.

Household Strategies for Climate Change Adaptation and Food Security Risk Reduction

Households in Nandi and Keiyo districts integrate both indigenous and modern coping strategies in mitigating the challenges of food security. As summarised in Annex 1, households in Lower Keiyo (22.5%) and Aldai (22.2%) go an entire day without a meal during times of acute food shortage, and more than 30% of households limit/reduce the portion of meals taken. A combined average of 33% reduces the number of meals eaten in a week and in other extreme cases, HH members, especially in Lower Keiyo (10%) are sent to seek food assistance from relatives, a traditional coping strategy known as *kesumet*. In the other ACZs, the proportions are much higher, representing 26.4% in Mid Keiyo, 22.2% in Upper Keiyo, 24.9% in Mosop, 31.4% in Tinderet and 25% in Aldai. In Upper Keiyo, Mosop and Tinderet, the number of HHs that often borrows food or relies on help from friends or relatives is 22.1, 18.9 and 14.4%, respectively.

Since climate emerged as the main factor affecting food production decisions, over time HHs have developed indigenous knowledge and information systems of weather patterns that enable them to prepare for abnormal events. Since most of these knowledge systems are based on environmental resources, increased environmental degradation and the death of older members is threatening the sustainability of IKS. Using intricate systems of gathering, prediction and interpretation, the HHs observe the changing behaviour and characteristics of animals, plants, celestial bodies and wind patterns (Annex 2) as important traditional ecological and climatic indicators that influence decision-making on the timing of planting dates, seed selection, harvest times, consumption patterns and HH asset management.

The onset of the rains was and is still decided using indicators such as the flowering or shading of leaves of certain indigenous tree species (e.g. *setiot*), wind flow direction observations, migration of birds and insects. This is normally accompanied by early dry sowing using seed broadcasting techniques in small kitchen gardens. Seed banking, a traditional practice that is still used, especially for crops such as millet, sorghum and maize seeds, is practised where the seeds are preserved by smoking them or covering them in ash, so that they cannot be used as food.

Due to the observed shortening of growing seasons and the irregular onset and amount of rainfall, most households are growing indigenous early-maturing crops that can evade drought or survive under limited rainfall. A significant finding is the re-introduction of traditional cereal crops such as finger millet (*Eleusine coracana)* and sorghum (*Sorghum bicolor*), and vegetable varieties such as spider plant (*Cleome gynandra*), black nightshade *(Solanum scabrum)*, amaranth (*Amaranthus blitum*), pig weed (*Amaranthus spp.)*, pumpkin leaves (*Cucurbita moschata*) and jute mallow (*Corchorus olitorius)*. These crops (Annex 3) are preferred due to their ability to

Annex 1 Household food security consumption coping strategies

Strategy	Frequency	Lower Keiyo (%)	Mid-Keiyo (%)	Upper Keiyo (%)	Mosop (%)	Tinderet (%)	Aldai (%)
Go entire day without a meal	Never	35.0	31.0	21.1	5.7	26.1	63.9
	Seldom (1–3 days/month)	22.5	1.1	5.6	2.9	4.4	22.2
	Sometimes (1–2 days/week)	10.0	17.2	31.1	9.1	25.4	8.3
	Often (3–6 days/week)	2.2	–	10.0	1.4	2.3	2.8
Limit portion at mealtime	Never	2.5	6.9	7.8	2.9	11.6	–
	Seldom (1–3 days/month)	37.5	3.4	10.0	3.9	11.7	19.4
	Sometimes (1–2 days/week)	47.5	58.6	30.0	47.3	38.3	41.7
	Often (3–6 days/week)	2.5	3.4	19.9	4.9	3.4	8.3
Reduce number of meals per day	Never	10.0	13.8	7.8	5.7	21.0	41.7
	Seldom (1–3 days/month)	35.0	4.6	12.2	13.7	11.7	2.8
	Sometimes (1–2 days/week)	47.5	28.7	30.0	23.6	29.9	38.9
	Often (3–6 days/week)	–	4.6	11.1	3.9	4.6	2.8
	Daily	–	–	2.3	–	1.1	2.8
Borrow food from friends or relatives	Never	47.5	12.6	16.7	6.3	10.4	50.0
	Seldom (1–3 days/month)	25.0	3.4	12.2	26.6	17.0	19.4
	Sometimes (1–2 days/week)	10.0	26.4	22.2	24.9	31.4	25.0
	Often (3–6 days/week)	–	4.6	22.1	18.9	14.4	–
	Daily	–	–	4.4	4.9	1.4	2.8
Purchase/borrow food on credit	Never	5.0	–	4.4	7.7	10.4	33.3
	Seldom (1–3 days/month)	27.5	1.1	7.8	7.9	9.5	13.9
	Sometimes (1–2 days/week)	35.0	28.7	26.7	26.9	19.9	44.4
	Often (3–6 days/week)	2.5	6.9	8.9	3.9	5.3	–
Living off savings	Never	2.5	8.0	2.2	2.9	19.9	58.3
	Seldom (1–3 days/month)	2.5	8.0	8.9	1.1	4.2	12.6
	Sometimes (1–2 days/week)	7.5	8.0	11.1	7.9	6.3	–
	Often (3–6 days/week)	–	2.3	2.2	2.9	–	–

(continued)

Annex 1 (continued)

Strategy	Frequency	Lower Keiyo (%)	Mid-Keiyo (%)	Upper Keiyo (%)	Mosop (%)	Tinderet (%)	Aldai (%)
Reduce social activities	Never	2.5	2.3	1.1	1.4	13.6	41.7
	Seldom (1–3 days/month)	7.5	1.1	5.6	4.4	6.3	25.0
	Sometimes (1–2 days/week)	2.5	2.3	10.0	5.4	8.7	11.1
	Often (3–6 days/week)	2.5	5.6	3.3	2.4	3.4	6.9
Reduce expenditure on clothing	Never	2.5	1.1	2.2	2.9	6.3	22.2
	Seldom (1–3 days/month)	25.0	1.1	5.6	6	10.6	38.9
	Sometimes (1–2 days/week)	10.0	20.7	10.0	14.4	8.5	19.4
	Often (3–6 days/week)	–	6.1	4.4	2.4	6.8	11.1
	Daily	–	6.6	–	–	1.1	–
Reduce health expenses	Never	5.0	20.7	8.9	5.7	21	58.3
	Seldom (1–3 days/month)	2.5	–	3.3	6.4	4.2	–
	Sometimes (1–2 days/week)	2.5	5.7	1.1	7.3	12.5	25.0
	Often (3–6 days/week)	–	–	4.4	1.4	2.5	–
Reduce education spending	Never	5.0	18.4	7.8	6.7	19.9	58.3
	Seldom (1–3 days/month)	12.5	2.3	1.1	2.0	2.1	8.3
	Sometimes (1–2 days/week)	2.5	1.1	7.8	3.0	5.5	11.1
	Often (3–6 days/week)	–	–	1.1	2.0	2.2	5.6
Children dropping out of school	Never	5.0	19.5	12.2	4.3	16.7	61.1
	Seldom (1–3 days/month)	10.0	–	5.6	1	4.2	16.7
	Sometimes (1–2 days/week)	2.5	–	2.2	5.7	11.9	5.6
	Often (3–6 days/week)	–	–	1.1	2	–	2.8
	Daily	–	2.3	–	1.4	1.2	–

(continued)

Annex 1 (continued)

Strategy	Frequency	Lower Keiyo (%)	Mid-Keiyo (%)	Upper Keiyo (%)	Mosop (%)	Tinderet (%)	Aldai (%)
Reduce pocket money for school children	Never	2.5	17.2	7.8	6.3	15.7	36.1
	Seldom (1–3 days/month)	25.0		1.1	6.9	4.2	13.9
	Sometimes (1–2 days/week)	5.0	3.4	6.7	4.4	2.3	11.1
	Often (3–6 days/week)	–	–	3.3	2.0	2.3	8.3
	Daily	2.5	–	2.2	–	–	5.6
Gather unusual types or amounts of wild food/hunt	Never	–	10.3	7.8	4.3	19.9	72.2
	Seldom (1–3 days/month)	22.5	1.1	3.3	4	2.1	–
	Sometimes (1–2 days/week)	2.5	6.9	4.4	12.3	–	–
	Often (3–6 days/week)	–	3.4	–	4	1.2	5.6
Harvest immature crops (green maize, beans, peas)	Never	2.5	3.4	3.3	7.1	12.7	19.4
	Seldom (1–3 days/month)	7.5	–	7.8	11.1	9.47	16.7
	Sometimes (1–2 days/week)	2.5	29.9	25.6	45.4	30.1	27.8
	Often (3–6 days/week)	–	–	6.7	2.4	3.2	19.4
Send household members to eat elsewhere	Never	–	21.8	10.0	5.7	24.1	66.7
	Seldom (1–3 days/month)	2.5	1.1	1.1	3.0	–	8.3
	Sometimes (1–2 days/week)	2.5	–	1.1	7.0	4.4	5.6
	Often (3–6 days/week)	–	1.1	1.1	1.0	–	–
Send household members to beg	Never	–	20.7	11.1	5.7	24.1	75.0
	Seldom (1–3 days/month)	12.5	–	2.2	1.0	–	–
	Sometimes (1–2 days/week)	–	1.1	2.1	2.0	4.6	2.8
Reduce adult consumption so children can eat	Never	–	1.1	10.0	2.9	7.4	61.1
	Seldom (1–3 days/month)	27.5	–	–	5	6.3	11.1
	Sometimes (1–2 days/week)	5.0	24.1	8.9	5.7	9.7	11.1
	Often (3–6 days/week)	2.5	–	1.1	–	1.1	8.4

(continued)

Annex 1 (continued)

Strategy	Frequency	Lower Keiyo (%)	Mid-Keiyo (%)	Upper Keiyo (%)	Mosop (%)	Tinderet (%)	Aldai (%)
Rely on casual labour for food	Never	–	2.3	4.4	4.3	4.17	61.1
	Seldom (1–3 days/month)	25.0	1.1	4.4	9.3	11.6	2.8
	Sometimes (1–2 days/week)	5.0	14.9	8.9	24.6	11.7	22.2
	Often (3–6 days/week)	–	5.7	6.7	1.4	3.41	–
	Daily	–	4.6	4.4	8.3	12.9	2.8
Rely on less preferred foods as substitutes for staples	Never	–	2.3	6.7	2.9	14.6	44.4
	Seldom (1–3 days/month)	27.5	1.1	2.2	13.9	10.4	5.6
	Sometimes (1–2 days/week)	5.0	2.3	14.4	6.9	7.58	33.3
	Often (3–6 days/week)	–	1.1	3.3	4.4	–	2.8
	Daily	–	–	1.1	–	4.55	–
HH members regularly eat meals of vegetables only	Never	–	11.5	10.0	4.3	25.0	58.3
	Seldom (1–3 days/month)	25.0	–	–	9.9	–	13.9
	Sometimes (1–2 days/week)	2.5	5.7	5.6	13.9	1.1	11.1
	Often (3–6 days/week)	–	1.1	–	1	2.3	–
Reduce expenditure on agro and livestock inputs	Never	–	2.3	1.1	1.4	10.4	27.8
	Seldom (1–3 days/month)	2.5	17.2	3.3	10.9	15.7	11.1
	Sometimes (1–2 days/week)	7.5	1.1	11.1	10.4	7.6	30.6
	Often (3–6 days/week)	–	1.1	2.2	1.4	2.3	5.6
	Daily	–	–	–	–	1.14	–
Sell other HH assets for food	Never	2.5	26.4	8.9	5.3	23.1	66.7
	Seldom (1–3 days/month)	20.0	4.6	13.3	15.9	–	2.8
	Sometimes (1–2 days/week)	2.5	–	7.8	34.1	6.4	11.1
	Often (3–6 days/week)	–	–	3.3	–	–	–
Distress migration to find food or work	Never	–	19.5	12.2	5.7	24.2	80.6
	Seldom (1–3 days/month)	25.0	1.1	1.1	2	–	–
	Sometimes (1–2 days/week)	5.0	3.4	–	3.4	–	2.8

(continued)

Annex 1 (continued)

Strategy	Frequency	Lower Keiyo (%)	Mid-Keiyo (%)	Upper Keiyo (%)	Mosop (%)	Tinderet (%)	Aldai (%)
Receive food aid from government or NGOs	Never	–	3.4	12.2	5.7	23.1	72.2
	Seldom (1–3 days/month)	2.5	3.4	–	2	–	5.6
	Sometimes (1–2 days/week)	–	27.6	–	4.4	–	–
	Often (3–6 days/week)	2.5	–	–	–	1.1	–
	Daily	–	–	–	2	–	–
Early marriage	Never	–	24.1	11.1	5.7	25.2	80.6
	Seldom (1–3 days/month)	2.5	–	–	0	–	–
	Sometimes (1–2 days/week)	–	–	1.1	1.0	–	–
	Often (3–6 days/week)	–	–	–	–	2.1	2.8
Postponement of marriage	Never	2.5	18.4	10.0	2.9	23.1	69.4
	Seldom (1–3 days/month)	–	1.1	–	7.9	–	–
	Sometimes (1–2 days/week)	–	–	–	1.4	–	2.8
Living with family (husband's family)	Never	2.5	1.1	8.9	7.7	12.5	58.3
	Seldom (1–3 days/month)	2.5	–	–	4	–	–
	Sometimes (1–2 days/week)	–	1.1	–	2	–	2.8
	Often (3–6 days/week)	–	2.3	–	–	7.8	–
	Daily	–	18.4	2.2	3.0	11.6	5.6
		N = 40	N = 87	N = 90	N = 85	N = 68	N = 36

Annex 2 Ranking of plant, non-plant, celestial and animal behaviour indicators of rainfall in Nandi and Keiyo districts

Name	Local name	Observable indicator	Time lapse	% score	Rank	Reliability	Main activity
Plant indicators of rainfall prediction and food production							
Flame tree	*Tepeng'wet*	Flowering (shading leaves)	1–2 weeks (1–2 months)	36.2	1	High	1st ploughing (for time lapse >1 month)
	Setiot	Flowering	2–3 weeks	35.9	1	High	1st ploughing (for time lapse >1 month)
	Kakorwet	Flowering—red (shade leaves/buds)	1–2 months (approx. 3 weeks)	23.4	2	High	Second ploughing (for time lapse <3 weeks)
	Ewaat	Flowering (shading leaves—no rain)	1–2 weeks (approx. 1 month)	5.0	3	High	1st ploughing (for time lapse >1 month)
	Chepng'atiatet	Budding (*near onset*)	2–3 weeks	4.3	4	High	Buy inputs
Non-plant indicators of rainfall prediction and food production							
Celestial bodies, name							
Sun	*Asista*	Intense, signifies near rainfall	Approx. 1 month	31.5	1	High	Primary/secondary ploughing
		Position: Tropic of Cancer indicates onset	3–4 weeks			Medium	Weeding, mulching, terracing and thinning
		Position: Tropic of Capricorn indicates cessation					Preparing granary for harvest
		Stars' appearance: east signifies short rains, west signifies long rains	3–5 weeks	28.1	2	High	Sowing seeds; monitor movement of stars (10 p.m) for six months and start sowing when stars reach the position of the sun at 3 p.m.

(continued)

Annex 2 (continued)

Name	Local name	Observable indicator	Time lapse	% score	Rank	Reliability	Main activity
Stars	*Kecheik*	Cluster surrounded by halo signify good rain season; dull clusters signify no rains or poor rains					
		Appearance indicates cessation	2 months	25.8	3	High	Early harvest and preparation of granary
Moon	*Arawet*	Waning-size/shape indicates onset				–	–
		Appearance in east in evening indicates cessation	Approx. 1–2 months			High	Prepare granary and start of harvest
		Disappearance at night indicates heavy rainfall				Medium	Top dressing
Weather parameters							
Clouds	*Bolik*	Clouds thicken at horizon and wing veers or breaks to east and darkens in colour, signifies near onset. Thick, dark clouds signify heavy rains.	2–3 weeks	22.8	1	Medium to high	Purchase inputs, sow seeds, Second ploughing
Temperature	*Lalangyet or kaititiet*	High (night) and low (evening) signify near onset	2–3 weeks	21.6	2	Medium to high	Sow seed
		Low in the morning signifies no or far rainfall	–			–	–

(continued)

Annex 2 (continued)

Name	Local name	Observable indicator	Time lapse	% score	Rank	Reliability	Main activity
Wind	*Koristo or usonet*	Blowing eastwards indicates rainfall is near	1–2 weeks	19.8	3	Medium to high	Second ploughing, buy inputs, sowing
		Blowing westwards indicates far rainfall, cessation	1–2 weeks			Medium to high	–
Humidity		Misty (morning and evening) indicates no or far rain	1–2 weeks	19.5	4	–	–
		Dew (morning), raising water levels in streams, wetness in the riverbanks, springs and wetlands signify near rains	1–2 weeks			–	–
Lightning	*Ilet or tiandab barak*	If strikes in near horizontal position, indicates no rainfall	Approx. 1 month	16.2	5	Medium to high	1st ploughing
		If strikes in near vertical position, indicates near onset	2–3 weeks			High	Purchase inputs, Second ploughing

(continued)

Annex 2 (continued)

Name	Local name	Observable indicator	Time lapse	% score	Rank	Reliability	Main activity
Animal behaviour							
Insects	Termites (*Toik*)	Appearance on land, roads and paths, signify near rainfall onset	2–3 weeks	35.9	1	High	Second cultivation and buy inputs
	Ants (*Birechik*)	Appearance on land surface signifies rising water table, hence near rainfall	2–3 weeks	10.3	2	High	Second cultivation and buy inputs
	Butterflies (*Taburburet*)	Migration to the east signifies rainfall is very near	1 month	4.3	3	Medium	Second cultivation and buy inputs
	Locust (*Chereng' endet*)	Appearance during early phenological stages of plant crops signify expected food shortage or food insecurity	–	0.9	4	High	Eaten as food
Amphibian	Frogs (*Mororojet*)	Croaking sound in rivers, streams, springs and wetlands signify near rainfall season. When frogs move to homesteads, signify time for harvest is near.	1 month	12.0	1	High	Start early harvest of early mature crops

(continued)

Annex 2 (continued)

Name	Local name	Observable indicator	Time lapse	% score	Rank	Reliability	Main activity
Birds	*Kapcheptalaminik*	Flying towards the north signifies near rainfall. When flight changes towards the south, signifies cessation	2–3 weeks	11.5	1	Medium to high	Sow seeds
	Chepkutkutiet	Sings and appearance along wetlands and streams signify near rainfall.	2–3 weeks	11.1	2	High	Second cultivation and buy inputs
	Kiptiltiliet	Sound and increase in population signify near rainfall	2–3 weeks	3.4	3	High	Second cultivation and buy inputs
	Terkekiat	Produces a sound that is interpreted to be telling farmers that it is time to start planting. This signifies rainfall is very near.	3–4 weeks	2.6	4	Medium to high	Second cultivation and buy inputs
Mammals	*Kipkamoiyet*	Sound and increase in population signifies near rainfall	2–3 weeks	1.7	1	High	–
	Kibworere	Sound and increase in population signifies near rainfall	2–3 weeks	0.9	2	High	Second cultivation and buy inputs
	Tabkosiet	Sound and increase in population signifies near rainfall	2–3 weeks	0.5	3	High	Second cultivation and buy inputs

Annex 3 Traditional and modern food crops and traditional storage equipment in Keiyo and Nandi

Name of crop	Type	Botanical name	Local name
Cereal crops			
Finger millet	Indigenous	*Eleusine coracana*	Kipsongik
Sorghum	–	*Sorghum bicolor*	Mosongik or kiplemiat
Maize	Modern	*Zea mays*	Bandiat, bandek
	Indigenous	*Zea mays*	Cheborosiat
	Indigenous	*Zea mays*	Kalambiliet (reddish-orange in colour)
	Indigenous	*Zea mays*	Chesamoiyot (*teta ne samoo*),
	Indigenous	*Zea mays*	Chelekeyiot (*Tap Rongoei*)
	Exotic	*Zea mays*	Number nane
Legumes			
Cowpeas	Modern	*Vigna unguiculata*	–
Beans	Modern	–	–
Peas	Modern	–	–
Green grams	Modern	–	–
Groundnuts	Indigenous	–	–
Vegetables			
Amaranth	Indigenous	*Amaranthus blitum*	–
Pig weed	Indigenous	*Amaranthus spp.*	–
Pumpkin leaves	Indigenous	*Cucurbita moschata*	Chebololet
Jute mallow	Indigenous	*Corchorus olitorius*	
Spider plant	Indigenous	*Cleome gynandra*	Isakiat
Black nightshade	Indigenous	*Solanum scabrum*	Isochot
Kale	Modern	–	Sukumawiki
Cabbage	Modern	–	–
Spinach	Modern	–	–
Fruits			
Molongonik	–	–	–
Muiyengiot	–	–	–
Siriat	–	–	–
Chobinek	–	–	–
Bekab tarit	–	–	–
Kimolwet	–	–	–
Roots and tubers			
Sweet potatoes	–	–	Rabwonik

(continued)

Annex 3 (continued)

Name of crop	Type	Botanical name	Local name
Cassava	–	–	–
Yams	–	–	–
Storage and preservation equipment			–
Granary	–	–	Choge
Pots	–	–	Terenik
Woven baskets for storing cereals	–	–	Indait
Woven baskets for storing milk	–	–	Chemalany busi
Calabash	–	–	Sotet/sotonik

Post-harvest and storage/preservation techniques for crop and animal products	
Crops:	*Animal products:*
Use of wild fruits, berries, roots of plants such as *tilingen* and *ng'oswet* (*Balanities aegyptiaca*) as a source of food during dry seasons. Some of these fruits are boiled for up to 12 h before being eaten.	Frying of animal fat, mixing with honey and its preservation for future use
Harvest was stored in two ways: *bekab lagok* (general granary) and *bekab boiyo/toek* (emergency food reserve) stored in traditional granary called *kitonget/indait* (woven basket from herbal reeds).	Milk preservation done using calabashes (*sotonik*)
Vegetable dried and ground and stored in calabashes	Indigenous tree species used for preparing (*kisute sotonik*) the calabash for storing milk (as preservatives, additives or flavour), such as *murkuiywet, tendwet, kesisitiet*
Honey production	Animals slaughtered and meat mixed with honey for preservation or sliced into long rolls (*sirigenik*) and smoked for future use
Morik (remains of *ugali*) stored in pots	Taboos—no taking of milk after eating meat

tolerate drought. In addition, a modern strategy that involves the cultivation of early-maturing crops such as cowpeas (*Vigna unguiculata)* and improved varieties of maize (*Zea mays)* is emerging. The production of millet and sorghum (traditional crops) in particular has also shifted from seed broadcasting to row cropping, an integration of modern cropping systems.

In Lower Keiyo, along Kerio River, small-scale irrigation farming is practised with the use of fertilisers. Similarly, modern crop varieties such as cowpeas, green grams and watermelon are intercropped with traditional varieties such as sorghum and groundnuts. In Nandi, indigenous crops such as millet, sweet potatoes (planted continuously throughout the year), an indigenous maize variety (*Chemborosiat*) and numerous wild fruits (*molongonik, muiyengiot, siriat, chobinek, bekab tarit, kimolwet*, etc.) are preferred. Most households supplement these food production systems by rearing goats for milk. This is usually used to feed the very young and old, especially in times of extreme drought events (indigenous). In addition, bee-keeping (indigenous and modern) and the sale of animals (cows and sheep) to buy food (modern) are also practised.

In Nandi and Keiyo districts, different traditional storage techniques are still used, both for harvested crops and animal products. Special pots (*terenik*) and woven baskets (*indait*), which were traditionally used to store shelled and ground cereals for use during times of HH food emergencies, has been abandoned by most HHs due to increased production that requires larger granaries. However, as observed in parts of Aldai and Tinderet, some households have modified the traditional granary using modern technology for it to accommodate larger quantities of millet and sorghum in particular. In keeping with tradition, most households still use calabashes (*sotonik*) to store sour milk (*mursik*), while fresh milk is kept in containers hanged on a traditionally woven rope (*chemalany busi*, which translates to "so that no cat can climb and drink the milk").

Among the Nandi, specialisation existed, with HHs categorised as either livestock keepers (*bikab tuga*) or farmers (*bikab bek, bikab rabuonik*, etc.). The two separate traditional practices have since been integrated and most HHs practised both dairy and crop farming. A highly regarded strategy, *kendoetab kesutik* (a disciplined way of consuming stored food) and which is still practised by some HHs is aimed at reducing misuse, overuse and wastage of food. In the past, social cohesion ensured control of consumption of what had been harvested. This has however been abandoned due to the increasing adoption of individualistic lifestyles and commercialisation of communal tasks. During seasons of poor harvest in Lower Keiyo, millet was stored in traditional granaries with the stalks to delay consumption and save for use in times of acute food shortage.

Reliability and Opportunities for Integrating IKS in CCA and Food Security

Observations over time indicate that changes in climate experienced since the late 1970s and 1980s are characterised by shifts in the seasonality, duration, intensity, amount and onset of rainfall, prolonged inter-seasonal dry spells, shifts in sowing dates of crops, varying intra-seasonal rainfall regimes, increased pest infestations and disease attacks of crops and livestock as well as increased temperatures. These observable climatic changes have affected food production by disrupting the timing of sowing, increasing inter and intra-seasonal dry spells, reducing seed germination rates at times, and additional economic costs associated with replanting.

Most respondents opined that indigenous practices were no longer reliable when used on their own, but can be enhanced if integrated with scientific techniques. This was demonstrated by existing practices already used by the communities in the two districts. Across the ACZs, households use integrated approaches that promote livelihood and crop diversification by planting indigenous drought-resistant crop varieties (groundnuts, cassava, traditional vegetables, millet, sorghum) alongside modern crop varieties such as maize, beans, cowpeas and bananas, using modern cropping techniques that increase yields.

In Aldai and Tinderet, for instance, farmers keep both indigenous cows (East African Zebu) and high-breed varieties in order to render dual benefits to the household by improving milk production, to cushion household assets and in some cases interbreed to increase resistance to diseases. In other areas of Keiyo and Nandi, traditional food crops are planted and/or intercropped with modern varieties using modern cropping systems and in some riparian areas, supplemented by irrigation.

However, for IKS to be reliable and fully integrate with modern scientific techniques of CCA and food security, the study emphasises three dynamic characteristics of IKS among the Nandi and Keiyo that should be carefully assessed. First is the replacement of prior ideas and values, where IKS is constantly being revised by outside influences, for instance with the introduction of markets and commercialisation of food crops, many traditional crops are no longer grown since monetary reasons prevail over subsistence needs. The second involves a reinterpretation of prior ideas, where indigenous use of slash-and-burn and oxen in land preparation is compared with, rather than replaced by the use of tractors. Finally, IKS is being modified with the fusion of modern ideas into the existing IKS.

Conclusions

Results from the study show that IKS plays a crucial role in rainfall prediction, food production, climate change adaptation and food security. Traditional indicators for determining rainfall onset are still used in food crop production among the Nandi and Keiyo communities. Based on the results of this study, it is recommended that the identified traditional indicators of rainfall onset forecasting and food security should be further monitored in the study area to provide more information on the performance and scientific interpretations.

There is also a need for policy interventions to integrate IKS and encourage the perpetuation of intra and intergenerational transfer of IKS practices related to food production, food security and CCA. The study reveals that Nandi and Keiyo communities use both traditional indicators and observable parameters (temperature, clouds, wind, humidity and lightning), which are also used for scientific weather forecasting in determining the onset of rainfall. It is therefore necessary for the communities to mobilise the available IKS and work closely with the Kenya Meteorological Department (KMD) to ensure the integration of the existing systems with the conventional weather forecasting techniques for easier dissemination. Likewise, traditional food crop varieties suited to local agro-ecological conditions should be identified and further research done to improve their tolerance to climate extremes. At the household level, options should be explored for livelihood diversification such as food crop farming, dairy production and bee-keeping.

Community food reserves should be established alongside government food strategic reserves in order to empower communities regarding the need to ensure food security at the local level and to improve opportunities for marketing crops. This should be accompanied, at the household level, by improved facilities for traditional food storage systems. Improved initiatives are required for environmental conservation in order to enhance strategies for climate protection at the farm level and to contribute towards the overall goal of reducing the impacts of climate change and enhancing adaptation options.

References

Adger WN (2006) Vulnerability. Glob Environ Chang 16:268–281

Adger WN, Huq S, Brown K, Conwaya D, Hulme M (2003) Adaptation to climate change in the developing world. Prog Dev Stud 3(3):179–195

Agrawal A (1995) Dismantling the divide between indigenous and scientific knowledge. Dev Chang 26:413–439

Anderson MD, Cook JT (1999) Community food security: practice in need of theory. Agric Hum Values 16:141–150

Briggs J (2005) The use of indigenous knowledge in development: problems and challenges. Prog Dev Stud 5:99–114

IPCC (2007) Climate change 2007: impacts, adaptation and vulnerability. In: Parry ML, Canziani OF, Palutikof JP, van der Linden PJ, Hanson CE (eds) Contribution of working group ii to the fourth assessment report of the intergovernmental panel on climate change, Annex I. Cambridge University Press, Cambridge

Fischer G, Shah M, van Velthuizen H (2002) Climate change and agricultural vulnerability. Int Inst Appl Syst Anal, Laxenberg

Food and Agriculture Organization (FAO) (2006) World agriculture: toward 2030/2050, Interim report (Food and Agriculture Organization, Rome, Italy

Kipkorir EC, Raes D, Massawe B (2002) Seasonal water production functions and yield response factors for maize and onion in Perkerra, Kenya. Agric Water Manage 56(3):229–240

Leary N, Conde C, Kulkarni J, Nyong A, Pulhin J (eds) (2008) Climate change and vulnerability. Earthscan, London

Mogaka H, Gichere S, Davis R, Hirji R (2006) Climate variability and water resources degradation in Kenya: improving water resources development and management. World Bank working paper No. 69. Washington, D.C.

Smit B, Pilifosova O, Burton I, Challenger B, Huq S, Klein RJT, Yohe G, Adger N, Downing T, Harvey E, Kane S, Parry M, Skinner M, Smith J, Wandel J (2001) Adaptation to climate change in the context of sustainable development and equity. Cited In: McCarthy JJ, Canziani O, Leary NA, Dokken DJ White KS (eds) Climate change 2001: impacts, adaptation and vulnerability. IPCC Working Group II. Cambridge University Press, Cambridge, pp 877–912

Stockholm environment Institute (SEI) (2009) The economics of climate change in Kenya, Stockholm environment institute, Oxford

Todisco F, Vergni L (2008) Climatic changes in central Italy and their potential effects on corn water consumption. Agric For Meteorol 148:1–11

UNEP (2008) Indigenous knowledge in disaster management in Africa. UNEP, Nairobi, Kenya, p 118

Warren DM (1991) Using indigenous knowledge for agricultural development. World Bank Discussion Paper 127. The World Bank, Washington, D.C.

Author Biographies

Charles Kipkorir Songok Charles Kipkorir Songok is a Junior Livelihoods Analyst at the Food Security and Nutrition Analysis Unit (FSNAU) Somalia and holds a Master of Philosophy degree in Environmental Information Systems from Moi University, Kenya (2009). His research focuses on GIS and remote sensing applications in land use/land cover mapping and watershed modelling; environmental risk, vulnerability and capacity assessment; climate change adaptation and disaster risk management. He is a member of the Ecological Society of East Africa (ESEA), Young Professional Involvement in Agricultural Research and Development (YPARD) and the Society for Conservation GIS (SCGIS) and has participated in research and consultancy projects in the fields of environmental management, climate change adaptation, food security risk and vulnerability assessment and watershed modelling. He has two scientific research project reports and five announcements at international conferences. Charles Kipkorir Songok is the corresponding author and can be contacted at: ngok4@yahoo.co.uk.

Emmanuel Chessum Kipkorir Emmanuel Chessum Kipkorir is a Senior Lecturer in the Department of Environmental Monitoring, Planning and Management at the School of Environmental Studies, Moi University, Kenya, with a Ph.D. in Applied Biological Sciences (2002) and an M.Sc. in Water Resources Engineering (1997) from KU-Leuven University (Belgium). His areas of expertise include water resource development, planning and management; water, health and environment; irrigation engineering; agro-meteorology, and climate change adaptation. He has been involved in a number of research and consultancy projects in the fields of water resource management and environmental monitoring, planning and management. He has already 14 scientific publications in international journals with reviewers, 15 announcements at international conferences and many articles published in journals and books.

Edward Musungu Mugalavai Edward Musungu Mugalavai is an Assistant Lecturer at the School of Environment and Earth Sciences, Maseno University, Kenya, with a Master of Philosophy degree in Environmental Earth Sciences from Moi University, (Kenya). He is a Ph.D. Student in Environmental Earth Sciences (Moi University, School of Environmental Studies). His areas of research interest are in agro-meteorology, hydrometeorology, oceanography, climatology, climate change assessment, water resources management and applied geography. He has been involved in major research projects in the fields of agro-climatology and climate change adaptation. In the above-mentioned fields, he has already four scientific publications in international journals with reviewers and eight announcements at international conferences.

Chapter 6
Climate Change and the Emergence of Helter-Skelter Livelihoods Among the Pastoralists of Samburu East District, Kenya

Eunice Ongoro Boruru, Edward Ontita, William Okelo Ogara and Nicholas Otienoh Oguge

Abstract Due to climate change, northern Kenya is experiencing intense droughts at shorter intervals with intermittent high precipitation and flooding. This has a major impact on livelihoods of the pastoral community, whose ways of life revolve around the availability of pasture and water for their livestock. Our study set out to determine the Samburu people's perception of climate change and how they cope with its effects. Such information is important in developing evidence-based policies on adaptation strategies for African drylands. Our study undertook a survey of knowledge, attitudes and life histories among the Samburu communities using semi-structured questionnaires and focus group discussions. Respondents indicated that climate change leads to soil erosion, local extinction of some faunal species, water and pasture scarcity, and intensification of conflicts over resources. This has negatively impacted on their

E. O. Boruru (✉) · E. Ontita
Department of Sociology and Social Work,
University of Nairobi, Nairobi, Kenya
e-mail: euniceboruru@yahoo.co.in

E. Ontita
e-mail: gizemba@yahoo.com

W. O. Ogara
Department of Public Health Pharmacology and Toxicology,
University of Nairobi, Nairobi, Kenya
e-mail: wogara@gmail.com

N. O. Oguge
Centre for Advanced Studies in Environmental Law and Policy (CASELAP),
University of Nairobi, Nairobi, Kenya
e-mail: ooguge@yahoo.com

W. Leal Filho (ed.), *Experiences of Climate Change Adaptation in Africa*,
Climate Change Management, DOI: 10.1007/978-3-642-22315-0_6,

livelihoods due to food shortage, human and livestock diseases, social and family instability, and general insecurity.

Keywords Climate change · Adaptation · African studies · Dry lands · Pastoralist economy · Natural resources

Introduction

Climate change has an adverse effect on livelihoods in the dry lands of northern Kenya in terms of longer and harsher droughts, shorter and more intense precipitation, and floods. The Intergovernmental Panel on Climate Change (IPCC) defines climate change as any change in climate over time, whether due to natural variability or as a result of human activity (IPCC 2007). The IPCC further argues that Africa is one of the most vulnerable continents to climate change and climate variability as a result of the interactions of multiple stresses, including land degradation and desertification, decreasing runoff from water catchments, high dependence on subsistence agriculture, HIV/AIDS prevalence, inadequate government mechanisms, rapid population growth occurring at various levels and low adaptive capacity due to factors such as extreme poverty, frequent natural disasters such as drought and floods, and dependence on rain-fed agriculture.

The likely impacts of climate change will add to these existing stresses and exacerbate the effects of land degradation; increased temperature levels, which are expected to cause diseases; additional loss of moisture from the soil; unpredictable and more intense rainfall, and higher frequency and severity of extreme climatic events such as floods and droughts. These factors are already leading to a loss of biological and economic productivity and putting population in dry lands at risk of short and long-term food insecurity. The Fourth IPCC African Assessment Report estimates that by 2020 between 75 and 250 million people are likely to be exposed to increased water stress and that rain-fed agricultural yields could be reduced by up to 50% in Africa if production practices remained unchanged.

Drought-prone areas are *inter alia* particularly likely to suffer complex localised impacts of climate variability or change. In the Sahel for instance, changes in temperature and rainfall patterns have reduced the length of the vegetative period and make it difficult to continue the cultivation of traditional varieties of long and short-cycle millets (Rosenzweig et al. 2007). Given the social, legislative, market and weather-based sources of vulnerability already prevailing in the region, reduction in agricultural productivity and land area suitable for agriculture due to even slight climate change will cause disproportionately large detrimental effects (IPCC 2007; Dietz and Geest 2004).

The most vulnerable communities to the impacts of climate change inhabit the dry land areas. For instance, the World Health Organization (WHO) states that changes in the patterns of the spread of infectious diseases are likely to be a major consequence of climate change in the dry lands. The WHO (2003) indicates that malaria represents a particular and an additional threat in Africa. There are between 300 and 500 million cases of malaria in the world each year, with a very high proportion of those occurring in Africa, largely among the poor. Malaria causes between 1.5 and 2.7 million deaths per year of which more than 90% are children under 5 years of age. In addition, malaria slows economic growth in Africa by up to 1.3% each year (UNEP 2005).

Climate change is almost certainly making an already bad situation worse and may already be contributing to the problem of poverty. In one highland area of Rwanda, malaria incidence increased by 337% in 1987. Some 80% of the increase could be explained by changes in rainfall and rising temperatures. Further, small changes in temperatures and precipitation could trigger malaria epidemics beyond current altitudinal and latitudinal disease limits. Global warming will increase the incidence of floods, warming and drought, all of which are factors in disease transmission. In South Africa, it is estimated that the area prone to malaria will double and that 7.2 million people will be at risk. Greater climatic variability will introduce the disease to areas previously free of malaria. Populations within these areas lack immunity, which will increase the impact of illness (Zhou et al. 2004). Pastoralist communities are the most vulnerable to climate change. This is because they tend to be located in geographically vulnerable areas, such as flood-prone Mozambique, drought-prone Sudan or cyclone-prone Bangladesh and in more vulnerable locations such as slums and informal settlements—in all cases heavily dependent on natural resources for their livelihoods (Jennings and McGrath 2009).

Vulnerability to climate change is not just a function of geography and dependence on natural resources; it also has social, economic and political dimensions, which influence how climate change affects different groups (Action Aid 2005). Pastoralist communities rarely have insurance to cover loss of property due to drought, storms or cyclones. They cannot pay for the healthcare required when climate change-induced outbreaks of malaria and other diseases occur. They have few alternative livelihood options when their only cow drowns in a flood or drought destroys their entire maize crop for the year and they do not have the political clout to ask why their country's early-warning system did not warn them of likely flooding. Climate change will also have psychological and cultural effects. For example, beliefs and traditions associated with the seasons are undermined by climate change (Jennings and McGrath 2009).

Therefore, climate change interlocks with people's life-worlds differently for different reasons. The geography of a people's location relative to other people may position them more acutely in harm's way when climate change ramifications roll out. The magnitude and impact of the resulting harm—however, small or big—depends on where the affected population is located in the politico-economic landscape of the overarching polity. The impact may also be further exacerbated by the cultural terrain when it occurs, deepening and broadening to the extent that the population is conscientised and prepared for the effects of climate change or to which they are left to conjecture their causes and effects on the basis of their local cultural repertoires. Understanding this can inform and make policy more relevant. This study is positioned within the aforementioned debate on climate change, dry lands and pastoralist livelihoods, by considering a case study of the Samburu pastoralist people of northern Kenya. Hence, this study poses and answers the question of what the Samburu people's perspectives of climate change are. It then identifies and explains the effects of climate change on Samburu livelihoods, and describes their coping and adaptation strategies.

Methods

The study was carried out in Samburu East district, northern Kenya. Save for the market centres, game reserve and other tourist installations, the district is predominantly occupied by Samburu pastoralists. The research project employed four methods: a household survey, focus group discussions, key informant interviews and life histories. Survey questionnaires were administered to 180 household heads in Waso and Wamba locations of Samburu East district.

Life histories were collected from old and resourceful people such as *kursa* (wise men) of the Samburu community. The aim was to locate junctures and processes of climate change in the respondents' life trajectories and explain them. Life histories were collected from four elderly people aged between 60 and 75 years and drawn from each of the four divisions in the district, namely Ngaruni, Soldoo, Matakwanii and Lengusaka.

Focus group discussions (FGDs) were conducted in each division in the district. Five FGDs were conducted with members as follows: twelve women aged between 30 and 50 years, twelve *ngolontoi* (girls) aged between 15 and 25 years, twelve *morans* (boys) aged between 15 and 25 years, twelve middle-aged men of between 30 and 45 years and twelve old men aged between 55 and 75 years. The discussions in the groups revolved around the perceptions of climate change; trends in weather patterns; challenges of climate change to their livelihoods and how each group was adapting and coping with the effects of climate change.

Key informant interviews were conducted with specific members of the community who were elderly and resourceful, regarded by the community as opinion leaders. These included chiefs, rainmakers, midwives, circumcisers, the manager of Westgate Conservancy, a retired Catholic priest and an indigenous veterinary worker. These individual interviews took an in-depth approach whereby the respondents freely discussed their perceptions of climate change, the challenges to their livelihoods and ways in which the community was adapting and coping with the challenges. A checklist of questions was used to guide and narrow the discussions' relevant questions in relation to the research question.

Data from the survey was analysed using the SPSS computer program and the resulting descriptive statistics, mainly frequencies, were tabulated and used in this report. Qualitative data from the life histories, focus group discussions and key informant interviews were transcribed and the key patterns, themes, trends and concepts were drawn out to illustrate and explain the various dimensions of the complexities of climate change and Samburu pastoralist livelihoods.

Results and Discussions

The Samburu of Samburu East districts are mainly pastoralists whose livelihoods rely on natural resources: primarily water and pasture for their livestock. These resources are sensitive to climate change. However, because of differences in wealth, power and natural resource endowments within Samburu society, their vulnerability to climate change and their capacity to adapt also vary over time and space. This section of the paper entails a discussion of the various climate change-induced livelihood dislocations and how the Samburu pastoralists confront and live with and in spite of them.

Drought and the Pastoralist

Drought is a major effect of climate change that impacts people severely. The Samburu people reported that drought was increasingly prolonged, frequent and severe. The effects of drought are felt by livestock, wildlife, vegetation and human beings. The pasture area dwindles, water sources run dry, most seasonal rivers dry up and the livestock are hard hit. In Samburu East district, the sources of water are mainly springs, seasonal rivers and streams, and boreholes. When severe droughts occur, including notable droughts in 1984, 1995, 2005, 2006 and 2009, the water sources dry up leaving the people in crisis. The boreholes in the district are few and located only in Wamba town. There is no borehole in the outskirts and the adjoining rural areas. Community members usually scoop sand from riverbeds to create shallow wells in order to access water (Fig. 1). The water is always inadequate. The few water pans available are shared by livestock, wild animals and people. The water is contaminated due to multi-sharing and its consumption can cause water-borne diseases such as diarrhoea and cholera. Overall, women trek long distances and spend valuable time fetching water both for domestic use and livestock. Girls, too, spend a good part of the day fetching water from long distances, hindering their personal development through school attendance and play.

During these times, the livestock species were not resilient enough, so cattle, sheep and donkeys died in large numbers. This had a direct impact on the people's livelihood because livestock is their main source of food and the mainstay of their economy. With their livestock dead, local purchasing power was wiped out and with that the local economy was in shambles. Samburu culture prohibits livestock sales except in extreme circumstances. Key informants reported that the Samburu

Fig. 1 A girl fetching water from a scooped shallow well on a dry riverbed

Table 1 Pastoralists whose livestock herds have shrunk over the past ten years

Herd sizes decreasing	Frequency	Percent
Yes	174	97
No	5	3
Total	179	100

generally lost livestock through deaths during droughts or cattle raids by neighbouring groups such as the Pokot and Borana. Both of these ways of losing livestock result from the aftermath of climate change. Therefore, as shown in Table 1 the majority of the Samburu pastoralists (97%) reported that their herd sizes had declined over the past ten years.

While it is evident that the decline in herd sizes was the result of climate change, this change may have affected the Samburu even when it occurred beyond their borders in southern Ethiopia and greater northern Kenya. This is because cattle raiders from the two regions were reported to be more vicious after prolonged droughts in their areas. The aim of the vicious cattle raids was to restock after losing livestock to the harsh droughts. It is for this reason that an agitated elderly respondent quipped that the colonial government was 'more relevant to the Samburu livelihoods than the independence government, because the former did not permit cattle raids across ethnic boundaries, while the latter has no control and is unable to stop foreign raiders from Ethiopia'.

Conflicts Across Ethnic and International Borders, and with Wildlife

The aforementioned international dimensions of climate change imply that while the triggers of change such as greenhouse gas emissions may be taking place on other continents such as Europe and the Americas and at a rather non-disruptive pace, the ramifications far away in northern Kenya and southern Ethiopia are disruptive, bloody and far-reaching. In northern Kenya in particular, there is an arms race pitting ethnic groups against each other and thus pushing the implications of climate change beyond livelihoods and economics to political stability across international borders. The arms race is always in preparation for conflict over water and pasture between the Samburu and their neighbours, particularly the Pokot and Turkana of Kenya and the Boran of northern Kenya and southern Ethiopia. These conflicts happen during the dry spells when contests over water and pasture spread beyond Samburu-land across ethnic and international borders. As demonstrated in Table 2 the main causes of conflicts with other ethnic groups include contests over pasture and water (83%) and cattle rustling attempts for purposes of restocking after droughts or after losing livestock to raiders (17%).

Water and pastures constitute a particularly important source of inter-ethnic conflict because these natural resources are limited and disputes arise over sharing, each community demanding priority over the others or exclusive rights over watering points. The conflicts get so intense that they lead to outright gun battles lasting days and claim casualties in human lives and lost livestock. Climate change policy will thus not only focus on slowing down climate change through measures such as growing more trees to soak up greenhouse gases, but also on stabilising livelihoods in marginal lands, mopping up small arms in the hands of herders and restocking across borders contemporaneously.

During the drought period, the wildlife move closer to human settlements for water from the earth dams and in the process they attack livestock. Hyenas were reported to attack goats and camels at night, baboons also turn into carnivores during the drought, period killing goats because they cannot get any fruit to eat. Wildlife such as elephants and baboons has a tendency to become violent in the face of water scarcity, resulting in destruction of property and human lives. Grevy zebras contaminate the watering points with their urine and livestock do not drink. In most cases, elephants destroy watering points, therefore lowering the quality of water both for livestock and domestic use. When the wild animals share water points with livestock, they infect them. This wildlife/livestock interface creates an environment for disease transmission. Some of the diseases reported were: anthrax (Lokoshum) from anthrax-positive cases, Brucellosis and Tuberculosis and Trypanosomiasis from milk; (Lodwa) from buffalo, heart water (poroto) from primates. Unfortunately, there is no mechanism to compensate Samburu pastoralists for the livestock lost to such disease outbreaks, which heightens hostility towards wildlife.

Table .2 Causes of conflicts between the Samburu and neighbouring ethnic groups

Causes of conflict	Frequency	Percent
Water and pastures	122	83
Cattle rustling	25	17
Total	147	100

The Aftermath of Floods

Samburu district experiences both short and long rains. The short rains are usually expected between March and April while the long rains usually occur between October and January. However, due to climate change, rainfall seasons have been shifting. The rains are unpredictable and erratic: when they come, it is for a short period but with so much intensity that they cause flooding. This is the time that the seasonal rivers flood and burst their banks, giving way to flooded landscapes. The flooding has a negative impact on people because it disrupts transport in the area, drowns people and animals, and causes extensive erosion (Fig. 2). Some homesteads that are built on the lowland get submerged. It generally causes a lot of damage, leaving the community vulnerable. During the floods, a lot of vector diseases emerge as a result of stagnant water that creates a breeding environment for the vectors. Human diseases such as malaria, cholera and dysentery (due to water contamination), and animal diseases such as East Coast fever, foot rot, and foot and mouth disease are common during the floods.

These diseases also become a burden to the community because the district lacks adequate capacity to effectively and efficiently respond to disease outbreaks due to poor infrastructure and there is also no compensation given to the people for their destroyed property. Overall, the floods and the accompanying soil erosion, as well as human and livestock diseases complicate post-disaster recovery because cropping and pasture recovery are hampered by the poor and compacted soils that result. These cycles of intense floods and prolonged droughts have led to the disappearance of certain species of fauna, especially fruit trees that some respondents said were readily available in the area when they were young in the 1940s. This has complicated livelihoods further as there are no berries and wild fruits collected with which to supplement livestock food products especially during droughts. Food shortages are thus routinely characteristic of the Samburu people during droughts and after the floods.

Family and Community Instability

The findings from the focus group discussions, particularly from women and girls, indicated that families were undergoing difficult times especially in searching for a sustainable livelihood. Droughts have killed many livestock and most families have fallen into extreme poverty and food insecurity. Poverty is also a reality

Fig. 2 Deep gully erosion, resulting from seasonal flash floods

because people die due to hunger and child mortality rates were reported as high because of the change in diet from animal protein to cereals and carbohydrates that relief agencies occasionally provide. There is not enough to eat, sharing and festivities are no longer practised in households because of low purchasing power, and the individualistic tendencies which have emerged in the Samburu society as people struggle as individuals for survival. This has led to a lack of social harmony and a general feeling of social insecurity. Community life has been shattered as sharing is limited and rituals are few and far between. With the sense of community disappearing, vulnerability increasingly becomes an individual burden and, with it, social disorganisation. Climate change policies should thus address issues of social harmony and community bonding in order to create viable communities that are strong enough to take collective actions to not only improve their environments and livelihoods but also provide a secure foundation for socio-economic change and growth.

Facing up to Climate Change: Adapting and Coping

As a people, the Samburu of Wamba live in constant contest with nature, with droughts, floods, wildlife and diseases on the one hand and aggressive neighbours training their eyes on their livestock on the other. Given this very complex and

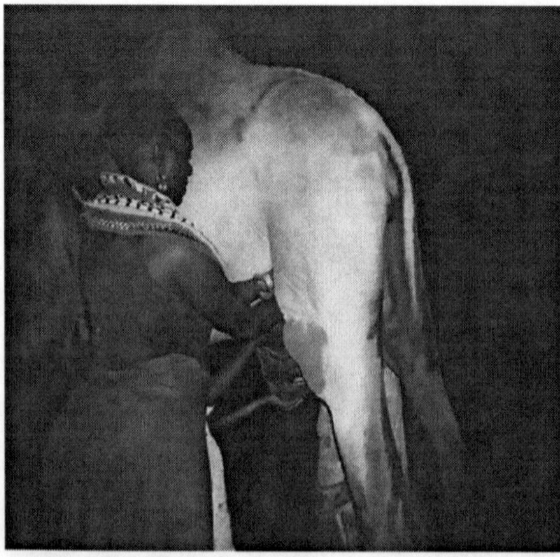

Fig. 3 A Samburu woman obtaining milk from a camel, a resilient animal able to withstand harsh and severe drought with outstanding milk production in harsh environmental conditions

ever-changing scenario, Samburu livelihoods are continually a work in progress. They try out many different ways of survival, sometimes sequentially and at other times haphazardly, in order to make ends meet. In the face of climate change, the Samburu are not oscillating their livelihoods with seasons ordered around well-known patterns of water and pasture availability—they walk several paths at the same time, all in the name of climate change-induced coping and adaptation mechanisms. There are a variety of such mechanisms.

Tapping into the Resiliency of the Camel

The Samburu have not always reared camels. Throughout their history, their main livestock have been goats and cattle. They have resorted to camel rearing in order to cope with persistent droughts, which their cattle and goats could not survive. They 'imported' the camels mainly from the Somali in the neighboring north-eastern province to save their livelihoods from the worrying effects of climate change. The study findings show that they use the camel in a multifunctional manner with the following general production aims: milk production (Fig. 3), meat production, provision of blood, transport, hides and skins. In the face of climate change, the Samburu have used the camel as an adaptation measure to sustain their livelihood in times of drought. The camel is by nature a resilient animal that is able to withstand the harsh and severe drought more easily. The camel has an outstanding milk production in harsh environmental conditions in which they are kept, producing more in the same circumstances compared to cattle and other small stocks. Its lactation persists well in the dry seasons and rarely ceases even

Table 3 Decline in importance of cropping in Samburu livelihoods in the past ten years

Importance of cropping declining	Frequency	Percent
Yes	14	87.5
No	2	12.5
Total	16	100.0

during extended dry spells. The Samburu also prefer camel milk to milk from other livestock because of its taste, nutritious value and for health reasons, arguing that camel milk prevents them from thirst even when walking for a long distance. However, they do not use camel meat as often as beef and other livestock. To them, the camel is a sure way of sustaining a livelihood in the difficult circumstances caused by drought.

Camels are also a source of blood that is mixed with milk to form a diet component that is highly nutrititious and which is reserved for children, expectant mothers and ailing older people. Culturally, the Samburu use camels as important gifts or loans to relatives and friends especially in times of need, such as during drought periods.

Turning to Agro-Pastoral Practices

The Samburu have responded to the precariousness of their livestock economy by cultivating small patches of land near the Manyatta for farming maize, beans and vegetables. But this adaptation mechanism is not sustainable because of greater unpredictability of the timing of rain seasons and the patterns of rain within seasons, making it more difficult to decide when to cultivate and plant. As shown in Table 3 the importance of crop production in people's livelihoods has been decreasing for 87.5% of the respondents involved in crop production over the past ten years invariably as climate change has intensified. This is in spite of the fact that only a minority (9%) of the 180 respondents interviewed reported any involvement in crop production.

The people also lack knowledge of the right seeds to plant, for example, early-maturing crop seed for the short rains. They also lack the resources to take advantage of the right time for planting and to maintain crops and animals through dry spells. The problem of heat stress, lack of water at crucial times and pests turns this into a precarious affair (Fig. 4). In any event, soil erosion has left behind poor and compact soils that make tilling and subsequent crop root development very difficult. Therefore, climate change constitutes a conspiracy of factors to undermine the livestock and crop economies of the Samburu people. Herein lies the challenge and starting point of devising effective policies to combat the impact of floods through soil and water conservation. Such a strategy might enhance short-rain crop production and livestock sustenance if water is conserved in large dams, for instance.

Fig. 4 Vegetable farming on a small patch of land near Manyatta

Livestock Mobility and Micro-Business Ventures

The Samburu have a culture and philosophy of movement. They believe that when it is dry in one place, it must be wet at another and so they keep moving, especially to areas where there are watering points and pastures. This is not working well for them because when drought comes the impact is felt all through the rangelands and resources dwindle. The conflicts with their neighbouring communities have also hindered these movements. Even in the face of these conflicts, the Samburu are not without arms and they March on other lands and communities to access pasture and water on the Pokot Hills and on the Boran wet valleys, often at the expense of huge losses of human lives and livestock. Their livelihoods are in this sense a series of trail, error, triumphs and failures here and there.

In addition to the movements with their livestock, they have introduced community groups that engage in small-scale businesses; these groups include women's groups and youth groups. They engage in income-generating activities such as bead-making, poultry-keeping and brick-making. Their products are sold to make income, which they use as collateral to get small loans from local banks for personal development.

Concluding Remarks

The Samburu have confronted climate change with all the ingenuity that they could summon. Climate change impacts include drought, floods and hence livestock deaths, conflicts with neighbours and wildlife, food shortages, and family and community instabilities. The resulting livelihood avenues are multiple, ranging from adopting the resilient camel, taking up agro-pastoral activities, nomadic practices and some small business activities. These multiple avenues to making a living conjure up an image of Helter-Skelter livelihoods.

The future of Samburu livelihoods in the face of climate change calls for concerted efforts by the government and other actors to enhance soil and water conservation efforts. Damming floodwater will slow soil erosion, create new microclimates and opportunities for small-scale irrigation activities as well as provide water for livestock during drought seasons. If linked with better livestock marketing strategies for the Samburu herders, such strategies will help them offload their livestock to the markets at good prices before they waste away during long drought periods. Policy efforts will thus bring order to Helter-Skelter livelihoods.

Acknowledgments The authors are grateful to the Earth watch Institute for funding the fieldwork. We also acknowledge the contributions of the Samburu community. We are thankful to Mr Nduhiu Gitahi and Alfred Mainga of the department of Public Health Pharmacology and Toxicology, University of Nairobi, for their contributions to this research.

References

Action Aid (2005) Module I: framework and approach; action aid adaptation toolkit: integrating adaptation to climate change into secure livelihoods; Action Aid

Dietz KT, Geest VD (2004) A survey on risk and vulnerability in drylands with a focus on the sahel. In: The impacts of climate change in drylands with a focus on West Africa

IPCC (2007) Climate change impacts, adaptation and vulnerability, contribution of working group ii to the forth assessment report of the intergovernmental panel on climate change. Cambridge University Press, UK

Jennings S and McGrath (2009) What happened to the seasons. Oxfam GB Research, UK

Rosenzweig C, Casassa G, Karoly DJ, Imeson A, Liu C, Menzel A, Rawlins S, Root TL, Seguin B, Tryjanowski P (2007) Assessment of observed changes and responses in natural and managed systems. Climate change 2007: impacts, adaptation and vulnerability. Contribution of working group ii to the fourth assessment report of the intergovernmental panel on climate change Cambridge university press, Cambridge, UK, pp 79–131

UNEP (2005) African environment outlook past, present and future perspectives. Impacts of the state of the environment. UNEP, Nairobi

WHO (2003) Climate change and human health–risks and responses; summary. WHO, Geneva

Zhou GN, Minakawa N, Githeko AK, Yan G (2004) Association between climate variability and malaria epidemic in east african highlands. Proc Natl Acad Sci U S A 101:2375–2380

Author Biographies

Eunice Ongoro Boruru is a Master of Arts student at the Department of Sociology and Social Work, University of Nairobi. Her research interests are climate change and dry-land livelihoods and she has completed an MA thesis entitled: 'Climate Change and Pastoralist Livelihoods Shifts in Northern Kenya: The Case of Samburu East District, Kenya'.

Dr. Edward Ontita is a lecturer at the Department of Sociology and Social Work where he teaches Social Policy, Rural Sociology and Program Monitoring and Evaluation. He specialises in rural resources and livelihoods and has conducted extensive research on climate change and rural livelihoods in the arid and semi-arid lands in Kenya, Uganda and Tanzania.

Dr. William Okelo Ogara is a lecturer in the Department of Public Health Pharmacology and Toxicology. He holds a PhD in Veterinary Public Health and has worked extensively in wildlife conservation and ecosystem health, knowledge management and technology transfer. He has supervised a number of students for Master's and PhD degrees, and has headed a number of professional bodies and organisations.

About the Authors Nicholas Otienoh Oguge is an associate professor of environmental policy at the Center for Advanced Studies Environmental Law and Policy (CASELAP), University of Nairobi, and has experience in teaching spanning over a period of 25 years and is currently the director of Earth watch Institute in Kenya. He has worked with scientists from universities and research institutions to develop environmental research projects to address various aspects of conservation. Over the years, he has been engaged by key conservation organisations to develop technical papers towards policy, dialogue and advocacy on protected areas management.

Chapter 7
Carbon Changes Following the Establishment of Exclosure on Communal Grazing Lands in the Semi-Arid Lowlands of Tigray, Ethiopia

Wolde Mekuria, Edzo Veldkamp, Marife D. Corre and Mitiku Haile

Abstract Most dryland ecosystems are degraded and are currently far from saturated with carbon. This indicates their potential to sequester a significant amount of carbon through sequestration measures. In this study, we: (1) investigated changes in ecosystem carbon stocks following establishment of exclosure on communal grazing lands, and (2) identified easily measurable biophysical and management-related variables which can be used to predict ecosystem carbon stocks restoration in the semi-arid lowlands of northern Ethiopia. We measured above-ground and soil carbon stocks, determined soil characteristics and collected climatic data and management information from replicated ($n = 3$) exclosures of 5, 10 and 15 years old which were paired with grazing lands. All exclosures displayed higher ecosystem carbon stocks than the adjacent grazing lands. Differences in ecosystem carbon stocks between exclosures and grazing lands varied between 18 ($\pm$2.8) and 40 ($\pm$6.3) Mg C ha^{-1} and increased with exclosure age. In exclosures and in communal grazing lands, much of the variability in ecosystem carbon stocks was explained by a combination of precipitation, silt and clay content, woody biomass, vegetation canopy cover and exclosure age.

W. Mekuria (✉) · M. Haile
Department of Land Resource Management and Environmental Protection, Mekelle University, Tigray, Ethiopia
e-mail: wolde_mekuria@yahoo.com

M. Haile
e-mail: gualmitiku@yahoo.com

W. Mekuria · E. Veldkamp · M. D. Corre
Buesgen Institute, Soil Science of Tropical and Subtropical Ecosystems, Faculty of Forest Sciences and Forest Ecology, University of Göttingen, Göttingen, Germany
e-mail: eveldka@gwdg.de

M. D. Corre
e-mail: mcorre@gwdg.de

W. Leal Filho (ed.), *Experiences of Climate Change Adaptation in Africa*, Climate Change Management, DOI: 10.1007/978-3-642-22315-0_7,

Our results help to establish baseline information for carbon sequestration projects such as the establishment of exclosures on communal grazing lands, and to predict the expected ecosystem carbon storage under exclosures.

Keywords Above-ground carbon · Ethiopia · Semi-arid lowlands · Soil carbon · Tigray

Introduction

Drylands cover about 40% of the Earth's land surface, and are home to more than two billion people (WRI 2002). Because of severe land degradation, dryland ecosystems contribute between 0.23 and 0.29 Gt of carbon a year to the atmosphere, which is about 4% of global emissions from all sources combined (Lal 2001). Since carbon losses from drylands are associated with loss of vegetation cover and soil erosion, management interventions that reverse these processes can simultaneously achieve carbon sequestration (Trumper et al. 2008). The estimation of carbon stock in each region of the world shows that dryland carbon storage accounts for more than one third of the global stock (Campbell et al. 2008). However, in Africa, a very high proportion of carbon (59%) is in drylands, so any sequestration measures there would need to address dryland ecosystems.

Arid or semi-arid lowlands below 1,500 m in altitude occupy about half of the total land area of Ethiopia. About 10% of the population of the country lives in these lowland arid or semi-arid areas (Tewoldeberhan 2006). Livestock rearing is an important occupation in the Ethiopian lowlands. However, owing to overgrazing, the natural vegetation in these areas has virtually disappeared, providing only few scattered grasslands with irregularly spaced trees and shrubs and vast areas of bare lands devoid of vegetation (Ayana and Fekadu 2003; Abdu and Schultz 2005; Solomon et al. 2007a).

Similar to other lowland parts of Ethiopia, the lowlands of Tigray are occupied by a farming pastoral population, which employs a communal resource system for livestock production and grazing exclosures. These exclosures are relatively new trends that are established on communal grazing lands with the goal of restoring degraded soils and vegetation. Within exclosures, grazing and other agricultural activities are not allowed. As the exclosures are not fenced, guards are hired by the local administration on a food-for-work basis (Yayneshet et al. 2009). According to local agricultural offices, exclosures now cover 6–21% of the total area of the districts covered in our study.

Exclosures have been established on communal grazing lands in the semi-arid lowlands of Tigray for about two decades. Since their establishment, the local communities are concerned that exclosures cause a shortage of grazing area (personal communication with the local people during fieldwork). This drawback of exclosures is important in the semi-arid lowlands of Tigray as most of the local people depend on livestock for their livelihood. This in turn calls for a careful evaluation of the benefits and costs of establishing exclosures on communal grazing lands. One aspect that needs to be addressed is the change in ecosystem carbon stocks following the establishment of exclosures on degraded communal grazing lands. Such information is necessary for: (1) evaluation of whether rehabilitation projects such as exclosure establishment should be expanded, and (2) incorporation of the economic and social consequences of such projects in policymaking. Although there are some studies that have shown the ecological and economic importance of exclosures (Tilahn et al. 2007; Bedru et al. 2009; Mekuria et al. 2010), all of these studies are located in the northern highlands of Ethiopia. The local land managers working in the lowlands of Ethiopia cannot solely depend on the outcomes of these studies to design and implement sustainable degraded land restoration projects as there are differences between the lowlands and highlands in biophysical and socio-economic settings.

We conducted the present study to: (1) investigate changes in ecosystem carbon stocks following establishment of exclosures on communal grazing lands in the semi-arid lowlands of Tigray, Ethiopia, and (2) identify which easily measurable biophysical and management-related factors can be used to predict ecosystem carbon stock in exclosures and in adjacent communal grazing lands. The easily measurable variables that we used in this study were variables related to climate (e.g. precipitation), soil (e.g. texture), and vegetation (e.g. vegetation cover, woody biomass) that had the potential to be obtained through remote sensing or soil survey. We formulated the following hypotheses: (1) soil and above-ground carbon increase after establishment of exclosures on communal grazing lands, and (2) soil and above-ground carbon in exclosures and in adjacent grazing lands, and the difference between exclosure and adjacent grazing land in soil and above-ground carbon stocks can be predicted based on soil, vegetation and climatic parameters. The observed differences in soil and above-ground carbon stocks following the establishment of exclosures on communal grazing lands were used as indicators of exclosures' effectiveness in restoring degraded soils and vegetation.

Materials and Methods

Study Area

Our study was conducted in three districts of Tigray (12°–15° N latitude and 36° 30′–40° 30′ E longitude), the northernmost region of Ethiopia (Fig. 1). All sites have a tropical semi-arid climate. Mean annual rainfall (for the years 2000–2006) varied between 488.2 and 644.5 mm/year^{-1} with an average of 562 mm/year^{-1}. The mean minimum temperature ranged from 11 to 17°C and mean maximum temperature ranged from 26 to 34°C (Ethiopian Meteorological Service Agency 2007). The altitude of the study sites varied between 1,400 and 1,600 m.a.s.l. The rainy season usually occurs between June and September, which

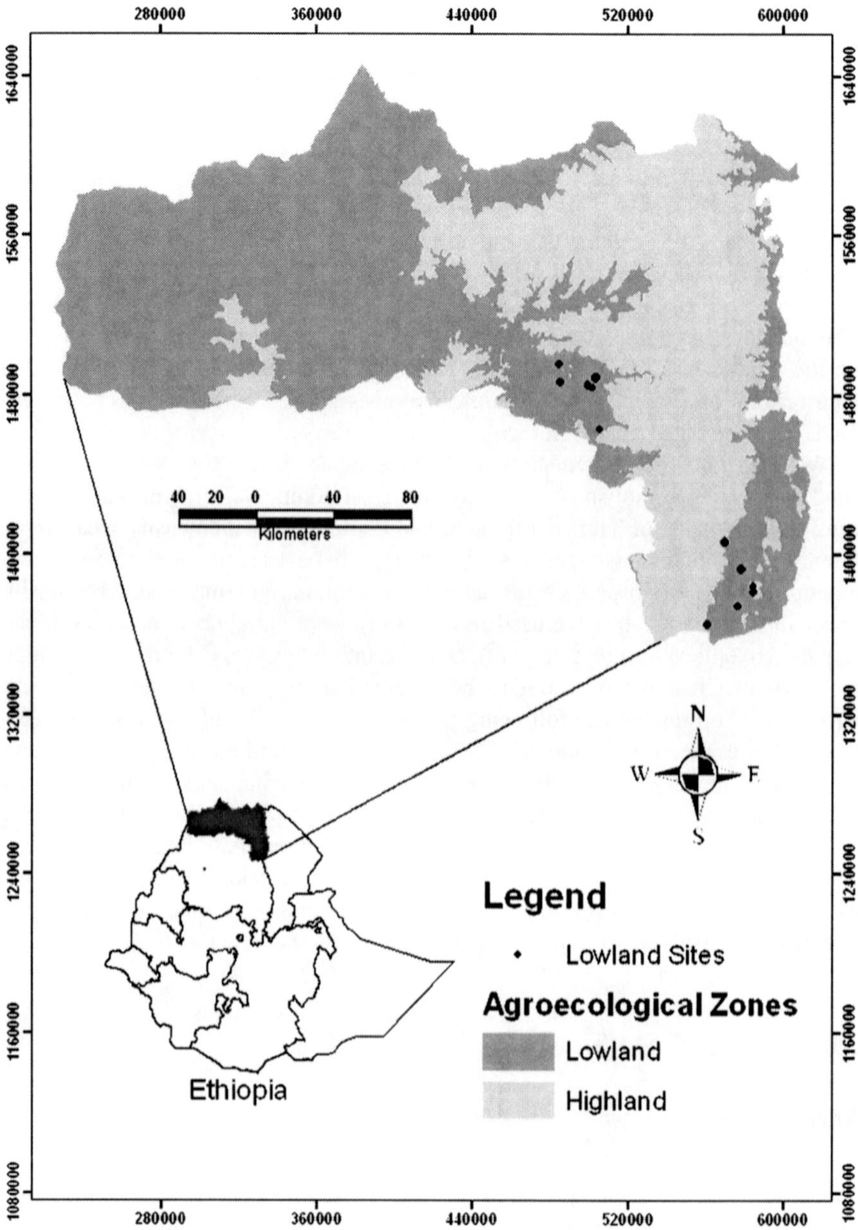

Fig. 1 Location of the study area in the lowlands of Tigray, northern Ethiopia, with the distribution of specific sites indicated with an asterisk (*)

is also the growing season (varying between 60 and 90 days). According to the district's Agricultural and Rural Development Office, average farm size ranges from 0.75 to 2.5 ha per household.

Soils in the study sites were classified into three major groups: Leptosol, Regosols, and Cambisols (WRB 2006). The most common woody vegetation species included *Acacia seyal, Acacia melifera, Acacia etbaica* and *Balanites aegyptiaca*. Diverse assemblages of grasses and herbs dominated the under-storey vegetation.

Major land uses in the study sites included cultivated lands (between 24 and 62% of the area), forest lands (0–14%), exclosures (6–21%), grazing lands (2–18%) and others (1–30%). Mixed crop-livestock farming is the backbone of households' livelihood in all of the study sites. Major cultivated crops include *Zea mays* (maize), *Eragrostis teff* (teff), *Sorghum bicolour* (sorghum) and *Arachis hypogea* (peanut).

In the study area, the first exclosures were established two decades ago and accordingly we selected replicated exclosures of 5, 10 and 15 years, which are paired with adjacent communal grazing lands. The area of the selected exclosures ranged from 7.1 to 85 ha, whereas that of the adjacent communal grazing lands ranged from 2.8 to 38.3 ha. In exclosures, grazing is not allowed, but grass harvesting (using a cut-and-carry system) is permitted once a year, typically after the seeding stage, starting three to five years after exclosure establishment. The main reason to restrict grass harvesting is to restore the soil seed bank. In the adjacent communal grazing lands, unlimited access for free grazing is practiced. In both exclosures and communal grazing lands, soil and water conservation structures such as hillside terraces, stone bunds and micro-basins have been constructed. However, the soil and water conservation structures are in better condition in exclosures (i.e. not damaged by livestock or humans) because livestock and human activities are restricted.

In addition to exclosures and the adjacent grazing lands, we selected three isolated forest fragments that remain around churches, also called "church forests". The Ethiopian Orthodox Tewahido Church has a long history of planting, protecting and preserving trees. In northern Ethiopia, the only areas where patches of forests can be found are areas surrounding churches. These church forests have survived as a result of the traditional conservation system and protective patronage of the Ethiopian Orthodox Tewahido Churches (Wassie et al. 2009). In the church forests studied, activities such as livestock grazing, collection of firewood and construction materials were allowed depending on the needs of the local community. The area of the church forests in our study varied between 5.9 and 15.3 ha.

Experimental Design

We used space-for-time substitution to monitor changes in ecosystem carbon stocks after conversion of communal grazing lands to exclosures with ages of 5,

10 and 15 years. The implicit assumption of this approach is that each paired grazing land and exclosure age should have comparable initial conditions such that changes in carbon stocks are a consequence of the change in land use (i.e. establishing exclosures). Thus, we selected sites with adjacent communal grazing land and exclosures, ensuring that soil and terrain conditions were as similar as possible between each pair. To cover the variability in soil and topography, we selected three replicates for each paired grazing land and exclosure age throughout the lowlands of Tigray (Fig. 1). We also selected replicated church forests ($n = 3$) to detect whether exclosures had reached the ecosystem carbon level of these forests. In each exclosure, grazing land and church forest, we randomly established two to four transects spaced at a minimum distance of 75 m. The number of transects per site was based on vegetation density, spatial heterogeneity of vegetation and area of the site (Tefera et al. 2005). Transects were parallel to each other and to the topography of the landscape. In each transect, we delineated three landscape positions (upper slope, mid slope and foot slope), and in each landscape position we established a sampling plot of 20 m × 20 m. In each 20 m × 20 m plot, five 2 m × 2 m subplots were set up (one at the centre and two along each of the two diagonal lines that cross at the centre of the plot). Landscape position was included in our sampling design in order to characterise the effects of topography-related processes on ecosystem carbon stocks. The upper slope position receives little or no overland flow but contributes runoff to downslope areas. The mid slope position receives overland flow from the upper slope and contributes runoff to the foot slope. The foot slope receives overland flow and deposition from the upper and mid slope positions (Mekuria et al. 2009).

Soil Sampling, Analyses and Carbon Stock Calculation

In each 2 m × 2 m subplot, soil samples from 0 to 0.2 m depth were collected at five random sampling points. One soil core sample was also taken from each plot for bulk density determination. The samples collected from each landscape position per replicate (i.e. 50–100 random sampling points from two to four transects) were mixed thoroughly in a large bucket to form one composite soil sample. In the entire study, we collected a total number of 63 composite soil samples [i.e. (2 paired exclosure and grazing land × 3 age classes × 3 landscape positions × 3 replications) + (church forest × 3 landscape positions × 3 replications) = 63]. Soil samples were air-dried, sieved through a 2 mm sieve and ground prior to analysis. Soil organic carbon was analysed using the Walkley–Black method (Walkley and Black 1934), bulk density was measured using the core method (Blake and Hartge 1986), and particle size was determined using the hydrometer method (Gee and Bauder 1982).

Soil organic carbon stock at the 0–0.2 m depth was calculated as follows:

SOC (Mg C ha^{-1}) = $(\%C \div 100) \times Bd\ (Mg\ m^{-3}) \times depth\ (m) \times 10000\ m^2\ ha^{-1}$, where *Bd* is the bulk density. The average bulk density of the oldest exclosure was used to calculate the soil organic carbon stock in all exclosures, grazing lands and church forests. We used this conservative approach to avoid overestimation of the soil organic carbon stock due to changes in bulk density (Veldkamp 1994).

Vegetation Inventory, Above-Ground Biomass and Carbon Estimation

In each 20 m × 20 m plot, we measured the following vegetation parameters: diameter at breast height (DBH), or for smaller and multi-stemmed shrubs, diameter at stump height or at the height of 30 cm (d_{30}) from the ground, crown diameter, total height and species identity. DBH and d_{30} were measured using calipers. Crown diameter was measured using a measuring tape. Total height was measured using either a clinometer or a measuring tape, depending on tree height. In the entire study, we examined 156 sample plots of which 75 were in exclosures, 57 in adjacent communal grazing lands and 24 in church forests.

To estimate above-ground biomass, we identified dominant woody species using our vegetation inventory data. The dominant woody species were determined based on their relative importance value (i.e. the sum of relative basal area, relative frequency and relative density) in each exclosure, adjacent grazing land and church forest. This approach ensures that species that are few in number but productive are not excluded. Species comprising 80–92% of the woody species composition were selected. The sum of the relative abundance of the dominant species ranged from 76–95% (average 89%) in exclosures, 70–100% (average 88%) on communal grazing lands and 87–94% (average 90%) in church forests. The number of woody species selected for biomass estimation varied between two and ten in exclosures, between one and eight on grazing lands, and between three and ten in church forests.

We used the methods of Hoff et al. (2002) and Snowdon et al. (2002) for above-ground woody biomass measurement. The dominant woody species were grouped into three diameter classes in order to minimise errors that can arise from variable sizes of individuals. In exclosures and grazing lands, selected individuals representing the dominant species were harvested and weighed. We harvested a total of 270 trees and shrubs (162 from exclosures and 108 from communal grazing lands). In the church forests, the partial harvest method was employed because tree felling is strictly forbidden; biomass was collected from 24 individuals. For estimation of under-canopy biomass, all vegetation (mainly grasses and herbs) in one subplot (located at the centre of the main plot) were harvested and weighed as the variability in grass cover among the subplots was less. Fresh mass of above-ground vegetation was adjusted to dry mass using the moisture content measured, determined by oven-drying sub-samples of stems, branches and leaves at 65°C until constant mass was attained (about 78 h).

From the selected individuals used for biomass measurement, organic matter and carbon content were estimated separately for the stem, branches and leaves based on the method shown in Armecin and Gabon (2008):

$$OM\ (\%) = \frac{DM - AM}{DM} \times 100, \text{ and}$$

$$OC\ (\%) = \frac{\%\,OM}{1.724}$$

where: OM = organic matter, DM = dry mass of the sample at 65°C, AM = ash mass of the sample after combustion in a furnace at 550°C for 8 h, OC = organic carbon, and 1.724 = van Bemmelen factor (i.e. OM contains 58% of OC; Armecin and Gabon 2008). The estimated OC of the selected woody species varied between 0.48 and 0.57. This estimated organic carbon value was subsequently used to calculate carbon stock of woody species. Carbon stock in under-canopy species was estimated by multiplying the oven-dried biomass by a factor of 0.5 (Snowdon et al. 2002).

Statistical Analyses

We first conducted tests for normality (Kolmogorov–Smirnov D statistic) and equality of variance (Levene statistic) of soil organic carbon and above-ground carbon stocks. We tested the differences between landscape positions in ecosystem carbon stocks for each exclosure age and communal grazing lands using one-way analysis of variance (ANOVA). The differences in carbon stocks between an exclosure age and adjacent communal grazing land were assessed using a Kolmogorov-Simirnov two-sample test (independent by groups). The patterns of changes in carbon stocks (i.e. difference between an exclosure age and the adjacent grazing lands) across exclosure ages (5, 10 and 15 years) were assessed using one-way ANOVA with Kruskal–Wallis test at $p < 0.05$. Spearman correlation tests were conducted to examine the relationships between carbon stocks and independent variables using the values of the three landscape position for each land use type ($n = 27$). Finally, step-wise multiple regression analyses were done to establish predictive relationships between carbon stocks and easily measurable independent variables.

Results

Ecosystem Carbon Stocks in Exclosures and Adjacent Communal Grazing Lands

We did not detect differences among landscape positions ($p > 0.05$) in soil and above-ground carbon stocks within any of the exclosure ages and adjacent grazing lands, indicating that landscape position was not the main driver of changes in ecosystem carbon stock. At each exclosure age, exclosures showed higher soil carbon concentration, soil carbon stocks, and above-ground carbon stocks than the adjacent grazing lands (Tables 1, 2). The difference between exclosures and adjacent grazing lands in soil carbon stock ranged from 13.6 to 34.1 Mg C ha^{-1} whereas above-ground carbon stock varied between 3.4 and 6.6 Mg C ha^{-1}. This means soil carbon stocks increased by 36–50% and above-ground carbon stocks increased by 44–49% following conversion of degraded grazing lands to exclosures. Soil carbon stock contributed most of the increase in ecosystem carbon stock (ranging from 75 to 91%) compared to above-ground carbon in exclosures. The influence of exclosure age on ecosystem carbon stocks was not linear with time as the increase in ecosystem carbon stocks was only significant between 5 and 15 year exclosures (Fig. 2).

Soil, Vegetation and Climatic Variables Explaining Ecosystem Carbon Stocks

Soil and above-ground carbon stocks in exclosures and their differences from adjacent grazing lands were positively correlated with woody biomass, vegetation canopy cover, silt content, clay content and exclosure age, and inversely correlated with bulk density and sand content (Tables 3, 4). In grazing lands, soil carbon stocks were positively correlated with woody biomass and silt content and inversely correlated with sand content. Above-ground carbon stocks in the adjacent grazing lands showed positive correlations with vegetation canopy cover and silt and clay content and showed negative correlation only with sand content.

Using multiple linear regression models, we were able to explain a considerable part of the variation in ecosystem carbon stored in exclosures and grazing lands (Table 5). Soil carbon stock in exclosures and their differences from grazing lands were predicted by woody biomass (ranging from 2 to 67 Mg ha^{-1}), precipitation (ranging from 488 to 644 mm/year^{-1}), silt content (ranging from 8 to 38%) and exclosure age (ranging from 5 to 15 years), whereas woody biomass carbon stocks were predicted by silt content, clay content (ranging from 6 to 35%) and exclosure age. In grazing lands, soil and woody biomass carbon stocks were predicted by woody biomass (ranging from 0.2 to 55 Mg ha^{-1}), vegetation canopy cover (ranging from 3 to 70%), silt content (ranging from 12 to 38%) and precipitation (ranging from 488 to 644 mm/year^{-1}).

Ecosystem Carbon Stocks and Site Characteristics in Church Forests

The environmental (e.g. precipitation) and topographic (e.g. slope) conditions of the church forests were within the same range as those of exclosures and the

Table 1 Mean values ($n = 3$, ± SE) of site characteristics, soil carbon concentration and soil carbon stocks in 0–0.2 m depth in paired exclosures and adjacent communal grazing lands

Exclosure	Exclosures								Adjacent communal grazing land								Difference[c]	
Age (year)[a]	Stone cover(%)	pH (1:5 H_2O)	Bd[b] (g/cm^{-3})	Sand (%)	Silt (%)	Clay (%)	C (%)	Soil C (Mg C ha^{-1})	Stone cover (%)	pH (1:5 H_2O)	Bd (g/cm^{-3})	Sand (%)	Silt (%)	Clay (%)	C (%)	Soil C (Mg C ha^{-1})	C (%)	Soil C(Mg C ha^{-1})
5	24	5.8	1.4	63.7	20.9	15.4	1.4	37.0	48	6.0	1.5	69.7	17.3	13.0	0.9	23.4	0.51**	13.6**
	(4.4)	(0.1)	(0.0)	(1.9)	(1.9)	(2.1)	(0.1)	(2.8)	(5.4)	(0.1)	(0.0)	(2.4)	(1.5)	(2.4)	(0.1)	(1.5)	(0.1)	(2.8)
10	41	6.4	1.3	57.4	30.0	12.6	2.6	67.4	53	6.4	1.5	57.4	28.4	14.1	1.3	34.7	1.24*	32.7*
	(7.8)	(0.2)	(0.1)	(3.8)	(2.1)	(1.9)	(0.4)	(10.6)	(7.7)	(0.2)	(0.1)	(3.3)	(2.2)	(1.6)	(0.1)	(3.7)	(0.26)	(6.9)
15	60	6.2	1.3	52.3	22.0	25.7	2.6	67.9	71	6.2	1.6	54.3	23.1	22.6	1.3	33.9	1.30*	34.1*
	(5.8)	(0.2)	(0.0)	(3.6)	(3.6)	(1.7)	(0.4)	(10.6)	(4.1)	(0.2)	(0.0)	(3.4)	(3.1)	(1.4)	(0.2)	(5.8)	(0.22)	(5.8)

[a] Time since the conversion of grazing lands to exclosures

[b] Bd bulk density

[c] Differences in carbon concentrations and stocks (calculated as: exclosure—adjacent grazing lands) were significant at $*p < 0.05$; $**p < 0.01$ (Kolmogorov–Smirnov two-sample test (group))

Table 2 Mean values ($n = 3, \pm$ SE) of vegetation characteristics and above-ground carbon stock in paired exclosures and adjacent communal grazing lands

Exclosure age (year)	Exclosures				Adjacent grazing lands				Differences[b]		
	Canopy cover[a](%)	Under-canopy cover (%)	Above-ground biomass		Canopy cover[a] (%)	Under-canopy cover (%)	Above-ground biomass		Above-ground biomass(Mg C ha^{-1})	Canopy cover[a](%)	Under-canopy cover (%)
			Mg ha^{-1}	Mg C ha^{-1}			Mg ha^{-1}	Mg C ha^{-1}			
5	38.3	60.9	9.9	5.2	12.8	6.3	1.4	0.7	4.5^{**}	25.5^{**}	54.5^{**}
	(5.1)	(4.7)	(1.2)	(0.6)	(3.2)	(0.7)	(0.4)	(0.2)	(0.6)	(7.3)	(4.7)
10	45.0	55.9	13.0	6.9	36.3	28.3	7.1	3.5	3.4^{*}	8.8	27.6^{*}
	(6.8)	(6.1)	(2.9)	(1.5)	(5.5)	(4.4)	(2.5)	(1.3)	(0.9)	(6.1)	(5.5)
15	74.2	43.9	29.0	14.8	25.5	7.0	16.3	8.2	6.6^{*}	48.7^{**}	36.8^{**}
	(5.4)	(6.8)	(8.4)	(4.1)	(7.6)	(3.6)	(7.3)	(3.7)	(1.5)	(9.0)	(8.2)

[a] Vegetation canopy cover was determined using crown diameter measurements and by assuming a tree or shrub of elliptical shape (Snowdon et al. 2002)
[b] Differences in above-ground carbon stock, vegetation canopy and under-canopy cover (calculated as: exclosure—adjacent grazing land) were significant at $*p < 0.05$; $**p < 0.01$ (Kolmogorov–Smirnov two-sample test (group))

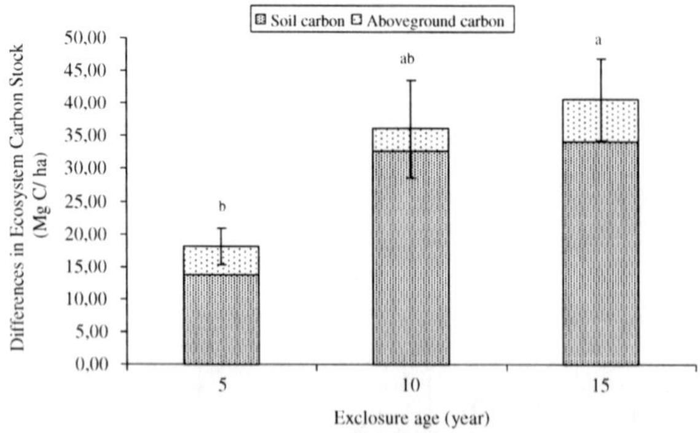

Fig. 2 Differences in ecosystem carbon stocks between exclosures and the adjacent communal grazing lands in relation to exclosure age. Means ($n = 3$; SE bars) followed by different letters indicate significant differences among categories of exclosure age (one-way ANOVA with Kruskal–Wallis test at $p < 0.05$)

Table 3 Spearman correlation coefficients ($n = 27$) between soil carbon and site and vegetation characteristics within exclosures and communal grazing lands

Site and vegetation characteristics	Exclosures soil C		Grazing lands soil C		Difference[a]	
	% C	Mg C ha^{-1}	% C	Mg C ha^{-1}	% C	Mg C ha^{-1}
Woody biomass (Mg ha^{-1})	0.64*	0.64*	0.46*	0.48*	0.58*	0.58*
Vegetation canopy cover (%)	0.53*	0.53*	0.36	0.37	0.54*	0.53*
Sand (%)	–0.52*	–0.51*	–0.41*	–0.43*	–0.48*	–0.48*
Silt (%)	0.42*	0.43*	0.59*	0.58*	0.33	0.33
Clay (%)	0.28	0.26	0.06	0.09	0.33	0.33
Bulk density (g/cm^{-3})	–0.49*	–0.49*	0.37	0.33	–0.53*	–0.54*
Precipitation (mm/year^{-1})	0.26	0.25	0.13	0.13	0.27	0.28
Exclosure age (year)	0.41*	0.43*	–	–	0.46*	0.46*

Significant at * $p < 0.05$

[a] Differences are calculated as exclosure minus adjacent grazing land and are correlated with site and vegetation characteristics in exclosures

adjacent grazing lands. Soils in the church forests had an average of 59.9% (±4.9%) sand, 26.0% (±4.4%) silt and 14.1% (±2.3%) clay content and an average bulk density of 1.34 (±0.06) g/cm^{-3}. The vegetation canopy and under-canopy cover of the church forest were 65.3% (±7.4%) and 34.2% (±9.9%), respectively. Soil carbon concentration, soil carbon stock in the 0–0.2 m depth and above-ground carbon stock of church forests were 1.7% (±0.1%), 44.8 (±2.2) Mg C ha^{-1} and 20.7 (±3.6) Mg C ha^{-1}, respectively. Although all of the exclosures had lower above-ground carbon stock than the church forests, the 10 and 15 year-old exclosures (Table 1) exceed the soil organic carbon levels of church forests.

Table 4 Spearman correlation coefficients ($n = 27$) between above-ground carbon stock and site characteristics within exclosures and communal grazing lands

Site characteristics	Exclosures	Grazing lands[a]	Differences[b]	
	Woody biomass (Mg C ha^{-1})	Grass biomass (Mg C ha^{-1})	Woody biomass (Mg C ha^{-1})	Woody biomass (Mg C ha^{-1})
Sand (%)	–0.43*	–0.37	–0.66*	–0.08
Silt (%)	0.22	0.61*	0.61*	0.05
Clay (%)	0.45*	0.08	0.38*	0.24
Vegetation canopy cover (%)	0.88*	–0.22	0.90*	0.55*
Bulk density (g/cm^{-3})	–0.36	0.30	0.32	–0.12
Precipitation (mm/year^{-1})	0.14	0.06	0.06	0.04
Exclosure age (year)	0.43*	–0.09	-	0.15

Significant at $^*p < 0.05$

[a] In grazing lands, correlation analyses were not conducted for grass biomass C because these areas did not have harvestable grasses

[b] Differences are calculated as exclosure minus adjacent grazing land and are correlated with site characteristics in exclosures

Discussion

Comparability of Exclosures and the Adjacent Grazing Lands

We tested our assumption (i.e. that the exclosures and adjacent communal grazing lands had similar conditions before exclosure establishment) using variables measured in the paired exclosures and grazing lands that are less dependent of land use (e.g. soil texture and slope). We observed no differences in soil texture and slope between any of the paired exclosures and adjacent grazing lands ($p > 0.05$). Significant differences were observed only for site characteristics that are dependent on land use. We conclude from these analyses that in general the paired sites were comparable and differences in soil and above-ground carbon stocks measured between the paired exclosures and adjacent grazing lands were caused by land use change (i.e. establishment of exclosures) and not by inherent site variability.

Variation of Soil and Above-Ground Carbon Stocks on Communal Grazing Lands

The variation in vegetation degradation among the adjacent grazing lands could be explained by the economic value of tree species that dominate the studied sites. In the communal grazing lands where economically important (e.g. *Boswellia papyrifera*) species dominate; the vegetation is less degraded and results in a high

Table 5 Multiple regression models for the prediction of soil carbon stock in the 0–0.2 m depth and woody biomass carbon in exclosures, communal grazing lands and their differences

Land use group	Model	p-value	R^2
	Soil carbon stock in the 0–0.2 *m depth*		
Exclosure	Mg C ha^{-1} = −68 + 0.8 × *woody biomass* (*Mg* ha^{-1}) + 0.1 × *precipitation* (*mm* yr^{-1}) + 0.9 × *silt* (%) + 1.5 × *exclosure duration* (*yr*)	0.00	0.59
Grazing land	Mg C ha^{-1} = −24.8 + 0.4 × *woody biomass* (*Mg* ha^{-1}) + 0.8 × *silt* (%) + 0.06 × *precipitation* (*mm* yr^{-1})	0.00	0.76
	Woody biomass carbon		
Exclosure	Mg C ha^{-1} = −17.2 + 0.57 × *silt* (%) + 0.63 × *exclosure duration* (*yr*) + 0.27 × *clay* (%)	0.00	0.44
Grazing land	Mg C ha^{-1} = −7.8 + 0.16 × *vegetation canopy* cover (%) + 0.35 × *silt* (%)	0.00	0.58
	Differences between paired exclosures and grazing lands in soil and above-ground carbon stock		
Soil carbon	(Mg C ha^{-1}) = −52 + 0.5 × *woody biomass* (*Mg* ha^{-1}) + 0.1 × *precipitation* (*mm* $year^{-1}$) + 1.2 × *exclosure duration* (*yr*)	0.00	0.39
Woody biomass	(Mg C ha^{-1}) = −0.95 + 0.09 × *vegetation canopy* cover (%)	0.00	0.36

We used adjusted R^2 values for predictive models with more than two independent variables

above-ground biomass. However, the soil carbon did not show the same trend as the communal grazing lands where free grazing took place, which results less under-canopy vegetation cover (Table 2) and consequently less biomass carbon input to the soil from easily degradable grasses and herbaceous species. Similar results were reported from studies conducted on communally owned grazing lands found in the semi-arid lowlands of Ethiopia (Solomon et al. 2007b; Angassa and Oba 2010).

The positive correlations of soil carbon with woody biomass and vegetation canopy cover (Table 3) suggest that reducing vegetation degradation in the communal grazing lands is key to restoring soil carbon. In addition to the increase in organic carbon input to the soil, restoring vegetation in communal grazing lands is critical to reducing soil erosion and subsequently the loss of sediment-associated soil carbon (Girmay et al. 2009). The negative correlation of soil and woody biomass carbon with sand content as well as the positive correlations of soil and woody biomass carbon with silt and clay content (Tables 3, 4) suggest the influence of soil texture on soil and woody biomass carbon storage. The fractions of coarse and fine-textured soil particles could affect soil and above-ground carbon stocks through their effect on soil moisture and nutrient retention.

We were able to explain a considerable part of the observed variation in soil and vegetation carbon stocks in communal grazing lands using multiple regression models (Table 5). Our field-based regression models open up possibilities to predict soil and vegetation carbon stocks in communal grazing lands before exclosures are implemented. Such baseline information will be critical if the positive effect of exclosures on ecosystem carbon stocks is to be evaluated. Furthermore, all the variables used in the prediction of soil and vegetation carbon stocks can easily be obtained from remote sensing or soil surveys. This increases the applicability of our field-based models for other restoration projects in the region.

Changes in Ecosystem Carbon Stock in Relation to Exclosure Age

The higher soil and above-ground carbon stocks in all exclosures indicate that exclosures are one of the viable options for restoring degraded soils and vegetation. Similar trends were reported from case studies conducted in the semi-arid lowlands of Ethiopia (Abebe et al. 2006; Angassa and Oba 2010). The substantial increase in ecosystem carbon sequestration from 5 to 15 year exclosures also signified that exclosure age strongly influenced carbon restoration. The relatively large increase in soil carbon stock in the first five years compared to above-ground carbon stock (Fig. 2) may have resulted from the large carbon inputs derived from grasses and herbaceous biomass. The large grass and herbaceous species' biomass is the outcome of a management practice that restricts grass harvesting for three to five years after exclosure establishment. The relatively slow increase in ecosystem

Fig. 3 Example of a church forest in the semi-arid lowland of Tigray region, northern Ethiopia (photo by Wolde Mekuria; June 2008)

carbon storage from 10 to 15 years of exclosure establishment was probably a manifestation of vegetation succession and the partial removal of organic matter input through the harvesting of grasses and herbaceous biomass. Other studies conducted in the southern lowlands of Ethiopia (Angassa and Oba 2010) and in South African savannah areas (Smit et al. 1999) have also shown that the age of exclosures decreases herbaceous and grass species' biomass and diversity with the increase in trees and shrubs. The 10 and 15 year-old exclosures reached the soil organic carbon level of church forests, suggesting that at least a decade is required to increase the soil organic carbon in the communal grazing lands to the level of the soil organic carbon in church forests through the establishment of exclosures. This pattern, however, does not mean that there is no more potential to sequester carbon after about 10 or 15 years. The relatively low soil organic carbon stocks in church forests compared to the 10 and 15 year-old exclosures could possibly be associated with the effects of livestock grazing in church forests (Fig. 3), which result in low under-canopy vegetation cover and organic matter inputs to the soil.

The positive correlations of soil carbon in exclosures with woody biomass and exclosure age (Table 3) indicate the influence of organic matter input in restoring the soil carbon stock, which increases with time. Other studies conducted in similar ecosystems also reported increasing soil organic carbon along with increasing above-ground biomass and vegetation diversity (Abule et al. 2007; Fantaw et al. 2007). The positive correlation of soil carbon with vegetation canopy cover could possibly be related to the effect of vegetation cover in reducing soil erosion and the loss of sediment-associated soil carbon (Valentin et al. 2005; Descheemaeker et al. 2006; Girmay et al. 2009; Mekuria et al. 2009). The negative correlation of soil carbon with bulk density could be an indication that land degradation in general, and soil erosion in particular, has a negative effect on soil carbon storage. The positive correlations of soil and above-ground carbon with silt and clay content suggest the influence of soil texture on restoring soil carbon through their effects on soil moisture and nutrient conservation as well as on vegetation growth. In the Ethiopian semi-arid lowland systems, soil water availability is critical to increasing woody, herbaceous and grass biomasses, which consequently have a direct impact on soil carbon storage (Abdu and Schultz 2005; Abule et al. 2005). The positive correlations of above-ground carbon with vegetation canopy cover and exclosure age is related to the productivity of the restored vegetations.

The high R^2 that we obtained with our multiple regression models using soil, climate, vegetation and management-related predictors (Table 5) illustrates that it is possible to predict changes in ecosystem carbon stocks after establishment of exclosures. These field-based models open up the possibility to evaluate changes in soil and above-ground carbon stocks after the establishment of exclosures. Such information is important for local land managers to plan and allocate places for future exclosures and to incorporate the economic value of exclosures in their land use and resource management-related decisions.

Potential of Exclosures to be Considered as a Clean Development Mechanism Project

The large surface area of drylands gives dryland carbon sequestration a global significance. Furthermore, the fact that many of the dryland soils have been degraded means that they are currently far from saturated with carbon and their potential to sequester carbon could be very high (Farage et al. 2007). In Tigray, the northernmost region of Ethiopia, about 53% of the land (2.7 million ha) is lowland (Kola), i.e. below 1,500 m.a.s.l. (Beyene et al. 2006) and about 20% of the 2.7 million ha of land is severely degraded so cannot be fully used for other agricultural production (TAMPA 2009). Part of this could be converted into exclosures.

In the present study, we showed that it is possible to make quantitative predictions of ecosystem carbon increments should exclosures be implemented on degraded communal grazing lands. Such predictions are critical in order to analyse whether exclosures are also an economically feasible alternative. In the case of the semi-arid lowlands of Tigray region, northern Ethiopia, we could make the following prediction: taking an average increment of 40.7 Mg C ha^{-1} in exclosures compared to the adjacent grazing lands in 15 years' time, the establishment of exclosures could increase ecosystem carbon stocks by 2.7 Mg C ha^{-1} a year. Using this annual increase in ecosystem carbon stocks, conversion of only 10% (50,000 ha) of the 0.5 million ha of degraded communal grazing lands in the region could sequester approximately 135,000 Mg C per year. The establishment of exclosures on degraded communal grazing lands thus has the potential to be considered in programmes such as Clean Development Mechanism (CDM) projects, which aim to generate financial compensation for increasing ecosystem carbon stocks and to enhance sustainable development in developing countries. With the present study, we cannot analyse whether such a project would be an economically feasible option. That would depend on opportunity costs and the social consequences of such projects. However, the information generated in this study would be critical for such an analysis.

Conclusions

We conclude from this study that exclosures are one of the viable options to restore degraded soils and vegetation in the semi-arid lowlands of Tigray region, northern Ethiopia. Our study showed that ecosystem carbon stocks in exclosures, in adjacent communal grazing lands and the change in ecosystem carbon stocks following the establishment of exclosures on communal grazing lands can be predicted using easily measurable biophysical and management-related indicators. The information generated by this study can be used by local land managers and policymakers to evaluate the benefits and costs of future expansion of exclosures before the establishment of exclosures, and may also be helpful in predicting the potential changes in ecosystem carbon stock following the establishment of exclosures. Although our study showed that exclosures are effective at restoring ecosystem carbon stocks, expansion of exclosures would increase grazing pressure on the remaining communal grazing area. Therefore, the decision to expand the establishment of exclosures should also include the economic and social impacts of future exclosures.

References

Abdu A, Schultz RC (2005) Water harvesting in a 'runoff-catchment' agroforestry system in the drylands of Ethiopia. Agroforestry Forum 63:291–298

Abebe Y, Assefa T, Lalisa A, Alemayehu N (2006) Soil carbon sequestration following bush encroachment in Afar and Borana rangelands, Ethiopia: implications for climate change and land management. Deserts and desertification challenges and opportunities 2006, Book of abstract of international conference in Sede Boqer Campus, Israel, p 97

Abule E, Snyman HA, Smit GN (2005) The influence of woody plants and livestock grazing on grass species composition, yield and soil nutrients in the Middle Awash valley of Ethiopia. J Arid Environ 60:343–358

Abule E, Snyman HA, Smit GN (2007) Rangeland evaluation in the middle awash valley of Ethiopia: III relationships among soil and vegetation variables. J Arid Environ 70:293–303

Angassa A, Oba G (2010) Effects of grazing pressure, age of enclosures and seasonality on bush cover dynamics and vegetation composition in southern Ethiopia. J Arid Environ 74:111–120

Armecin RB, Gabon FM (2008) Biomass, organic carbon and mineral matter contents of abaca (*Musa textilis* Nee) at different stages of growth. Ind Crops Prod 28:340–345

Ayana A, Fekadu B (2003) Current range condition in southern Ethiopia in relation to traditional management strategies: the perceptions of Borana pastoralists. Trop Grasslands 37:53–59

Bedru B, Muys B, Fredu N, Tollens E, Nyssen J, Deckers J, Mathijs E (2009) The economic contribution of forest resource use to rural livelihoods in Tigray, Northern Ethiopia. For Policy Econom 11:109–117

Beyene A, Gibbon D, Haile M (2006) Heterogeneity in land resources and diversity in farming practices in Tigray, Ethiopia. Agric Syst 88:61–74

Blake GR, Hartge KH (1986) Bulk density. In: Klute A (ed) Methods of soil analysis: part 1, physical and mineralogical methods, ASA monograph, vol 9. Soil science society of america, Inc, Madison, pp 363–375

Campbell A, Miles L, Lysenko I, Hughes A, Gibbs H (2008) Carbon storage in protected areas: technical report. UNEP World Conservation Monitoring Centre

Descheemaeker K, Nyssen J, Rossi J, Poesen J, Haile M, Moeyersons J, Deckers J (2006) Sediment deposition and pedogenesis in exclosures in the Tigray Highlands, Ethiopia. Geoderma 132:291–314

Ethiopian Meteorological Service Agency (2007) Annual climate bulletin for the year 2007. Ministry of Water Resources, Addis Ababa, Ethiopia, pp 15

Fantaw Y, Ledin S, Abdelkadir A (2007) Changes in soil organic carbon and total nitrogen contents in three adjacent land use types in the Bale Mountains, south-eastern highlands of Ethiopia. For Ecol Manag 242:337–342

Farage P, Ardö J, Olsson L, Rienzi E, Ball A, Pretty J (2007) The potential for soil carbon sequestration in three tropical dryland farming systems of Africa and Latin America: a modelling approach. Soil Tillage Res 94(2):457–472

Gee GW, Bauder JW (1982) Particle size analysis. In: Klute A (ed) Methods of soil analysis: part, physical and mineralogical methods, ASA Monograph, vol 9. Soil science society of america, Inc, Madison, pp 383–411

Girmay G, Singh BR, Nyssen J, Borrosen T (2009) Runoff and sediment-associated nutrient losses under different land uses in Tigray, Northern Ethiopia. J Hydrol 376:70–80

Hoff C, Rambal S, Joffre R (2002) Simulating carbon and water flows growth in a Mediterranean evergreen Quercus ilex coppice using the FOREST-BGC model. For Ecol Manag 164:121–136

Lal R (2001) Potential of desertification control to sequester carbon and mitigate the greenhouse effect. Clim Chang 51:35–72

Mekuria W, Veldkamp E, Haile M, Kindeya G, Muys B, Nyssen J (2009) Effectiveness of exclosures to control soil erosion and local community perception on soil erosion in Tigray, Ethiopia. Afr J Agric Res 4:365–377

Mekuria W, Veldkamp E, Tilahun M, Olschewski R (2010) Economic valuation of land restoration: the case of exclosures established on communal grazing lands in Tigray, Ethiopia. Land Degrad Dev 21:1–11

Smit GN, Richter CGF, Aucamp AJ (1999) Bush encroachment: an approach to understanding and managing the problem. In: Tainton NM (ed) Veld management in South Africa. University of Natal Press, Pietermaritzburg, South Africa, pp 246–260

Snowdon P, Raison J, Keith H, Ritson P, Grierson P, Adams M, Montagu K, Huiquan BB, Eamus D (2002) Protocol for sampling tree and stand biomass. Technical report, Australian Greenhouse Office, No. 31

Solomon TB, Snyman HA, Smit GN (2007a) Cattle-rangeland management practices and perceptions of pastoralists towards rangeland degradation in the Borana zone of southern Ethiopia. J Environ Manag 82:481–494

Solomon TB, Snyman HA, Smit GN (2007b) Rangeland dynamics in southern Ethiopia: (1) botanical composition of grasses and soil characteristics in relation to land-use and distance from water in semi-arid Borana rangelands. J Environ Manag 85:429–442

TAMPA (Tigray Agricultural Marketing Promotion Agency) (2009) Jatropha–a bio-fuel plant: an opportunity for Tigray Regional State, Bulletin, No. 68, 1–15 March 2009

Tefera M, Demel T, Håkan H, Yonas Y (2005) The role of enclosures in the recovery of woody vegetation in degraded dryland hillsides of central and northern Ethiopia. J Arid Environ 60:259–287

Tewoldeberhan GE (2006) The role of forest rehabilitation for poverty alleviation in drylands. J Drylands 1(1):3–7

Tilahun M, Olschewski R, Klein C, Kindeya G (2007) Economic analysis of closing degraded Bosewellia papyrifera dry forest from human intervention: a study from Tigray, Northern Ethiopia. For Policy Econom 9:996–1005

Trumper K, Ravilious C, Dickson B (2008) Carbon in drylands: desertification, climate change and carbon finance. UNEP-UNDP-UNCCD technical note for discussion at CRIC 7, Istanbul, Turkey

Valentin C, Poesen J, Li Y (2005) Gully erosion: impacts, factors and control. Catena 63:132–153

Veldkamp E (1994) Organic carbon turnover in three tropical soils under pasture after deforestation. Soil Sci Soc Am J 58:175–180

Walkley A, Black IA (1934) An examination of the Degtjareff method for determining organic carbon in soils: effect of variations in digestion conditions and of inorganic soil constituents. Soil Sci 63:251–263

Wassie A, Sterck FJ, Teketay D, Bongers F (2009) Effects of livestock exclusion on tree regeneration in church forests of Ethiopia. For Ecol Manag 257:765–772

WRB. World reference base for soil resources (2006) 2nd edn. World Soil Resources Reports, No. 103, FAO, Rome

WRI. World Resources Institute (2002) Drylands, people, and ecosystem goods and services: a web-based geospatial analysis. http://www.wri.org. (Accessed on 19 Feb. 2010)

Yayneshet T, Eik LO, Moe SR (2009) The effects of exclosures in restoring degraded semi-arid vegetation in communal grazing lands in northern Ethiopia. J Arid Environ 73:542–549

Author Biographies

Wolde Mekuria is a lecturer in the Department of Land Resource Management and Environmental Protection at Mekelle University, Tigray, Ethiopia. He is a PhD candidate at Buesgen Institute, Soil Science of Tropical and Subtropical Ecosystems, University of Göttingen, Germany. His research focuses on evaluating the recently introduced degraded land restoration effort in northern Ethiopia in relation to ecosystem carbon sequestration, soil nutrient and vegetation restoration. He has participated in a number of projects in the field of restoration ecology and land resources management. He has four publications in international peer-reviewed journals and seven publications at international conferences.

Edzo Veldkamp is a professor and section head of the Soil Science of Tropical and Subtropical Ecosystems, Buesgen Institute, Faculty of Forest Sciences and Forest Ecology, University of Göttingen, Germany. His research focuses on understanding how natural and managed (forest and agricultural) ecosystems are affected by 'global change' processes such as: changes in temperature and rainfall, increased N deposition, and land use changes. His research projects focus on the effects of human influences on carbon and nitrogen cycles and on interactions between water and matter fluxes. In his research projects, he applies a wide range of methods: stable isotopes as tracer (^{13}C) and in pool dilution approaches (^{15}N), radioactive isotopes at natural levels (^{14}C and ^{222}Rn), chambers techniques with gas chromatographs and chemiluminescence techniques for field measurements of trace gas fluxes, measurements of soil nutrient leaching, and physical, chemical and biological indicators of C and N cycling in combination with modelling techniques. He has 60 publications in international peer-reviewed journals.

Marife D. Corre is a research scientist and a junior research group leader hosted by the Soil Science of Tropical and Subtropical Ecosystems, Buesgen Institute, Faculty of Forest Sciences and Forest Ecology, University of Göttingen, Germany. Her research focuses on: (1) impact of global change processes (e.g. increased nitrogen deposition, land use change in the tropics, changes in rainfall levels) on soil carbon and nitrogen cycling, net primary production, leaching losses, and biogenic trace gas fluxes, (2) landscape-scale assessment of crop yield and of trade-offs between crop production and associated environmental effects, (3) landscape-scale quantification of soil carbon changes and greenhouse gas fluxes, and (4) applications of stable isotopes, e.g. ^{13}C and ^{15}N, for quantifying carbon and nitrogen fluxes among soil, plant, water and air. She has 20 publications in international peer-reviewed journals, and at present five more papers are being submitted to international journals with peer-review systems.

Mitiku Haile is a professor and president of Mekelle University, Ethiopia. His research focuses on the effects of human influences on the use, management and protection of natural resources and on designing new approaches for natural resource conservation in the highlands of Ethiopia. He has participated in a number of projects in the field of restoration ecology and land resource management. He has more than 50 publications in international peer-reviewed journals.

Chapter 8
Coastal Forest Buffer Zones and Shoreline Change in Zanzibar, Tanzania: Practical Measures for Climate Adaptation?

Johanna Mustelin, Miza Khamis, Robert G. Klein, Abbas J. Mzee, Tahir Abbas Haji, Bakar Asseid and Taimi Sitari

Abstract Coastal environments worldwide are going to be affected by climate change impacts in the future. These environments are densely populated especially in Africa and thus such impacts as rising sea level, increase in extreme storms and natural hazards, will have a bearing on local, regional and national levels. This research looks at environmental change, both past and current, on the east coast of Zanzibar, Tanzania in terms of the implications of sea level rise, extreme storms and the current vulnerability of coastal settlements. Current vulnerabilities in the area relate to deforestation, coastal erosion, and differential access to and ownership of natural resources. The creation and strengthening of coastal forest buffer

J. Mustelin (✉)
Griffith Centre for Coastal Management and Urban Research Program, South East Queensland Climate Adaptation Research Initiative, School of Environment, Griffith University, Southport 4222, Queensland, QLD, Australia
e-mail: j.mustelin@griffith.edu.au

J. Mustelin · R. G. Klein · T. Sitari
Department of Geography, University of Turku, Turku, Finland
e-mail: springbokvlakte@yahoo.com

T. Sitari
e-mail: taisit@utu.fi

M. Khamis · A. J. Mzee · T. A. Haji · B. Asseid
Department of Commercial Crops, Fruits and Forestry, Zanzibar, Tanzania
e-mail: mizakhamis@gmail.com

A. J. Mzee
e-mail: abassmzee@yahoo.co.uk

T. A. Haji
e-mail: twahirhaji@yahoo.com

B. Asseid
e-mail: b.s.asseid@redcolobus.org

W. Leal Filho (ed.), *Experiences of Climate Change Adaptation in Africa*,
Climate Change Management, DOI: 10.1007/978-3-642-22315-0_8,

zones is one of the suggested approaches for proactive adaptation, which aims to increase the resilience of the natural environment, while producing multiple benefits for multiple stakeholders. The study suggests that anticipatory adaptation needs to be first and foremost incorporated into the local realities in order to be effective and valued. Considering and understanding the value judgements and existing priorities among multiple stakeholders is crucial since otherwise adaptation efforts will remain conflicting and separate processes.

Keywords Climate change adaptation · Zanzibar · Coastal erosion · Multiple stakeholder cooperation

Introduction

Africa is projected to suffer the most from the impacts of climate change (Brown et al. 2007). This is because many African countries depend on climate-sensitive sectors (agriculture, fisheries, tourism), which are forecasted to face significant changes in the future (UNECA 2008; Osman-Elasha 2007). In addition, human population density is extremely high on coastal areas throughout the continent (Nicholls 2000) and the projected impacts will affect large numbers of populations (Technical Recommendations 1993). Furthermore, changing climatic conditions affect people's livelihoods and can lead to differential gender impacts. In Africa, women often depend on livelihoods directly connected to the natural environment, whereas men often seek market-based opportunities such as those found in tourism (Paavola 2006). However, whether climate change is treated and understood as an additional factor influencing processes and changes taking place in Africa (Boko et al. 2007) or as the sole cause of change, its impacts will significantly alter the outcomes (Dubi 2000). Nevertheless, communities in Africa have been adapting to changing conditions for decades, and there exists a range of indigenous adaptation and coping strategies, which should be considered when planning for adaptation (Boko et al. 2007; Downing and Osman-Elasha 2007).

It is clear that not all countries are affected the same way and there will be disparity in the scale and nature of impacts even within regions. For example, some countries can even profit from higher rainfall patterns, whereas other countries face increase in drought conditions (Boko et al. 2007). This is a case in point with Tanzania where the projections suggest a decrease in rainfall and higher temperatures in inland areas and increases in rainfall on coastal areas (Mwandosya et al. 1998), which could indicate more severe flooding and increased erosion on the coast (Paavola 2003). National variations thus will also differ and lead to multiple adaptation responses, which are context- and place-specific.

Tanzania's National Adaptation Plan for Action (NAPA 2006) focuses strongly on the mainland but leaves Zanzibar (an island state in union with Tanzania) out of the main discussion and the projections. This is perhaps not surprising since the climate projections will differ for Zanzibar as it belongs to the category of small islands and thus will be affected in different terms than the mainland (Paavola 2006). Such issues as size and geographical setting make small islands more limited in developing responses to climate change impacts (Pelling and Uitto 2001; Nicholls and Lowe 2004; Mimura et al. 2007). Sea level rise is most likely to be the most considerable impact for all small island states (Ragoonaden 2006) impacting on subsistence farming areas, villages and infrastructure.

If we understand current vulnerabilities and problems, there is a better chance to formulate such responses to climate change, which fit the particular context and produce multiple benefits (van Aalst et al. 2008). It is therefore crucial to examine the current environmental changes and the factors and processes creating change and the implications these have regarding adaptation to climate change. Furthermore, adaptation involves individuals, communities, governments and NGOs (Hay and Mimura 2006; Paavola and Adger 2006) with diverse values, which affect the decision-making processes (Adger et al. 2005). Therefore, the examination of the multiple stakeholders in a certain context has high relevance in adaptation research, as also Adger et al. (2005) have remarked that adaptation is often not equal and individual adaptation might even trigger negative outcomes on others' part. However, such measures as coastal forest buffer zones can be seen as responses to current experienced problems simultaneously addressing also future vulnerabilities and aiming to provide benefits or solutions to multiple problems. For instance, functional bio-zones on coastal areas act as natural protection against storms, erosion and saline water intrusion.

The purpose of this chapter is to describe a recent research project that took place on the east coast of Zanzibar, Tanzania. The project focused on examining coastal governance and the linkages between implications of climate change impacts (sea level rise, erosion, extreme events) and coastal stakeholders in the area, and what climate adaptation would mean in that particular context. The chapter is organised as follows: firstly, we provide a short background on Zanzibar. We then discuss the project background, aims and methods. Secondly, we will present project results in terms of vulnerable areas, shoreline change and sea level rise scenarios. Thirdly, we will discuss the practical issues in planning coastal forest buffer zones through collaborative planning and fourthly, we make recommendations and observations in regards to future research.

Zanzibar

Zanzibar is located forty kilometres west of the Tanzanian mainland and consists of numerous islands of which Unguja and Pemba are the largest. Unguja, often called Zanzibar, is the best known of the islands. Since the early 1970s tourism has been growing as a sector and the 1983 Zanzibar Development Plan already recognized tourism as an integral part of the economy (Gillies 2001). Foreign investors have been able to construct the tourism sector due to a lack of government regulation (Rutherfoord 1992). Although international tourism has provided increased opportunities for employment, the actual earnings for coastal local communities have still remained relatively low (Mustelin 2007).

The increased tourism has brought with it conflicts regarding land-use, ownership and land rights (Gössling 2001; Mustelin 2007), which in turn affect communities' opportunities for natural resource management, access and use (Dolan and Walker 2004). Moreover, as entitlements to land, the environment, social capital and institutions determine the existing possibilities and dependencies in a location (Barnett and Adger 2007), these should be examined in order to understand the context in which adaptation discussions take place. As tourism in Zanzibar is very focused on the beach landscapes, there is also an incentive for the tourism industry to try to maintain the exotic image that they sell (Mustelin 2007). In addition, coastal erosion is already a problem in Zanzibar (Dubi 2000), which relates closely to questions of environmental governance and resource entitlements.

The coastal zone in Zanzibar is highly contested in terms of multiple stakeholders' claims on the use rights to the beach environment, the intertidal zone and adjacent areas (Gössling 2001; Mustelin 2007). The communities utilize the coast for diverse livelihood activities including seaweed farming, fishing, firewood and building pole collection, subsistence farming and for cemeteries (Sitari 2005; Mustelin 2007); in other words, the coastal zone is a livelihood space for the communities (Yap 1996). Seaweed farming has in particular provided better income opportunities and enhanced living standards for women (Tobey and Torell 2006) but simultaneously it has come in conflict with the tourism industry as it utilizes the same intertidal zones (Mustelin 2007; Mustelin et al. 2009). In addition, increased erosion, loss of coastal vegetation and deforestation are some of the current trends, which play into the context in which communities and other stakeholders, such as the tourism industry, exist. Issues of coastal governance thus are very much part of what the context for adaptation looks like and what opportunities and barriers exist within that context (Adger et al. 2005; Barnett and Adger 2007).

In Zanzibar, the identified factors that accelerate beach erosion are sand extraction from the beach, cutting of mangroves, building seawalls and jetties, and building infrastructure too close to the high watermark (Nyandwi 2001). Sand mining is a common feature in both Tanzania and Zanzibar, which increases coastal erosion. Beach sand is used for construction purposes such as roads and infrastructure and is often illegally mined (Masalu 2002). These are human-induced problems, which most often have led to aggravated erosion rates instead of protecting the shore.

Building regulations especially have an important role in regulating the zones appropriate for infrastructure development in coastal environments. In Zanzibar, there is a rule not to built housing infrastructure nearer than 30 m from the high watermark; however, this rule is not in any legislation but is more of a preferred guideline regarding good practice on coastal infrastructure building (Mustelin et al. 2009).

Although there has not been much climate change specific research conducted in Zanzibar, there is an emerging concern particularly among the institutional stakeholders in what should or can be done to adapt to the projected impacts (Mustelin et al. 2009, 2010). Most of the climate data available is produced by the weather stations under the Tanzania Meteorological Agency's Zanzibar Section. The Institute of Marine Science (University of Dar es Salaam) collects also data on sea level change along the coasts of Zanzibar. What is partly causing concern is the range of extreme weather phenomenon that has occurred in recent years. In March 2007, Zanzibar experienced the highest rise in sea-level, which caused flooding of the old town. The warmest temperature recorded in Zanzibar in the last 68 years was in February 2007 when the Zanzibar International Airport station recorded a temperature of 39.4°C. The highest rainfall for the last 50 years was recorded in April 2005 in Stone Town Victoria Garden (471 mm in 24 h) (Suleiman Khamis 2008). However, much of the responses regarding natural resource management and extreme events have been rather reactive than proactive (Mustelin et al. 2009).

Project Background, Methods and Research Area

The research project was a collaboration between the Department of Geography, University of Turku, and the Department of Commercial Crops, Fruits and Forestry, Government of Zanzibar and funded by the Ministry for Foreign Affairs of Finland as part of its development policy research. A research gap was identified in regards to the planning and usage of coastal forest buffer zones as a climate change adaptation response, which simultaneously would address current vulnerabilities, and questions of coastal governance such as resource use and rights among multiple stakeholders. The research agenda was jointly negotiated between the institutions.

The objectives of the research were (a) to examine the environmental changes in the shoreline and coastal strip of the chosen area over the past 50 years and to examine the implications to local livelihood security, coastal biodiversity and infrastructure development, and (b) to propose a model for the use of coastal forest buffer zones development as a climate change adaptation tool.

The methods used in the research included key informant discussions with local communities, hotels and different institutional actors, vegetation surveys, GIS methods, aerial photograph analysis, stakeholder workshops and literature surveys.

Fifty persons have been discussed with in the communities, twelve hotels were directly visited and interviewed and sixty hotels were contacted through the email questionnaire. In addition, thirty-five persons were interviewed regarding firewood collection areas in the research villages and various institutions and organisations, such as NGOs and business associations, have also been discussed and collaborated with (Mustelin et al. 2009). Workshops were used throughout the project to facilitate greater involvement of multiple stakeholder groups. Stakeholder workshops with government and academic representatives were convened in the beginning and end of the project to make sure the project was relevant, and to raise awareness of climate change issues. An additional workshop was held with community members, hotel managers, NGOs and government officials to discuss the results, and to negotiate the way forward. This provided also an opportunity for interaction between these groups to discuss and voice their particular concerns. English and Kiswahili were used both in the workshops and interviews to guarantee that all participants could participate effectively.

GIS methods have included change analysis (aerial photographs between 1953 and 2004), wave power calculations, modelling of sea level rise (according to IPCC 2007), and visualisation of firewood collection points. GPS coordinates have been taken on recently cleared areas, stretches of beach with natural vegetation, hotel areas and demarcated areas for construction, and points of significant erosion on the shoreline. Based on the outcomes of the methods, highly vulnerable areas have been mapped along the coast. The research area is in the northeast coast of Zanzibar and includes the administrative units of Kiwengwa, Pwani Mchangani and Matemwe (Fig. 1).

Aerial photographs and GIS methods were utilized to analyse changes in the coastal area over the last 50 years and to identify areas vulnerable to the possible effects of climate change. By studying aerial photographs of 1953, 1978, 1989 and 2004 changes along the shoreline in particular were analyzed. The accurate detection of shoreline position through remote sensing methods is problematic due to the dynamic nature of the coast. This is because of the short term influence of tides and the long term influence of relative sea level rise on the shoreline position (Appeanding Addo et al. 2008). For this reason to study the dynamics along the shoreline, the changes along the beach edge were examined against a fixed shoreline obtained from the 1:10,000 topographic map of 1986. A shift in the beach edge could be a result of deforestation/reforestation processes caused by wave action and humans and hotel construction along the shoreline. The distances of the beach edge from the fixed shoreline were plotted on a graph for better visualization of the changes between the different temporal layers.

Sea level at 2,100 according to the Intergovernmental Panel for Climate Change (2007) is predicted to increase by 18–59 cm. This does not take into account uncertainties in climate-carbon cycle feedbacks nor the effects of changes in ice sheet flow. Considering a linear growth from the 1993–2003 rates, sea level rise is predicted to increase by an additional 10–20 cm. The effect of such increases in sea level within the chosen area was modeled by means of a Digital Elevation Model (DEM) showing areas flooded by 20, 40, 60 and 80 cm sea level rise. The input data for the DEM included contour lines and point elevation data from the 1:10,000 topographic map of 1985 as well as point elevations manually selected from digital stereo models of the 2004 aerial photographs (Mustelin et al. 2009). The aim with the sea level rise modeling was to give an increased understanding of what the implications would be for the area. It was understood that sea level rise is not static phenomenon but its impacts depend also on other factors, such as storm surge, extreme events and tidal activities.

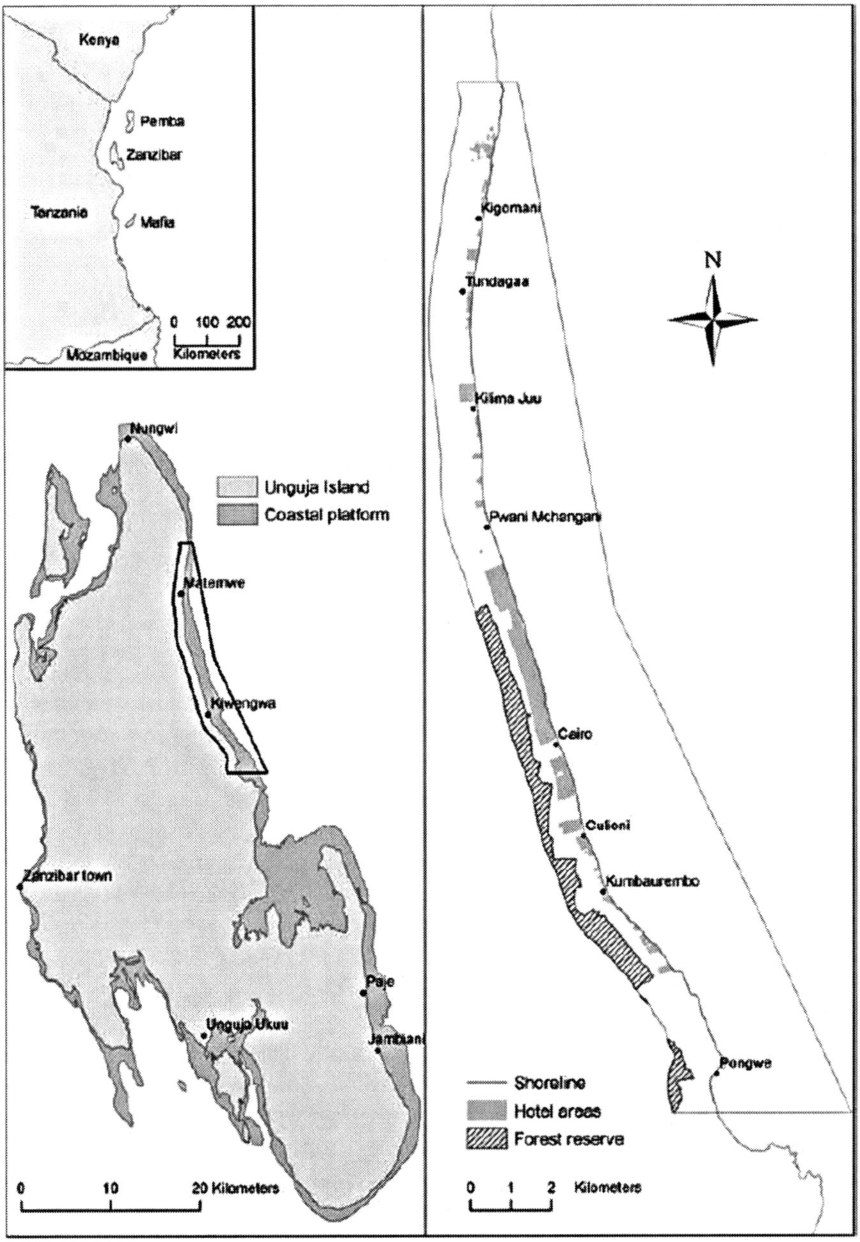

Fig. 1 The research area on the northeast coast of Zanzibar

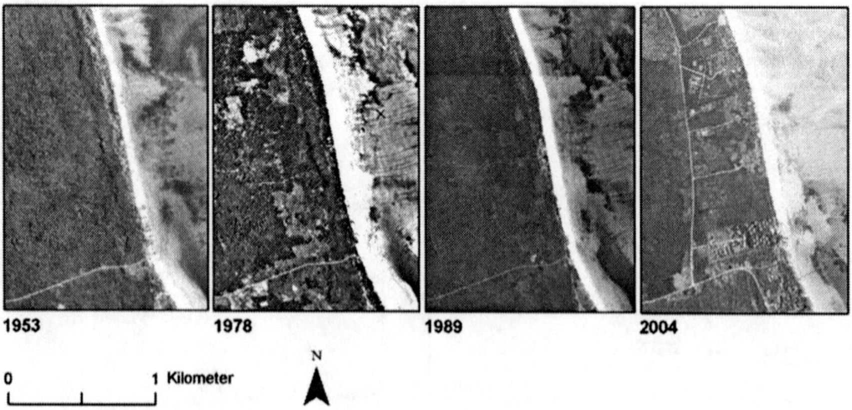

Fig. 2 Changes in Kiwengwa between 1953 and 2004

Coastal Erosion and Environmental Change

Physical Changes Over the Last 50 Years

Land-use changes have been significant in the research area mostly due to the growing tourism industry but also because of the utilization of natural resources by the communities and increase in population. A net expansion of deforestation is most evident in Kiwengwa, which also has had the highest tourism development (Fig. 2). Between Pongwe and Cairo most of the vegetation clearance has taken place over the last 20 years in the narrow strip between the Kiwengwa-Pongwe forest reserve and the coast. This strip of land seems to experience considerable pressure due to restrictions on wood collection in the adjacent forest reserve. The vegetation in Matemwe is fairly open of fragmented shrubland containing very little woody vegetation suitable for firewood and construction. As a result villagers from Matemwe often have to collect wood elsewhere, such as Kiwengwa, which adds to the environmental pressure.

During the past 50 years there have been extensive changes in the beach edge in Kiwengwa and Matemwe areas (Fig. 3). Each line in the graph represents the beach edge of the respective aerial photographs between 1953 and 2004. By studying these lines, one can see how the beach edge has shifted over the years. The graph also shows the presence of villages, hotels and vegetation along the beach in order to relate beach changes with nature of the adjacent land. Beaches in this region lie adjacent to villages, hotel areas, natural beach vegetation, or other vegetation cover such as coconut and Casuarina plantations. It is common for hotels to clear natural beach vegetation in front of the hotel area causing the beach to encroach inland as can be seen at hotel areas in Kumbaurembo, Gulioni, and the area between Cairo and Pwani Mchangani. Where natural beach vegetation grows along the beach the beach does not encroach inland and can be seen along the

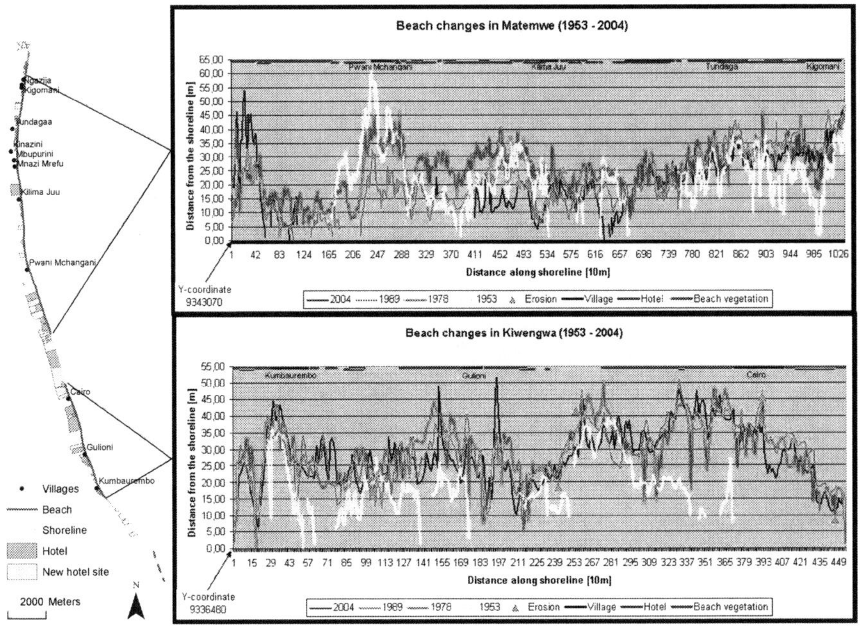

Fig. 3 Changes along the beach edge in Kiwengwa and Matemwe areas during the last 50 years

beach in Matemwe. The same can be observed at beach areas in front of villages (Gulioni, Cairo and Pwani Mchangani), where the boundary line between beach and land moves seaward over time. This could imply a problematic situation in which villages may expand onto the beach. With an increasing tourist industry, shops are constructed on the beach, and in recent years many have been washed away by the sea.

Physical Impacts of Sea Level Rise

Different scenarios of sea level imply different outcomes along the shoreline of Pongwe, Kiwengwa, Pwani Mchangani and Matemwe (Fig. 4). With the current coastline topography, Pongwe, Kiwengwa and Pwani Mchangani beach areas will be affected even by small increases in sea level. According to Fig. 4, a sea level rise of 20 cm will flood the entire village of Pongwe and beach fronts along the Kiwengwa and Pwani Mchangani shoreline. The accuracy of the DEM is, of course, a limitation when modelling changes of tens of centimeters. However, low lying beach areas are generally at risk as even minor changes in sea level can have significant local effects. During field work in May–June 2008 it was observed that the beach slope in Matemwe is steeper than the beach slope in Kiwengwa. Furthermore, signs of significant erosion, such as the exposure of tree roots, damages to constructions and remnants of washed-away constructions were far less in Matemwe than further south. Therefore, despite the inaccuracies in the DEM, the sea-level rise prediction model nevertheless presents some realistic outcomes. Although the maximum sea level rise prediction of the IPCC (2007) rarely extent more than 100 m inland, flooding of this extent will cause major implications to the local village communities and tourist industry. The beach sand lost by sand mining as well as excessive erosion caused by deforestation only exacerbates the future impacts of sea level rise.

Current Pressures and Local Perceptions

Hotels' impacts on the environment are considered to be mostly positive by the managers, especially regarding tree planting and landscaping activities. The positive actions include replanting of trees, maintaining the existing vegetation, distributing environmentally friendly information to tourists, using ecological technologies (grey water treatment, solar energy and recycling) and cooperation with the village communities (Mustelin et al. 2009). The majority of the informants are personally concerned of climate change and recognize its impacts on the tourism industry in the future in Zanzibar. However, hotels enhance coastal erosion by clearing vegetation adjacent to the beach and by taking sand, according to both community and expert interviews. Nevertheless, some hotels have left indigenous vegetation buffer zones and have experienced less erosion problems. Eaton (2008) also reports similar observations in the southeast coast of Zanzibar, where those hotels, which have planted indigenous vegetation, such as Goat's Foot Creeper (*Ipomoea pes-caprae)* in order to stabilise the beach, experience less erosion.

In Kiwengwa, some hotels badly affected by coastal erosion have begun building seawalls out of palm tree poles or cement. However, beach erosion prevention walls might not be a functioning solution as these often cause large-scale erosion somewhere else instead (Nyandwi 2001; Cooper and McKenna 2008; Eaton 2008). This corresponds with Adger et al. (2005) notion that some adaptation measures in general might benefit an individual but harm other stakeholders. Some hotels in Kiwengwa note that walls made of palm poles deteriorate quite quickly, which means that the wall has to be constantly restored. A manager notes that as his hotel has began using palm poles for shoreline protection, the erosion rate has slowed down. In community discussions, seawalls are not highly valued as these are perceived to shift erosion elsewhere along the shoreline.

The communities all imply that significant environmental changes have taken place in the last 20–30 years. For instance, before the villages could not be seen from the beach due to the density of beach vegetation. The factors causing the changes, as implied in the community discussions, relate to population growth, which causes the most pressure as people need to use the forests for their livelihoods. The community interviews for instance revealed that many of the villagers

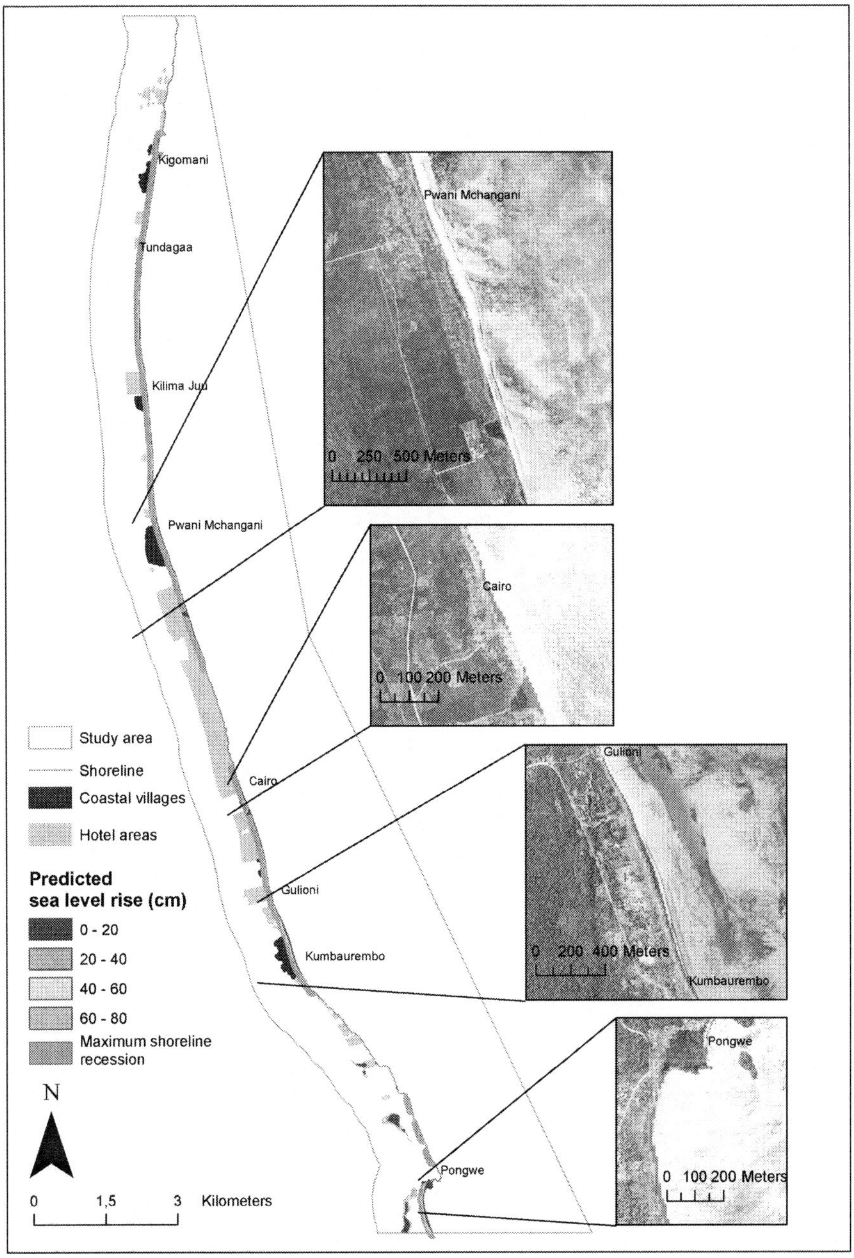

Fig. 4 Sea level rise scenarios according to the IPCC (2007) predictions in the research area

fetch their firewood from the Kiwengwa-Pongwe forest reserve, which implies an increasing pressure on the surrounding natural resources. Deforestation thus occurs both on the coast and in the inland areas. Vegetation loss is attributed to people cutting trees for firewood, fish traps and clearing the land for farming. As more tourism development is planned to the area, the benefits of tourism at local scale are somewhat unevenly distributed as local villagers have difficulty in obtaining jobs due to a lack of education, foreign language skills and hospitality skills.

The loss of beach vegetation is perceived to be a problem also on social grounds as now the villages are visible to all beach users. This is partly due to the cultural clash in a highly Islamic island where Western tourists beach attire is in contradiction to the prevailing cultural norms. Moreover, less vegetation and trees means also that there is less protection from high winds and extreme storms, and there is now no natural buffer to stop the seawater entering the villages during high tides and storm conditions; some of the villages already experience flooding once or twice a year.

Other changes in relation to environment were the rain seasons: these were experienced to have changed in both in length and intensity. The spring rain season (masika) had become more irregular and shorter, while the fall rain season (vuli) had almost disappeared. As the majority of the population depends on subsistence farming, change in rain seasons is a serious issue in terms of livelihoods. Some informants implied that they already have observed that the plants do not grow in their usual phase but rather slower due to the decrease in rainfall.

The measures to strengthen the coastal area received somewhat mixed perceptions within the communities. The proposed measures range from raising awareness to placing debris on the beach. However, the most popular suggested measure is vegetation planting including indigenous trees, shrubs and creepers. This correlates positively with the idea of coastal forest buffer zones as planting as an activity is valued by the communities.

Coastal Forest Buffer Zones: Planning and Issues

Vulnerable areas were mapped together with the villagers, while also identifying areas for planting coastal forest buffer zones (Fig. 5); this also included identification of the preferred species. The number of seedlings for the total area is 41,450 and the area to be planted is approximately 30, 5 hectares. The buffer zone is rather fragmented as the areas with most erosion and decreased vegetation cover are found in patches rather than as large areas. Furthermore, most of the planting areas are near communities as hotel areas have been excluded from the plan. The planting areas are situated only on the shoreline although deforestation is also a problem further inland. However, the plan is focused specifically on the strengthening of the shoreline and therefore planting in inland areas is not included, although the coastal environment obviously extends further than just the shore.

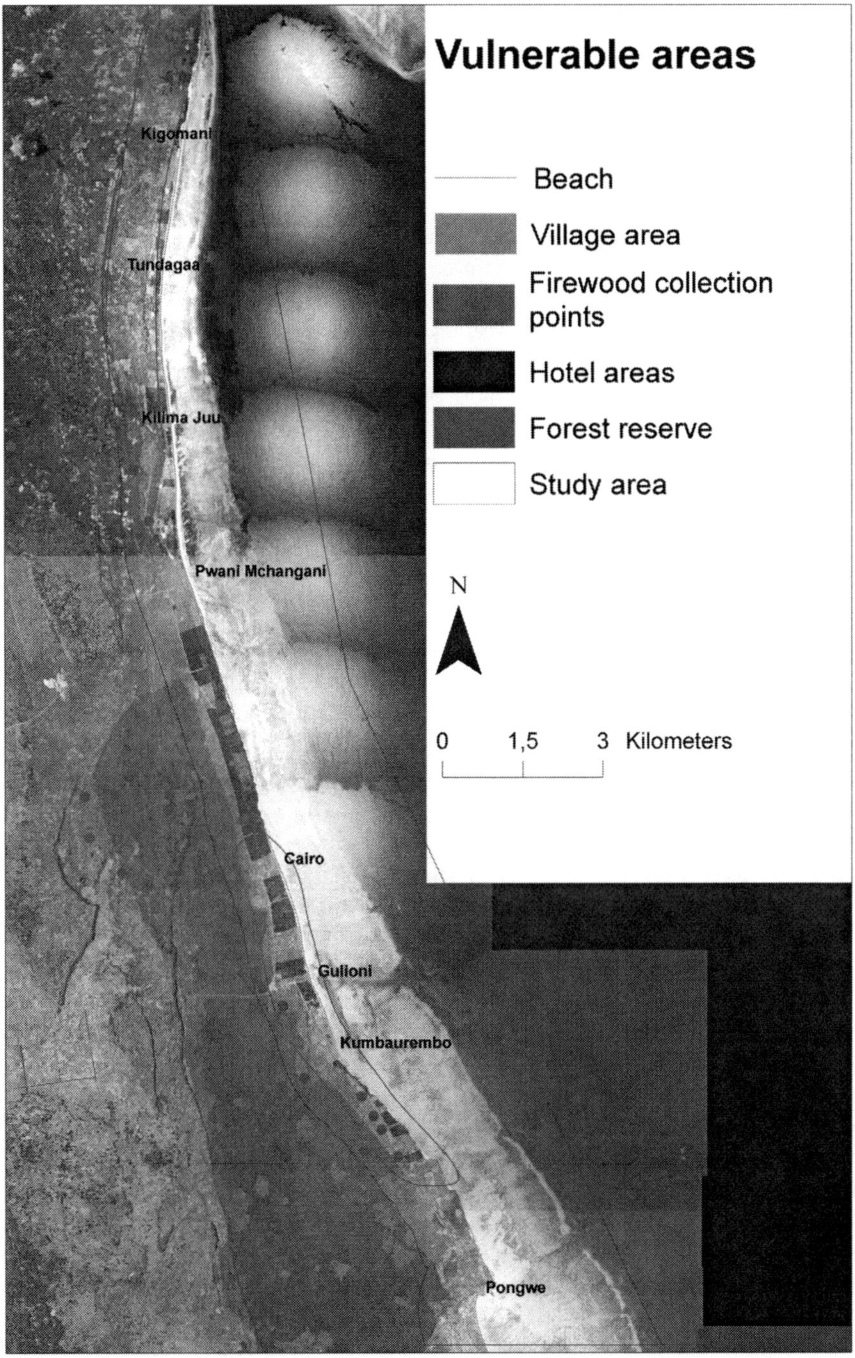

Fig. 5 Vulnerable areas along the coast and areas where most of the firewood is fetched from

The plan also includes training and monitoring activities, which have been set in place in order to guarantee the survival of the species both during the growth period and after the planting. Monitoring will take place once a month as a follow-up on the survival rate of the seedlings. Training is planned to take place several times in all of the involved communities in how to take care of the seedlings.

Alien species are disadvantageous for coastal buffers due to their high risk of invasion. Casuarina has a low invasive potential and is resistant against strong wave action, however its fibrous roots prevent undergrowth to establish, resulting in erosion (Avis 1989), and is therefore an unsuitable species as buffer on the beach. Furthermore, *Casuarina equisetifolia*, as well as *Lantana camara*, have a suppressing effect on indigenous species (Kombo 2003). Indigenous vegetation is not only adapted to local conditions, but also creates habitat and/or migration corridors for indigenous fauna (Van der Merwe and Lohrentz 2001). The species should thus reflect the coastal ecosystem's complexity and therefore should include both trees and shrubs but also creepers, which are often the first species on the beachfront.

The communities defined approximately twenty-five different species that could be used for the buffer zone. Some of the species can be obtained through making cuttings of the existing species such as the goat foot creeper; in addition, the communities emphasise that the best season for planting is during the *masika* rain season in the spring. The species need to be protected from the livestock, which is somewhat easier with trees but more difficult with creepers; the goats eat all the planted species as they often roam free. However, this takes place mostly during the rain season as the goats have little fodder during that time, which coincides with the proposed planting time. Some of the species could be grown in communities but prior training is needed in how to take care of the seedlings in an appropriate way. Being aware of these "everyday" issues is crucial as the some of the plants have long growing times during which they need to be protected. Since the proposed tree species are indigenous and there are not ready seedlings available, the activities will also include seed gathering and seedling production in a nursery.

In the discussions with the communities, it has been agreed that the planted trees would fall under community ownership and thus remain under community management. This is especially important since the planted species need to be also looked after in order to remain. Women in particular have been eager to participate in the planting efforts.

Some of the hotels have indicated a willingness to cooperate especially regarding environmental issues. As the local management plan is accepted and supported by the communities, it would be crucial to involve the hotels in the activities as well. Some of the hotels already have a green team, which looks after the coastal environment in the hotel area. However, stronger cooperation between the stakeholder groups is needed as the planting activities will not have a long-term effect if other activities, such as sand mining and seawall infrastructure, keep affecting the shoreline.

Conclusions

This study addressed the implications of climate change impacts on coastal communities in the north-east coast of Zanzibar, Tanzania. The results imply that while most of the current vulnerabilities are due to anthropogenic factors, such as increased pressure on natural resources and modifications of the coastline, climate change will aggravate and increase these vulnerabilities. The multiple stakeholders sharing the same coastal space have differing needs and aims for the maintenance and use for the coast, which seem to be often in conflict particularly in relation to land use rights (including the intertidal zone and the beach). Increased erosion is already a strong indicator for current change processes, which all stakeholders view as negative. Although most of the literature offers hard structures as possible adaptation strategies (NAPA 2006), this study has looked at the feasibility and the aspect of practical planning of coastal forest buffer zones in highly complex coastal governance systems. This is on the premise that natural coastal vegetation is a very effective buffer against excessive coastal erosion and flooding, while improved aesthetic values and protection from the storms also accrue from such activities. Although the booming tourism industry has contributed to "coastal squeeze" of the villages and the available livelihood spaces, simultaneously some villagers benefit from new income opportunities in tourism. Training and educating workforce for tourism could offer larger variety of possibilities for villagers in the future if the cooperation with the hotels would become more structured.

Adaptive planning and management approaches are crucial as new research and changing situations compel new strategies appropriate for that time and context as suggested also by Tompkins et al. (2008). Precautionary approaches in natural resource management are therefore valuable as they provide multiple benefits (Tri et al. 1998). The key issue in adaptation is willingness to adapt and to have the resources and skills available. However, given that often in developing countries there are much more urgent issues, such as poverty and everyday needs, any attempt for adaptation has to also include the notion of development and enhanced well-being.

Although it is highly unlikely that improved coastal vegetation alone can combat such large-scale phenomena as sea level rise and increased erosion, they provide a good base for environmental governance, which addresses multiple issues. Coastal forest buffer zones can become effective measures to strengthen the coastal environment and increase its resilience to face future change and even impacts of climate change. These zones do not only give environmental benefits but also respond to trends in aesthetics (improved visual landscape) and increased protection from the storms. Nevertheless, what needs to happen is the incorporation of multiple stakeholders, more stringent policy enforcement of building and environmental regulations, and improved and more equal access to the existing natural resources.

What is required for enhanced climate adaptation in Zanzibar are more resources among local institutions who have the official responsibility for natural resource management. This could be done in the form of training courses, increased research funding and expertise in climate modeling, social science approaches in climate change adaptation and through multi-sectoral collaboration towards a national adaptation policy specifically for Zanzibar. There exists vast and rich knowledge among local institutions, government officers, NGOs, and communities what the current issues are. These different kinds of knowledge make an important contribution to understanding what and how adaptation could (or should) take place in Zanzibar. However, what is crucial is that adaptation is locally driven and not merely formulated by external agencies. In addition, it also needs to be discussed what levels of thresholds are acceptable and when improving the current resilience actually increases future vulnerabilities. That discussion is first and foremost local.

Acknowledgements This research project is a collaborative project between the Department of Geography, University of Turku and the Department of Commercial Crops, Fruits and Forestry, Zanzibar. The project is funded by the Ministry for Foreign Affairs of Finland. The views in this article, however, are of the authors and do not necessarily represent the views of the Ministry.

References

Adger WN, Arnell NW, Tompkins EL (2005) Successful adaptation to climate change across scales. Global Environ Change 15:77–86

Appeaning Addo K, Walkden M, Mills JP (2008) Detection, measurement and prediction of shoreline recession in Accra, Ghana. ISPRS J Photogramm Remote Sens 63(5):543–558

Avis AM (1989) A review of coastal dune stabilization in the cape province of South Africa. Landsc Urban Planning 18:55–68

Barnett J, Adger WN (2007) Climate change, human security and violent conflict. Political Geogr 26:639–655

Boko M, Niang I, Nyong A, Vogel C, Githeko A, Medany M, Osman-Elasha B, Tabo R, Yanda P (2007) Africa. In: Parry ML, Canziani OF, Palutikof JP, van der Linden PJ, Hansen CE (eds) In climate change 2007: impacts, adaptation and vulnerability. contribution to the working group II to the fourth assessment report on the intergovernmental panel for climate change. Cambridge University Press, Cambridge, pp 433–467

Brown O, Hammill A, McLeman R (2007) Climate change as the 'new' security threat: implications for Africa. Int Affairs 83(6):1141–1154

Cooper JAG, McKenna J (2008) Social justice in coastal erosion management: the temporal and spatial dimensions. Geoforum 39:294–306

Dolan AH, Walker IJ (2004) Understanding vulnerability of coastal communities to climate change related risks. J Coastal Res 39:2

Downing TE, Osman-Elasha B (2007) Lessons learned in Africa. pp 19–21, Tiempo 65 Oct 2007. www.tiempocyberclimate.org

Dubi A (2000) Coastal erosion. In: Ngusaru AS (ed) Tanzania coastal management partnership: the present state of knowledge of marine science in Tanzania synthesis report. Tanzania Coastal Management Partnership Support Unit, Science and Technical Torking Group, Tanzania

Eaton L (2008) A historical case study of beach erosion in Paje, Bwejuu and Jambiani: possible methods of prevention. SIT Zanzibar, Coastal ecology, Zanzibar

Gillies J (2001) A revised political economy of tourism: a case study of the effect of tourism on rural livelihoods on Zanzibar's northeast coast. 17 dec 2001, Lancaster University, Department of geography, a dissertation submitted in partial fulfilment of the requirements of the degree MRes in environment and development (overseas placement)

Gössling S (2001) Tourism, economic transition and ecosystem degradation: inter-acting processes in a Tanzanian coastal community. Tourism Geogr 3(4):430–453 (Taylor & Francis, Routledge)

Hay J, Mimura N (2006) Supporting climate change vulnerability and adaptation assessments in the Asia-Pacific region an example of sustainability science. Sustain Sci 1:23–35

IPCC (2007) Climate change 2007: synthesis report. An assessment of the intergovernmental panel on climate change

Kombo YH (2003) The alien invasive species and their potential threats to the fragile ecosystems of Zanzibar. Forestry technical paper no 149, p 10

Masalu DCP (2002) Coastal erosion and its social and environmental aspects in Tanzania: a case study in illegal sand mining. Coastal Manage 30(4):347–359

Mimura N, Nurse L, McLean RF, Agard J, Briguglio L, Lefale P, Payet R, Sem G (2007) Small islands. pp 687–716. In: Parry ML, Canziani OF, Palutikof JP, van der Linden PJ, Hansen CE (eds) In climate change 2007: impacts, adaptation and vulnerability. contribution to the working group ii to the fourth assessment report on the intergovernmental panel for climate change. Cambridge University Press, Cambridge

Mustelin J (2007) Tourism, resource access and power in Zanzibar, Tanzania: examining the linkages between tourism, coastal communities and the coastal environment. Turku University Department of Geography Publications B 10, Digipaino, Turku, p 108

Mustelin J, Klein R, Asseid B, Sitari T, Khamis M, Mzee A, Haji T (2009) Practical measures to tackle climate change: coastal forest buffer zones and shoreline change in Zanzibar Tanzania. Turku University Department of Geography Publications B: 13, Turku, p 106, January

Mustelin J, Klein R, Asseid B, Sitari T, Khamis M, Mzee A, Haji T (2010) Understanding current and future vulnerability in coastal settings: community perceptions and preferences for adaptation in Zanzibar, Tanzania. Popul Environ 31(5):371–398

Mwandosya MJ, Nyenzi BS, Luhanga ML (1998) The assessment of vulnerability and adaptation to climate change impacts in Tanzania. Centre for Energy, Environment, Science and Technology (CEEST), Dar-es-Salaam, Tanzania

National adaptation programme for action (NAPA) for Tanzania (2006) Division of the environment. April 2006, p 48. www.napa-pana.org/private/modules/knowledgebox/io/file.php?entry=681&field=22. Accessed 30 Apr 2008

Nicholls RJ (2000) Sea-level rise and Africa: an outsider's view. In: de la Vega-Leinert AC, Nicholls RJ, Nasser Hassan A El-Raye M (eds) Proceedings of the SURVAS Expert Workshop on: African vulnerability and adaptation to impacts of accelerated sea-level rise (ASLR). www.survas.mdx.ac.uk/pdfs/cairopro.pdf. Accessed 15 July 2008

Nicholls RJ, Lowe JA (2004) Benefits of mitigation of climate change for coastal areas. Global Environ Change 14:229–244

Nyandwi N (2001) Survey of the extent of human-induced beach erosion problems in Tanzania. gridnairobi.unep.org/chm/EAFDocuments/Tanzania/Human_induced_beach_erosion.pdf. Accessed 3 July 2008

Osman-Elasha B (2007) Africa's vulnerability. pp 3–9, Tiempo 63 April 2007. www.tiempocyberclimate.org. Accessed 2 July 2008

Paavola J (2003) Vulnerability to climate change in Tanzania: sources, substances and solutions. a paper presented at the inaugural workshop of southern Africa vulnerability initiative (SAVI) in Maputo, Mozambique, June 19–21 2003. www.gechs.org/savi/workshop/maputo/papers/paavola_tanzania.pdf. Accessed 2 May 2008

Paavola J (2006) Justice in adaptation to climate change in Tanzania. In: Adger WN, Paavola J, Huq S, Mace MJ (eds) Fairness in adaptation to climate change. p 319. Massachusetts Institute of Technology, The MTI Press, USA, pp 201–222

Paavola J, Adger WN (2006) Fair adaptation to climate change. Ecol Economics 56:594–609

Pelling M, Uitto JI (2001) Small island developing states: natural disaster vulnerability and global change. Environ Hazards 3:49–62

Ragoonaden S (2006) Sea level activities and changes on the islands of the Western Indian ocean. West Indian Ocean J Mar Sci 5(2):179–194

Rutherfoord R (1992) Tourism and development: a village in Zanzibar. University of Wales, University College Swansea, Centre for development studies, A dissertation submitted in partial fulfillment for the degree of Bachelor of science

Sitari T (2005) Forestry in the community lifeworld in Unguja, Zanzibar. 37–58. In: Sitari (ed) Forestry, community and biodiversity in Zanzibar. Turku University Department of Geography Publications B: 3, pp 90, Newsprint, Raisio

Suleiman Khamis A (2008) An overview on climate change-Zanzibar perspective. Presentation at the department of commercial crops, Fruits and forestry, 16 May 2008

Technical recommendations of the workshop on integrated coastal zone management in Eastern Africa including the Island states (1993) In: Linden O (ed) Proceedings of the workshop and policy conference on integrated coastal zone management in Eastern Africa including the Island states, Swedish agency for research cooperation with developing countries, University of Dar es Salaam & Ministry of Tourism, Natural Resources and Environment, Tanzania, pp 315–323

Tobey J, Torell E (2006) Coastal poverty and MPA management in mainland Tanzania and Zanzibar. Ocean Coast Manag 49:834–854

Tompkins EL, Few R, Brown K (2008) Scenario-based stakeholder engagement: Incorporating stakeholders preferences into coastal planning for climate change. J Environ Manage 88:1580–1592

Tri NH, Adger WN, Kelly PM (1998) Natural resource management in mitigating climate impacts: the example of mangrove restoration in Vietnam. Global Environ Change 8:49–61

UNECA (2008) Climate change: African perspectives for a post-2012 agreement. p 17, United Nations Economic and Social Council, Economic Commission for Africa, African Union Commission. E/ECA/COE/27/8//AU/CAMEF/EXP/8(III). www.uneca.org/cfm/2008/docs/ClimateChange.pdf. Accessed 2 July 2008

Van Aalst MK, Cannon T, Burton I (2008) Community level adaptation to climate change: The potential role of participatory community risk assessment. Global Environ Change 18:165–179.

Van der Merwe JH, Lohrentz G (2001) Demarcating coastal vegetation buffers with multicriteria evaluation and GIS at Saldanha Bay, South Africa. Ambio 30(2):89–95

Yap HT (1996) Attempts at integrated coastal management in a developing country. Mar Pollut Bull 32(8/9):558–591 (Elsevier Science Ltd)

Author Biographies

Johanna Mustelin is a doctoral researcher with Griffith University. Her research interests include public policy in climate change adaptation and the role of public participation in such policy processes with specific interest in coastal communities. She worked as a project coordinator and researcher for the Department of Geography, University of Turku, prior to taking up a doctoral research position in Australia with the Griffith Centre for Coastal Management and Urban Research Program, and the South East Queensland Climate Adaptation Research Initiative.

Miza Khamis has a Master's degree in Natural Resource Management with specialisation in Forestry for Sustainable Development from the International Institute for Geo-information Science and Earth Observation, Enschede, the Netherlands. She also studied GIS methods in Turku. She is a member of the research team in Sustainable Landscapes on the Zanzibar project and for climate change issues in the department.

Robert G. Klein has a wide background in environmental and geosciences. He completed his MSc in Environmental Sciences, majoring in Geoinformatics, at the University of Turku. He is currently a PhD student in Geophysics at the University of Helsinki. Apart from his work in Zanzibar related to landscape and edaphic issues, Klein also worked at the National Botanical Institute (South Africa) for the Millennium Seed Bank Project.

Abbas J. Mzee BSc, is a forest officer from Zanzibar and works in the DCCFF. He studied GIS methods in Turku. He is a member of the research team in sustainable landscapes and climate change on the Zanzibar project. He is the GIS specialist in the department.

Tahir Abbas Haji Diploma Botanist, works at DCCFF and has experience in research concerning community development.

Dr Bakar Asseid is the Director of the Department of Commercial Crops Fruits and Forestry (DCCFF).

Dr Taimi Sitari has over 30 years of research and work experience from different parts of Africa, and she has worked particularly closely with development issues. Her research expertise relates to community development and natural resource management sectors in Africa.

Chapter 9
Conflicting Policies: Institutional Approaches Towards Decentralisation and Promoting Climate Change Resilience in Kenya

Samuel Munyua Kimani, Emily Obonyo Kamau and Paul Othim Ongugo

Abstract Decentralisation refers to 'any act by which a central government cedes rights of decision-making over resources to actors and institutions at lower levels in a politico-administrative and territorial hierarchy'. Kenya's history of a highly centralised forest governance regime has recently seen a shift in policy and legislation authorising decentralisation in the sector. But what is it that gets decentralised in the forestry and natural resources sectors? Is decentralisation effective in meeting the goals of equity, sustainability, poverty reduction and climate change in an environment characterised by conflicting policies? This paper attempts to answer these questions. To understand the resource management outcomes of decentralised programmes, the rights and capacities that are transferred to actors at lower levels were examined. Using both primary and secondary data from two Kenyan forest resources, an analysis was done to find out key roles played by relevant institutions in understanding what is expected to be decentralised; and what policy environments are required to ensure the effectiveness of a decentralised forest resource management system. Results indicate that despite their similarities in ecology, prominence of both forest in local and national economies and conservation of biological diversity, there are some clear differences in institutional regimes for their management. The study concludes that heterogeneity of community stakeholders as government institutions (ministries), parastatals (KWS and KFS), international organisations and

---

S. M. Kimani (✉) · E. O. Kamau · P. O. Ongugo
Kenya Forestry Research Institute IFRI- Kenya CRC,
PO Box 20412,Nairobi, 00200, Kenya
e-mail: kimanisam86@yahoo.com

E. O. Kamau
e-mail: emily.obonyo@gmail.com

P. O. Ongugo
e-mail: paulongugo@live.com

W. Leal Filho (ed.), *Experiences of Climate Change Adaptation in Africa*,
Climate Change Management, DOI: 10.1007/978-3-642-22315-0_9,

NGOs have overlapping mandates and policies that affect common pool resource management. The overlaps should be reduced to provide clear jurisdiction of governance and to enhance transparency in decision-making and equitable benefits distribution, which has long been wanting.

Keywords Decentralisation · Common pool resources · Policies · Mandates · Climate change

Introduction

Increases in emissions of CO_2 gas and other greenhouse gases (GHG) such as methane, nitrous oxide, chlorofluorocarbons (CFC), hydrochloroflourocarbons (HCFC) and halogens into the atmosphere has led to the overall rise in mean global temperature over the years and the resultant climate change. Key anthropogenic activities assumed responsible include fossil fuel combustion and land-use changes, particularly tropical deforestation. For most developing countries, the dependence on agriculture has remained high. Therefore, land-use changes have devastating effects on agricultural production and consequently on livelihoods. Kenya is currently experiencing some major effects of climate change: of special consideration is the variation in weather patterns. Prolonged drought and famine has left over 10 million people faced with starvation (Barrow et al. 2002), while floods and resurgence of pests and diseases have been noted in other parts of the country. Widespread poverty, inadequate socio-economic resources and a large climate-dependent agricultural sector make the country vulnerable to the consequences of climate change. The country is also ill-equipped to adapt to the long-term changes associated with climate. In spite of this, Kenya has embarked on various measures to mitigate climate change, such as the adoption of clean development mechanisms, reafforestation and the spread of green technology. Forest conservation and management in Kenyan history was guided by the forest policy of 1957, which was revised in 1968, and then again in 1994. This later draft formed the basis for policy and legislative reform a decade later. The main legislation is the Forest Act Cap 385 of 1962 that was revised three times: in 1982, 1992 and 2005. It was drafted in support of the 1957 policy and covers a broad range of activities from the gazettement/degazettement of forests and nature reserves, licensing of use, prohibitions of certain activities and imposition of penalties, etc. Subsidiary regulations cover the rights of forest-adjacent communities to utilise specified resources in specific ways. This Act had several crucial shortcomings. It covered only gazetted forest reserves, did not provide sufficient safeguards against forest excisions, provided only use rights to a narrow set of resources for communities, and did not recognise the importance of forests for environmental conservation.

A large number of reasons have been given for shifting the focus of decision-making and resource management away from central states to local governments or communities. State control was found to be largely unsuccessful, costly and financially unsustainable (Meinzen-Dick and Knox 2001; Shackleton 1999). Local communities, on the other hand, have been shown to be effective managers of local resources (Arnold 1990; Ostrom 1993; Bromley et al. 1992; Berkes 1989). Local communities not only have greater knowledge of local resources but are also better able to monitor resource use and rule compliance (Meinzen-Dick and Knox 2001; Gibson 2001; Peters 1994; McKean 1992). They are often directly dependent on the resource and are assumed to have the greatest incentives to maintain the resource base over time. The policy move towards greater local control is reflected in a wide range of community-based arrangements in the natural resources sector over the past decade (Barrow and Murphree 2000; Barrow et al. 2002; Hulme and Murphree 2001; Shackleton 1999).

However, the involvement of forest-adjacent communities and other stakeholders in forest management and conservation in Kenya is a key feature of the 2005 Forests Act. The main objective of wider stakeholder participation in forest management as captured in the draft Forest Policy (Sessional Paper No. 7 of 2007) is "to promote the participation of the private sector, communities and other stakeholders in forest management to conserve water catchments areas, create employment, reduce poverty and ensure the sustainability of the forest sector". For effective participation of communities in forest management, it became imperative that capacity building be done among the forest-adjacent communities, key stakeholders and forest authority staff. There are other national laws and regulations that refer to the forestry sector, including the Environmental Management and Coordination Act (EMCA) 1999; the Water Act 2002; the Wildlife (Conservation and Management) Act Cap 376; the Agriculture Act Cap 318; the Antiques and Monuments Act Cap 215; the Local Government Act Cap 265, and the Fisheries Act Cap 378.

The objective of this study was to investigate:

1. What is it that gets decentralised in the forestry and natural resources sectors?
2. Is decentralisation effective in meeting the goals of equity, sustainability, poverty reduction and climate change in an environment characterised by conflicting policies?

It is assumed that all lead institutions have a common goal in the management of common pool resources based on different policies and mandates aimed at both the protection and conservation of resources. However, this is not usually the case, especially in Kenya since the holistic approaches of management have no clear jurisdiction of governance or transparency in decision-making and equitable benefits distribution from natural resources.

Study Area and Methods

Study Country

The Republic of Kenya is a country in East Africa lying along the west of Indian Ocean and bisected by the equator. Kenya is bordered by Ethiopia (north), Somalia (north-east), Tanzania (south), Uganda plus Lake Victoria (west), and Sudan (north-west) (Fig. 1). The capital city is Nairobi. The population has grown rapidly in recent decades to nearly 38 million. Kenya has numerous wildlife reserves, containing thousands of animal species (MENR 1994).

Meteorologically, Kenya lies in one of the most complex sectors of the African continent. Its climate is influenced by large-scale tropical controls which include several major convergence zones, including the Inter-Tropical Convergence Zone (ITCZ) that are superimposed upon regional factors associated with lakes, topography and the maritime influence. Thus, the climatic patterns within the country are markedly complex and change rapidly over short distances (Wandiga 2006a). The annual temperature range is 2–35°C with the lowest value in March and April and the highest in July and August. Diannual temperature range is in the order of 10–20°C, far exceeding the annual temperature range. Mean annual net radiation received on a horizontal surface is between 450 and 550 cals/cm^2/day. Mean annual bright sunshine amounts to over 7–8 h per day in the highlands and 8–9 h per day in the lowlands (ibid). Rainfall is distributed in short and long rainy seasons with the former received in October–December and the latter in April–June, while July and August are the coldest months. Rainfall is influenced by conventional and relief microclimates depending on location. In addition, the rainfall variability is closely linked to the El Niño Southern Oscillation (ENSO) phenomenon and the sea-surface temperature (SSTs) fluctuations in the equatorial Indian and Atlantic Oceans. The rains are normally enhanced during the ENSO years that occur every 5–6 years. Thus, such high climate variability is likely to enhance due to climate change, in turn enhancing climate change impacts both at regional and local levels.

From the late 1970s to early 1980s, there was an unprecedented acceleration in the destruction of forests in Kenya, which to a large extent was blamed on lack of appropriate and all-inclusive forest policy and legislation. The policy and legislation used to manage forest resources were developed in 1957 by the colonial government, and changed only slightly in 1968 after independence. Even though it was expected that the new policy and law would be implemented then and followed up quickly in order to halt forest degradation, it took another ten years before a new policy was put in place and a further three years before the Forests Act came into being in 2005. Article II Sect. 4 of the new Forests Act requires the Kenya Forest Service (KFS), the new parastatal that has replaced the Forest Department, to enter into agreement with CFAs to manage natural forests. The service, contrary to expectations from civil society organisations, is involved in the formation of the CFAs. It is from these changes after involving other stakeholders in forest conservation that, in particular, necessitated the study Fig. 1.

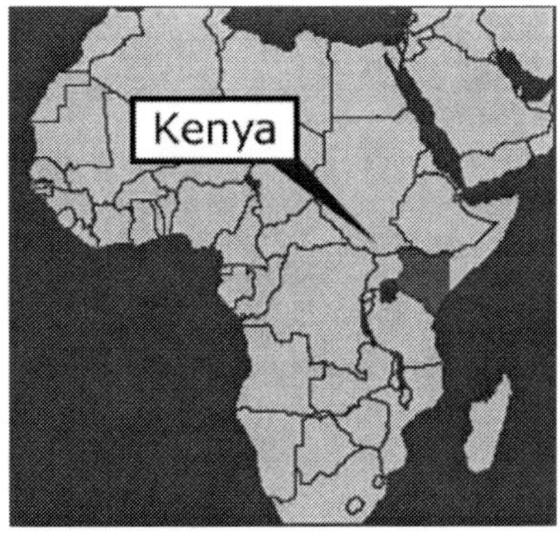

Fig. 1 The location of Kenya in Africa

Methodology

This study is part of a global multi-disciplinary team, the International Forest Resources and Institutions research programme (www.indiana.edu/~ifri). The IFRI research protocol was used to collect data from the forest site, from settlements around each forest and from among user groups and local organisations involved in forest management. Focus group discussions and key informant interviews were used to glean information on forest product use, actors and interactions in forest management, etc. Resource and resourse diagrams were additional tools developed to link and illustrate the flow of information and decisions across and between actors in management and conservation of forest. For each of the forest sites, researchers, user groups and others worked to illustrate the flow of decisions, information, authority, responsibility, and financial and human resources before and after the decentralisation reform. The aim is to help identify process weaknesses and to evaluate impacts on user group incentives. In general, data collection was conducted by research teams together with individuals drawn from communities living in the forest sites. Participatory rural appraisal (PRA) tools including mapping, semi-structured interviews and historical institutional profile analysis were also used.

Results and Discussion

Decentralisation of the Forest Sector

Kenya's reform of the forestry sector recognises the roles of local communities in resource management. The Forest Act of 2005 provides for the development of new institutions and partnerships for improved forest management. Kenya Forests Service is in the process of developing participatory forest management plans for all forest areas in the country in close consultation with Forest Associations and

Table 1 The roles played by different institutions

Institutions/groups	Roles
Government organisations	Direct management
Parastatals	Research, training and policy
International organisations and NGOs	Fundraising and information sharing
Local organisations	Conservation issues

gender-balanced local forest management committees. This is a major break from prior policy and practice, which was authoritarian and failed to recognise the roles of different actors. The Act provides for involvement of other stakeholders in forest conservation and protection. Table 1 shows various institutions involved in forest conservation and the roles they play therein.

Although KFS has not signed any agreement with any community forest association in Kenya, there have been some milestones in the devolution of forest activities in all natural forest in Kenya. To play a substantial role in decentralisation, it ratified the Kyoto Protocol of the United Nations Framework Convention on Climate Change (UNFCCC) in August 1994 and has over the years actively participated in and hosted the Conference of Parties. The country has also operationalised the implementation of target objectives and agreements albeit on a lower scale due to socio-economic challenges. In this regard, relevant ministries and state corporations in Kenya are in charge of major components of the CDM through licensing, inspection, monitoring and approval of related projects.

Thus:

1. The National Environment Management Authority is the Designated National Authority of the CDM.
2. The Kenya Forest Department is in charge of all CDM reafforestation and agroforestry projects.
3. The Ministry of Energy is in charge of all energy conservation and alternative energy projects.

However, institutions in Kenya have been operating on crossroads in resources conservation and curbing climate change. For example, NEMA, a government parastatal established to exercise general supervision and coordination of all matters relating to the environment, has done little in terms of enforcement and compliance of the Kyoto Protocol, which is aimed at reducing the levels of carbon emissions. The ultimate objective of UNFCCC Convention is the "stabilisation of the greenhouse gas concentration in the atmosphere at a level that would prevent dangerous anthropogenic interference with the climate system" (IPCC 2001). The Kenya Forestry Research Institute mission is to contribute, together with its partners, agricultural innovations and knowledge towards improved livelihoods and commercialisation of agriculture by increasing productivity and fostering value-chains while conserving the environment.

The Kenya Agricultural Research Institute (KARI) brings together research programmes in food crops, horticultural and industrial crops, livestock and range management, land and water management, and socio-economics. KARI promotes sound agricultural research, technology generation and dissemination to ensure food security through improved productivity and environmental conservation.

The Kenya Forest Working Group is a gathering of individuals and organisations (government and non-government, local, national and international) concerned with forests, their conservation and management. KFWG was formed in 1995 to provide a forum for exchanging and sharing information and experiences among members. Although it exists as a sub-committee of the East African Wild Life Society, it has done a commendable job by trying to bring on board various stakeholders in curbing climate change through forestry. This has been achieved through improving the status of Kenya's forests.

Equity, Sustainability, Poverty Reduction and Climate Change

Kenya has declared its national priority as poverty reduction. Incidentally, there are various sectoral programmes that support sustainable livelihoods, local food security and healthcare, especially for poor people. These include the promotion of indigenous food crops and traditional herbal medicine. There are institutions such as the Kenya Industrial Property Institute and the National Museums of Kenya that are mandated to protect traditional knowledge, innovations and practices. Regulations on access and benefit-sharing mechanisms have been drafted and are awaiting adoption by the government.

Aside from increasing the coverage of protected areas and establishing new special-status sites, Kenya also intends, through its Strategy for Revitalising Agriculture, to achieve by 2014 comprehensive development of the agricultural sector at all levels for the benefit of the population. On the subject of conservation of species, specific targets and programmes have been established regarding, among others, mangroves, coral reefs, turtles, and black rhinos. There are closed fishing seasons for some fish taxa to avoid overexploitation of certain species. Furthermore, Kenya states that there is increased public awareness and access to information on habitat conservation. The Third National Report also declares that the Kenyan Plant Health Inspectorate Service has a system for controlling and monitoring the introduction of invasive alien species, including tighter control and surveillance at ports of entry (some programmes on specific species, such as the water hyacinth, are mentioned).

Anthropogenic Impacts Contributing to Climate Change in Kenya

Major impacts contributing to climate change in Kenya include over-utilisation and degradation of natural resources; reduction in tree cover on farmlands, soil erosion and deforestation; industrialisation; rapid urbanisation with a projected

Fig. 2 Forest plantation infected with pests and diseases

60% of total population destined to live in cities by the year 2030 (Kenya Press Release 2007); and all the foregone products and services are driven by rapid population increase. As shown in Fig. 2, the impacts on the deforestation and wetland degradation are further illustrated since their conservation and management act as major opportunities for carbon sinks.

Impacts of Climate Change in Kenya

Africa is one of the most vulnerable continents to climate change—a situation aggravated by the interaction of multiple stresses occurring at various levels and as a result of its low adaptive capacity (IPCC 2007). Kenya is already experiencing the impacts of climate change and these are likely to worsen with the anticipated increase in temperatures. Direct impacts include changes in weather patterns with decreased rainfall, increased temperatures and higher evaporation rates in the dry areas.

It is estimated that rainfall will increase by 5–20% during the months of December–March while a decrease of between 5 and 10% will occur in June–August (meteorological department). Under more rapid warming scenarios, Kenya and other East African countries may receive up to double the usual precipitation, while arid areas are likely to receive even less than they receive now (IPCC 2001). Indirect impacts of climate change concern socio-development strategies such as health, livelihood support, education and conflict. Frequent drought spells over the years have led to severe water shortage, increased risk of food shortage and expansion of aridity and desertification into marginal lands and changes in the planting dates of annual crops. There will be notable increases in fungal outbreaks and insect manifestations due to changes in temperature and humidity, along with reductions in ecosystem integrity, its resilience and decline in biodiversity.

Fig. 3 Melting glacier in Mt. Kenya

Fig. 4 Flooding of Bundalangi

Other impacts include increases in human, crop and animal vector-borne diseases such as malaria, cholera and Rift Valley fever; sea level rise resulting in inundation of low-lying areas along the coast and islands while increased ocean acidity will result in coral reef bleaching along the Kenyan coast. The melting of glaciers on Mount Kenya is already occurring (Fig. 3), while extreme weather events will increase; *inter alia* (Case 2006; Githeko et al. 2000; IPCC 2007a; NEMA 2008; UN 2001; Wandiga 2006b). The impact on the tourism sector (Kenya's second foreign exchange earner) will also be high due to loss of biodiversity; spread of disease margins and inundation of low-lying coastal areas (Fig. 4).

Impact on Development and the Economy

The economic sector was found to be vulnerable to climate change. This vulnerability is exacerbated by existing developmental challenges of endemic poverty; complex governance and institutional dimensions; limited access to capital

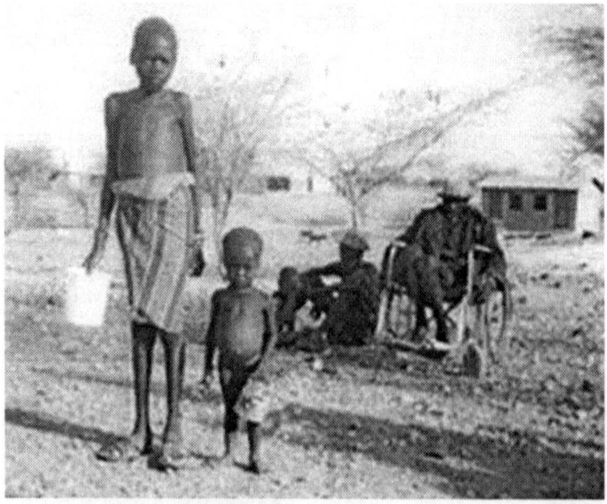

Fig. 5 Impact of drought as a result of climate change

including markets; inadequate infrastructural and technological development; ecosystem degradation and complex natural disasters (IPCC 2007). In Kenya, the economic sector is mainly agriculture-driven thus sensitive to change in this sector. Climate change will continue to negatively affect food production, especially tea production in the Kenyan Highlands as it is sensitive to temperature and rainfall. This will affect income from the leading foreign exchange earner.

Extreme Events and Natural Disasters

Over 70% of natural disasters in Kenya are weather-related and their frequency has increased over the years, with drought and floods being the main disasters Fig. 5. The drought oscillation period in past years recurred every five years but has now reduced to every two to three years. The worst drought since independence was in 1991–1992, while the 1997–2000 drought was the worst in the past 40 years. During the latter drought, nomadic communities incurred over 50% losses in livestock while food shortages were felt countrywide Fig. 6.

In 2008/2009, Kenya faced one of its worst droughts, which left over 10 million people without food and access to drinking water. This disaster was exacerbated by a combination of other factors including the 2007/2008 post-election violence that affected agricultural production in Kenya's bread basket areas, global economic recession that affected the purchasing power of the people, as well as inadequate disaster response mechanisms, which often take the form of disaster response rather than prevention and other governance issues. The worst floods in Kenya occurred in 1997/1998 with El Niño rains that resulted in displacement of persons, damage of physical infrastructure such as roads, bridges and railways, spread of human and livestock diseases *viz* cholera, malaria and Rift Valley diseases respectively, damage

Fig. 6 Loss of livestock in North-eastern province due to extended drought

Fig. 7 Dry kapiti plains

to agricultural produce and economic fiscal inflation (Kenya Press Release 2007). Climate change is predicted to enhance drought, flood, fire, ENSO and tropical storms within the country and along the coastal zone Figs. 7 and 8.

Initiatives in Protected Areas and Curbing Climate Change

Kenya has expressed its commitment to progressively increasing forest cover from the current 2–10% of the country under a protected area system (this target is not time-bound). There is also a generalised focus to increase the coverage of different biomes under protected areas to include spots not currently covered (including forests, heritage sites, national parks, game reserves, and marine parks and reserves). Several marine parks and reserves have been established but the

Fig. 8 Mau forest destruction from agricultural cultivation, fires and charcoal making

coverage for inland waters is still lacking. The integrated coastal zone management strategy may result in more marine parks and reserves being established.

State actions relating to climate change are growing in sophistication and importance. They are redirecting their economies towards new energy development and job creation, using comprehensive climate change action plans to identify and implement cost-effective measures that reduce greenhouse gas emissions and create other co-benefits. Some are requiring the direct reduction of greenhouse gas emissions while others see greenhouse gas reduction as a by-product of other policies.

Conclusions and Policy Recommendations

It is understandable that institutions have different strengths and capacities, and this calls for networking. Although multifaceted strategies should be pursued to ensure collaborative interventions by the various institutions, efforts should also be made by the government and other concerned institutions to strengthen linkages and networks to improve resilience.

The other challenge is for Kenya to develop strategies that would promote sustainable development without resulting in increased emissions of GHGs. It is necessary therefore to develop appropriate policies and response strategies to manage GHG emissions. Policies and strategies must be based on reliable inventories of GHG emissions and sinks. The First National Communication of Kenya is the main output of the enabling activity project. The Ministry of Environment and Natural Resources through the National Environment Secretariat (now NEMA) executed the Climate Change Enabling Activity. Four thematic areas need to be identified in accordance with the UNFCCC guidelines for the preparation of non-Annex 1 national communication. These include:

1. National GHG inventory, abatement, sinks and sequestration
2. Vulnerability and adaptation to climate change
3. Research and systematic observation

4. Awareness and understanding of climate change issues
5. Clean Development Mechanisms
6. Transfer of environmentally sound technologies
7. Synergy with other MEAs
8. Climate change mitigation strategies and policy frameworks
9. Convention negotiation capacity including issues of equity

Finally, more resources and actual budgets for disaster management have to be put in place, and this necessitates policy support.

References

Arnold JEM (1990) Social forestry and communal management in India: ODI: rural development forestry network (RDFN)

Barrow EJC, Isla G, Kamugisha-Ruhombe J, Tesse Y (2002) Analysis of stakeholder power and responsibilities in community involvement in forest management in Eastern and Southern Africa. Forest and social perspectives in conservation No. 9. IUCN Eastern Africa Program

Barrow E and Murphree M (eds) (2000) Community conservation from concept to practice. James Currey: Oxford. Berkes, Fikret, (ed) 1989. Common property resources: ecology and community-based sustainable development. London; 2001. Barrow, Edmund, Vogt and Bahati J. Decentralized governance and ecological health: why local institutions fail to moderate deforestation in Mpigi district of Uganda. Scientific Research and Essay (10):434–445

Berkes C (1989) Cultivating peace: conflict and collaboration in natural resource management. IDRC/World Bank, Ottawa

Bromley DW, David F, Margaret M, Pauline P, Jere G, Ronald OC, Ford R, James T (eds) (1992) Making the commons work: theory, practice, and policy. ICS Press, San Francisco

Case A (2006) Natural resource conflict management case studies: an analysis of power, participation and protected areas. FAO, Rome

Forest Act (2005) Change in Kenya forest landscape and vision 2030

Gibson J (2001) Legitimacy and disappointment in forest management. Environ Policy 24:141–148

Githeko E, Schroeder-Wildberg E, Carius A (2000) Confronting conflict: a first-aid kit for handling conflict. Hawthorne Press, Stroud

Hulme D and Murphree M (eds) (2001) African wildlife and livelihoods: the promise and performance of community conservation

IPCC (2001) Third assessment report on climate change

IPCC Climate Change (2007): Impacts, adaptation and vulnerability. Contributions of working group II to the fourth assessment report of the intergovernmental panel on climate change. Parry ML, Canziani OF, Palutikof JP, van der Linden PJ and Hanson CE (eds) Cambridge University Press, Cambridge

Kenya Press Release (2007) Managing tragedies: understanding conflict over common pool resources. Science, 1998

McKean J (1992) Changing to gray: decentralization and the emergence of volatile socio-legal configuration in central Kalimantan, Indonesia. World Dev 32(7):1199–1223

Meinzen-Dick R and Knox A (2001) Collective action, property rights, and devolution of natural resource management: a conceptual framework. In: Collective action, property rights and devolution of natural resource management: exchange of knowledge and implications for policy. Meinzen-Dick R, Knox A and Di Gregorio M (eds) pp. 117–126

MENR (1994) Kenya forestry master plan: development programs. Ministry of Environment and Natural Resources, Nairobi

NEMA (2008) Environmental mediation for environmental problems

Ostrom E (1993) Governing the commons: the evolution of institutions for collective action. New York: Cambridge University Press; 1990. Ostrom E, Schroeder L and Wynne S. Institutional Incentives and Sustainable Development, Infrastructure Policies in Perspectives

Peters W (1994) The dynamics of decentralization in the forestry sector in South Sulawesi: the history, realities and challenges of decentralized governance. CIFOR, Bogor

Shackleton K (1999) Globalized conflicts of discourse: the type of conflict of proxies. Geo J 52:45–58

United Nations (2001) The Africa water vision for 2025: equitable and sustainable use of water for socio-economic development. Economic Commission for Africa, Addis Ababa

Wandiga D (2006a) Observations on resource use in Kakamega forest reserve, Kenya. Afr J Ecol 45(Suppl. 1):2–6

Wandiga S (2006b) Conflict escalation and coping with workplace bullying: a replication and extension. Eur J Work Organizational Psychology 10(4):497–522

Author Biographies

Samuel Munyua Kimani is currently an environmentalist and GIS technologist at the Kenya Forestry Research Institute. The institute was established in 1986 under the Science and Technology Act (Chapter 250) to carry out research in forestry and allied natural resources. The institute has a role to play in influencing polices on forest resource management. Mr Kimani graduated with a Diploma in Environmental Resource Management in 2007 and is currently a third-year student undertaking Geo-informatics at Kenya Polytechnic University College. Mr Kimani has been involved in intricate relationships, policies and mapping of common pool resources.

Emily Obonyo Kamau is currently a PhD student at Bonn, Germany, and a Senior Research scientist at the Kenya Forestry Research Institute. She has worked as a socio-economist at various levels. Mrs Kamau holds a Bachelor's degree in sociology from the University of Nairobi, 1993. She attained her Master's degree in Sociology and Gender in 1996.

Paul Othim Ongugo is currently the Assistant Director for Natural Forest Programme at Kenya Forestry Research Institute. He is also the Coordinator of International Forestry Research and institutions in Kenya looking at decentralisation of common pool resources and policy. Dr Ongugo holds a Bachelor's degree in Forestry, 1980. He also holds a Master's degree in Forestry and Agriculture from the University of Helsinki, specialising in business and social economics, 1985. He did his PhD at the University of East Anglia's School of Development Studies in 2000.

Chapter 10
Climate Shocks, Perceptions and Coping Options in Semi-Arid Kenya

Barrack Okoba, A. Almeneh Dejene and Meshack Mallo

Abstract The purpose of this study was to assess farmers' awareness, perceptions and coping options to deal with climate change shocks in semi-arid Kenya. Household heads who have been farming in the study areas for the last 15–20 years were identified by village leaders. By random selection, 100 households (in Gachoka division–Mbeere district) and 69 households (in Lare division–Nakuru district) were interviewed using a structured questionnaire. The household data was analysed using SPSS—a statistical package to establish levels of responses on each issue of climate change. The results indicated that over 90% of respondents have noticed changes in climatic patterns over the last 20 years. They have experienced increasing average temperatures, decreasing average annual rainfall and variable rainfall patterns, while incidences of malaria and animal diseases are on rise. These events are closely linked to the severe droughts of 1984 and 2006–2009 and too much rainfall in 1997/1998 and 2004. These had a negative impact on the livelihood systems of the vulnerable farming community. This data indicated that most farmers tended to adopt technologies/strategies that are related to crop production than to livestock farming. Few respondents did not perceive the critical role of soil and water management as a coping option for

B. Okoba (✉)
Kenya Agricultural Research Institute (KARI)-Kabete,
PO Box 14733, 00800 Nairobi, Kenya
e-mail: okoba200@yahoo.com

A. A. Dejene
Environment and Climate Change Adapatation, Food and Agricultural Organization of United Nations, 00100 Rome, Italy
e-mail: alemneh.dejene@fao.org

M. Mallo
Food and Agricultural Organization of United Nations,
PO Box 5536, Addis Ababa, Ethiopia
e-mail: meshack.malo@fao.org

W. Leal Filho (ed.), *Experiences of Climate Change Adaptation in Africa*,
Climate Change Management, DOI: 10.1007/978-3-642-22315-0_10,

climate shocks. Failure to do nothing to climate shocks by over 30% of respondents in Mbeere district is worrying. Inadequate technical information and capital/credit sources hindered widespread adoption of different types of soil-improving fertilisers, change in planting dates and taking crop/livestock insurance against climate variability.

Keywords Climate change · Shocks · Coping options · Adaptation · Smallholder farmers · Kenya

Introduction

According to the IPCC (2007), sub-Saharan Africa is going to experience warming that is greater than the global average and in parts of the region rainfall will decline. However, some model predictions indicate that the East Africa region is going to have increased rainfall events due to climate change. Other recent research suggests that local circulation effects will result in decreased precipitation instead (Funk et al. 2008). This anticipated climate variability and climate changes pose great threats to food and water security, public health, natural resources and biodiversity (McCarthy et al. 2001). The farmers in the region are going to be most vulnerable to climate variability and climate change due to their heavy dependency on rain-fed agriculture (Tschakert 2007; IPCC 2001). This is because the majority of farmers have low adaptive capacity to anticipated increases in extreme events, resulting from widespread poverty, lack of economic and technological resources, and insufficient safety nets and educational progress. With long-term changes in rainfall patterns and shifting temperature zones, changes in water availability and carbon dioxide fertilisation, climate change is expected to increase the frequency of climate-related shocks in these regions. Climate change adaptation should be a major priority because of the region's vulnerability. Several studies have indicated that with the anticipated long-term climate changes, African farmers, with well-developed coping responses to short-term climate variability events, would not be able to tame the future/present adverse effects (Smithers and Smit 1997; Mortimore 1998; Mortimore and Adams 2001; Ziervogel et al. 2008).

Kenya's land mass has only 16% land that could be described as having high to medium potential for rain-fed agriculture. This area supports about 80% of the country's population in food/livestock production. The remaining 20% of population live in other remaining 84% of land mass, whose agro-ecological system is described as arid and semi-arid land (ASAL). Kenya experiences major droughts every decade and minor ones in three to four years, with the exception of the arid northern part where drought is experienced regularly with varied consequences. In recent years, critical drought periods in the country were experienced in 1984, 1995, 2000 and 2005/2006 (UNEP/GoK 2000). The likelihood of an upward trend in rainfall under global warming is high. Wet extremes (defined as high rainfall events occurring once every ten years) are projected to increase during both the

September–December and the March–May rainy seasons, locally referred to as the short and long rainfall seasons, respectively. Dry extremes are projected to be less severe in the northern parts of the region during September–December, but the models do not show a good agreement in their projected changes of dry extremes during March–May (Thornton et al. 2006). The impacts of these dry extremes or droughts on the population are increasing exponentially as a consequence of high population growth and increasing encroachment of agricultural activities in the ASAL. Due to frequent drought events, declining crop yields and the migration of livestock keepers in search of pasture on one hand but also experiences of landslides and floods on the other are becoming common. Such occurrences are greatly undermining food security efforts. Effective adaptation options need to be sought and must therefore address both the adaptation and mitigation of climate change. Agricultural adaptation is of higher interest to smallholder farmers than mitigation. The manner in which farmers update their expectations of the climate in response to unusual weather patterns defines how they adapt. According to Maddison (2006), farmers learn about the best adaptation options in three ways: (1) learning by doing, (2) learning by copying, and (3) learning from instruction. There is recognition that farmers' responses vary when faced with the same stimuli. Such varied responses, even within the same geographic area, are partly related to the variety of agricultural systems involved and the different market systems in which farmers operate (Bryant et al. 2000). A more important factor of varied farmers' responses is the differences between farmers in terms of personal managerial and entrepreneurial capacities and family circumstances. Also, farmers can be influenced by their peers' perceptions and by values present in their communities as well as their professional associations. A review of literature on adoption of new technologies identified farm size, tenure status, education, access to extension services, market access and credit availability, agroclimatic conditions, topographical features, and the availability of water as the major determinants of the speed of adoption (Maddison 2006). Examples of adaptation options likely to be taken by most farmers are not only those that ensure food security and livelihood but also those that will address conservation of natural resources.

Looking at farmers' perception is critically important if constraints to adaptations have to be identified. Gbetibouo (2009) observed that perceptions of climate variability by South African farmers in the Limpopo basin were in line with long-term climatic data records, implying that their memory of rainfall and temperature changes in past seasons was accurate and could be relied on to make necessary adjustments. More importantly, the study also observed that farmers with access to extension services were likely to perceive changes in the climate because extension services provided information about climate and weather changes. Having access to water for irrigation increased the resilience of farmers to climate variability; therefore, they did not need to pay as much attention to changes in the patterns of rainfall and temperature. In some cases, although farmers may be well aware of climatic changes, few would take steps to adjust their farming activities. These could be due to lack of credit and information on climate and appropriate technological options.

In Kenya, most research work on climate change impacts and adaptation has been based on analysis of individual crops, analysis of 'representative farms' or modelling exercises using aggregated data, with little or no participation of farmers and other stakeholders in analysing climatic variability and constraints to adaptation. The objective of this study was to assess the perceptions of communities living in semi-arid areas of Kenya with respect to understanding their local knowledge of climatic trends and related shocks, and the adopted coping options to different shocks. Through this, constraints to adaptation and appropriate recommendations to policymakers would be made.

Materials and Methods

The study was undertaken in two semi-arid areas of Kenya. In the Rift Valley province, the Lare division of Nakuru district was selected while in the Eastern Province, Gachoka division in Mbeere district was identified for the study (Fig. 1). These sites represent areas with relatively high population density compared to other semi-arid areas of Kenya, with approximately 95 persons per km^2. The study was conducted between January and April 2009. Both study areas experience a bi-modal rainfall pattern (500–800 mm per year), which is variable in time and space—typical of semi-arid areas of Kenya. The long rainfall season comes in March–May while the short rainfall season comes in October–November. According to the local experts, a successful cropping season always comes once every five years, the rest of the seasons' rainfall is below the ideal amount or there is the occurrence of droughts causing crop failure and lack of fodder. Grazing areas are similarly insufficient for the livestock population in both studied divisions, especially in periods prior to the wet seasons. The landscape is characterised by over-grazed shrubs/grasslands and smallholder agriculture plots. The expansive land has rolling hills of moderate gradients and degraded land with sandy/stony soils in most areas except in valley bottoms, whose ephemeral streams are becoming permanently dry in most years.

The majority of households are smallholders who own 1.2–2.6 ha of agricultural land. They are involved in both crop production and livestock keeping, especially for beef and dairy products. Because of the multiple challenges and constraints faced by smallholder farmers to cope with climate variability and the effects of climate change, the study was conducted to assess their knowledge and perceptions of these changes. We also wanted to establish which methods they were employing in order to cope with these long-term changes. Constraints to adaptation were similarly listed. The questionnaire had several sections. They included a section on the socio-economic characteristics of the household; household expenditures; shocks experienced by the household; perceptions of climate change; adaptation responses; and constraints to adaptation; whether they had noticed any long-term changes in mean temperature and rainfall, and the direction of the change; adjustments made in response to perceived changes in

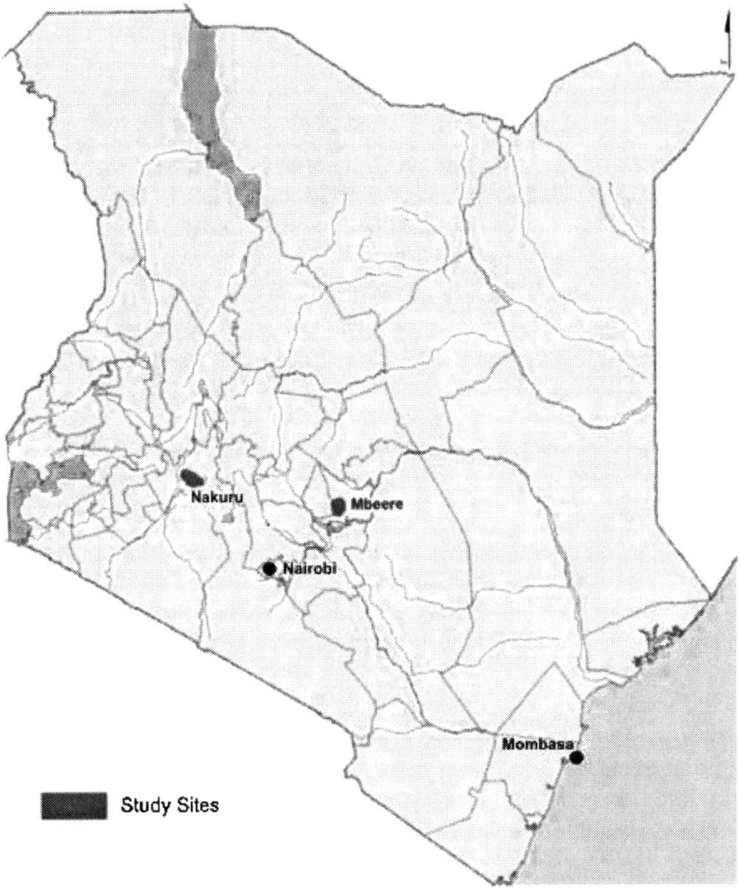

Fig. 1 The map of Kenya, showing the study sites

temperature and rainfall and the constraints to adaptation. The questions were structured to allow easy selection of right answers to the posed question but also to allow respondents to add other comments. Most farmers interviewed had primary level of education (7–10 years spent in school) and had an average household size of six members. Random sampling of farmers to respond to the questionnaire was done from a list of household heads (either men or women) generated by village elders in 60% of the villages in each division. All interviewees were household heads and must have had farming experience of between 15 and 20 years. Sixty nine farmers were selected in the Nakuru–Lare division and 100 farmers in Mbeere–Gachoka division were purposely selected using the set criteria. They proportionally represented 30% of the 15–20 years' farming experience by household heads. Analysis of data was done using SPSS, a statistical package program. Only some sections of the questionnaire are discussed in this chapter.

Results and Discussion

Climate Change Perceptions

While past studies have linked farmers' perceptions of climate change to the real climate data (Vedwan and Rhoades 2001; Hageback et al. 2005; Gbetibouo 2009), this study was not able to access local meteorological data. It therefore relies on regional model predictions to then establish what the farmers think about climate variability and climate change. It was on the basis of age or farming experience (not less than 15 or 20 years) that the perceptions are considered reliable. The farmers in both districts perceive that over the last 20 years the average temperatures and rainfall have changed. Over 92% of respondents ($n = 169$) from both districts observed that the average temperatures are increasing while rainfall is decreasing. They had also observed long-term change in rainfall timing (variability). Despite differences in magnitude of responses in the two study areas on various issues related to rainfall variability, it was clear to farmers that rains are becoming both more erratic and coming later in the cropping seasons. These changes are recent and more intense than they used to be. Traditionally, both study areas have bi-annual rainfall patterns. The first rainy season is from April to June (long rainy season) while the second rainy season is expected from October to November (short rainy season). Reliance on such a range of dates/months that change the timing of when to prepare land for planting is a big challenge. Also longer periods of drought were observed by majority of the respondents (30–84%) (Fig. 2). These findings agree with model predictions on the regional scale, which have stated that with climate change some regions of Eastern Africa will receive such climatic variability (KNMI 2006; Thornton et al. 2006). These variabilities will be more common in vulnerable regions like semi-arid and arid lands.

Asked when they experienced droughts and too much rainfall, farmers in both districts seem to have been experiencing consistent drought periods from 2004 to 2008 but worst experience was in 2008, especially for the Mbeere farmers (Fig. 3). Droughts were however, experienced in 1984 and 2008 by farmers in both districts except in 2009 by Nakuru farmers. With regard to too much rain, most farmers reported that 1997/1998 (El Niño rains) was their wettest season. Otherwise, periods between 2004 and 2007 were also mentioned as wet seasons (Fig. 4).

Both the drought and too much rainfall phenomenon are becoming common in most parts of Kenya both in perceived high rainfall areas (humid wet highlands) and low and erratic rainfall areas (the dry lowlands). As such, the likelihood of experiencing floods in drier areas or crop failure in traditionally wet highlands is not a surprise. Differences in the year when drought or too much rainfall occurred may be a matter of localised variabilities or farmers' own perceptions of what qualifies as drought or too much rainfall. In most cases, farmers in both study areas observed too much rainfall in 1997/1998, during the El Niño. The perceptions of farmers are much linked to how successful they were in their crop or livestock production in a particular year. Also, announcements on government radio and

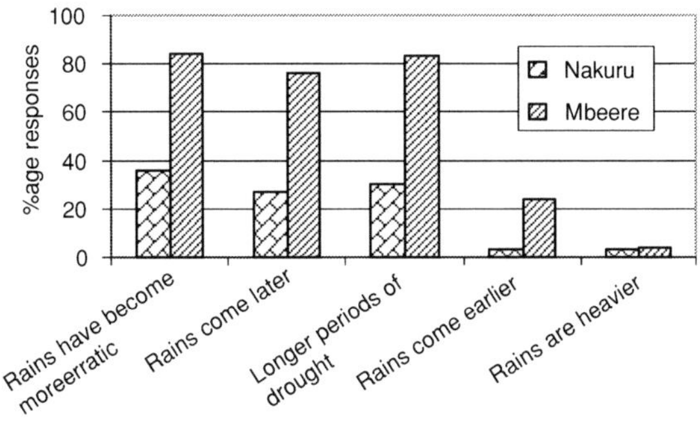

Fig. 2 Perceptions of rainfall variability in Nakuru–Lare and Mbeere–Gachoka

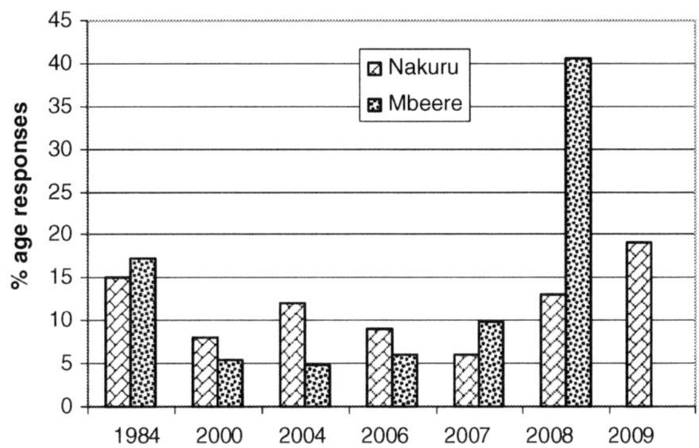

Fig. 3 Years when drought was experienced in Nakuru–Lare and Mbeere–Gachoka

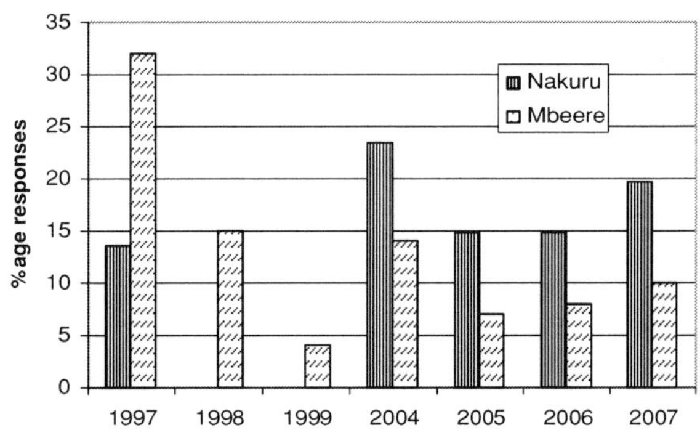

Fig. 4 Years when too much rainfall was experienced in Nakuru–Lare and Mbeere–Gachoka

television may also have had a strong influence on farmers' responses or expectations of the annual or seasonal rainfall pattern. In an expected dry year, food relief and other farm input aids accompany government warnings. So it was easy for farmers to recall years when it was dry.

Table 1 Types of major climatic shocks encountered by farmers in Nakuru–Lare and Mbeere–Gachoka

Type of major shocks	Nakuru–Lare		Mbeere–Gachoka	
	Responses	Percent of farmers	Responses	Percent of farmers
Drought	101	56	148	56
Erratic rainfall	52	29	87	33
Floods	14	8	11	4
Hailstorms	5		5	2
Pests and diseases	7	4	11	4
Total	179		262	

Climate Shocks and Outcomes

Several climate shocks were experienced as a result of the prevailing climate changes in the country. The outcome of major climatic shocks affected farmers differently. The most frequently mentioned climate shocks (Table 1) by respondents were drought (56%) and erratic rainfall (29–33%). Other shocks included floods (4–8%), hailstorms (2% only in Lare) and outbreaks of pests and diseases (4%) affecting humans, livestock and crops. Though many African farmers are said to have developed coping strategies to most of these shocks (Mortimore 1998; Mortimore and Adams 2001), there is still a high level of vulnerable community members due to diverse socio-economic and technological reasons.

These shocks resulted in various outcomes. Table 2 shows responses to several outcomes that farmers have experienced as a result of the climate shocks. Whereas all outcomes were mentioned in both study areas, some had higher frequency than others in respective districts. Droughts affected more farmers in Mbeere–Gachoka than in Nakuru–Lare. These resulted in declines in crop yields (47–82%), loss of income (11–88%), loss of entire crop yield (11–77%), food shortage/insecurity (11–76%), food price increase (10–86%) and death of livestock (8–100%). Whereas drought and erratic rainfall pattern were responsible for most of these outcomes, farmers perceived that the floods, hailstorm and pest and diseases were least responsible for these outcomes. These findings are consistent with the widely experienced climate shocks and impacts in ASAL in the region. The drought phenomenon is common and besides affecting crop and livestock production, it causes farmers to lose income from farm production, forcing them to sell household assets to meet immediate household needs. The farmers of Mbeere districts were more vulnerable to drought and erratic rainfall than those of Nakuru. Hailstorms were common in Nakuru district probably due to effect of the Rift Valley geographical systems on rainfall patterns. Due to the concavity of landscape and low tree densities in the Rift Valley, the wind speeds are high and these could contribute to causing hailstorms.

Specific pests and diseases mentioned by respondents were cases of malaria and ticks. Farmers observed different trends in the number of malaria cases (Table 3)

Table 2 Percentage responses to outcome of climate shocks in Nakuru–Lare and Mbeere–Gachoka

Shock outcome	Drought		Erratic rainfall		Flood		Hailstorm		Pests/disease	
	M*	N*	M	N	M	N	M	N	M	N
Decline in crop yield	82	47	14	45	82	64	2	63	2	41
Loss of income	88	11	6	16	0	19	3	16	3	29
Loss of entire crop	77	11	21	9	2	10	0	16	0	6
Food shortage/insecurity	76	11	24	9	0	5	0	0	0	12
Food price increase	81	10	19	11	0	2	0	5	0	6
Death of livestock	100	8	0	6	0	0	0	0	0	3

* *M* refers to Mbeere–Gachoka and *N* to Nakuru–Lare

Table 3 Noticed change in the number of malaria cases in the last 20 years in Nakuru and Mbeere

	Nakuru		Mbeere	
	Number of responses	Percent of farmers	Number of responses	Percent of farmers
Increased	52	51	30	31
Decreased	43	42	47	49
Stayed the same	3	3	7	7
Don't know	5	5	12	13
Total	103		96	

and incidences of diseases related to livestock ticks (Table 4) for the last 20 years. While 31–51% of farmers in the study areas perceived that there was an increase in malaria cases, between 42–49% perceived decreasing trends in the incidences of malaria cases. The perceived decrease by part of the respondents was possibly due to a government programme of supplying free mosquito nets to all pregnant mothers in the country and to families living in malaria-prone environments. Whereas it has been traditionally known that outbreaks of malaria cases were synonymous with lake or coastal regions, due to rising temperatures accompanied by too much rainfall mosquito breeding sites are growing outside these ranges, resulting in these findings.

Livestock diseases related to outbreaks of ticks were noted to be increasing and at the same time decreasing by a section of farmers in both study areas (Table 4). While some farmers had access to community cattle dips, others could afford their own pesticides to keep off the ticks but also practised livestock confined management i.e. a zero-grazing system. The other reason that some farmers observed an increase was because of free-range grazing tendencies in search of pasture during dry seasons or during drought events. Few farmers thought the disease incidences had stayed the same or were not aware of changes in the trends.

Table 4 Noticed change in the incidence of livestock diseases from ticks over the last 20 years in Nakuru and Mbeere

	Nakuru		Mbeere	
	Number of responses	Percent of farmers	Number of responses	Percent of farmers
Increased	42	41	33	35
Decreased	44	45	45	47
Stayed the same	2	2	2	2
Don't know	13	13	15	16
Total	103		95	

Coping Options and Knowledge of Climatic and Agricultural Information

Households hit by a climatic shock tend to develop strategies to cope with the shock, which is an important indication of adaptation. As a result of long-term shifts in temperature, rainfall amounts and variability, farmers have adjusted their farming strategies through several coping options. Figure 5 shows some of the adopted coping options. The results indicate that farmers (20–29%) from both districts tend to invest their resources in adapting to the shocks by focussing on crop-related adjustments. Adoption of crop varieties and types, and changing the planting dates were mentioned by 10–35% of respondents as some of the adaptation packages for climate change. Few farmers mentioned soil and water management or better management of livestock as strategies to adapt to climate change. A large number of farmers (40%), especially from Mbeere district, did nothing to counter the challenges brought about by climate change. Failure by farmers to take action against these shocks was mostly linked to widespread lack of access to technologies, credit or lack of information on climate change. There are two possible reasons for failing to take action. First, it could be that they had already adjusted to the frequent shocks in the region by growing drought-resistant/tolerant crop varieties or keeping local livestock breeds besides other measures deemed suitable. And second, it could be that their old coping strategies were by far superseded by current adverse climate changes shocks.

Another aspect the study undertook to find out was how receiving agricultural extension information through available media channels influenced adoption of coping measures to climate change shocks. Farmers adjusted to climate shocks by adopting crop management-related measures, changing areas of land under use, water management schemes and changing livestock management systems, among others. Table 5 shows the percentage responses of farmers who had adopted the listed measures and the proportion of those who had adopted them because they had been reached by extension agents with information on crop and livestock production. In both districts, of those who adopted coping measures about half had received information on crop and livestock production. More adoption measures

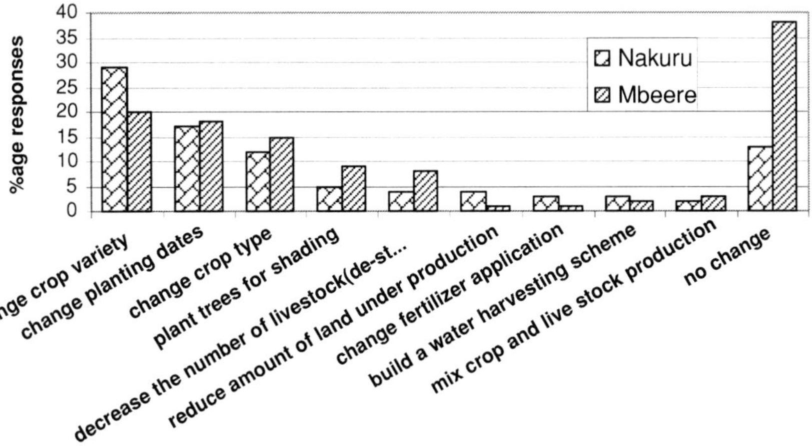

Fig. 5 Coping options adopted by farmers in Nakuru–Lare and Mbeere–Gachoka

Table 5 Interaction between adopted coping options with/without information on crop and livestock production from extension agents in Mbeere and Nakuru

Coping options	Mbeere			Nakuru		
		Percent of responses with/ without information			Percent of responses with/ without information	
	No. of responses	Yes	No	No. of responses	Yes	No
Change crop variety	42	45	55	56	54	46
Change crop type	32	56	44	23	52	48
Change planting dates	37	54	46	32	50	50
Reduce amount of land under production	3	33	67	7	57	43
Change fertiliser application	3	33	67	5	100	0
Build a water harvesting scheme	5	80	20	7	57	43
Plant trees for shading	19	58	42	8	63	37
Mix crop and livestock production	7	71	29	4	50	50
Decrease the number of livestock (de-stocking)	16	25	75	7	71	29
No change	8	50	50	23	52	48

involved the changing of crop varieties and types, and planting dates; planting of trees for shedding and on de-stocking. The first three coping measures were adopted by the majority of farmers in both Mbeere and Nakuru districts. Where a large proportion of responses is ‘Yes’, then the information may have influenced the adoption levels too. There are those who may have adopted measures not

Table 6 Interaction between adopted coping options with/without information on expected temperature and rainfall in Mbeere and Nakuru

Coping options	Mbeere			Nakuru		
		Percent of responses with/without information			Percent of responses with/without information	
	No. of responses	Yes	No	No. of responses	Yes	No
Change crop variety	42	64	36	55	82	18
Change crop type	32	66	34	23	74	26
Change planting dates	37	76	24	31	87	13
Reduce amount of land under production	3	100	0	6	100	0
Change fertiliser application	3	67	33	4	50	50
Build a water harvesting scheme	5	80	20	7	71	29
Plant trees for shading	19	79	21	8	88	12
Mix crop and livestock production	7	71	29	4	75	25
Decrease the number of livestock (de-stocking)	16	76	25	6	83	17
No change	8	63	37	23	65	35

because they had had any information/advice from local extension agents but from other sources. However, such responses imply that receipt of information on crop and livestock production from extension agents or other types of service providers could influence adoption of coping measures by a greater extent. It also implies that the level of vulnerability can be reduced greatly by providing critical farm management information to the farming communities.

We also assessed whether having knowledge of expected temperature and rainfall made farmers implement adaptation measures or not (Table 6). These results show that the majority of farmers adopted the listed coping measures because of the information on the expected temperature and rainfall received from the government climate prediction services. These findings show how critical accurate provision of seasonal weather information could be in facilitating widespread countermeasures to climate change shocks. All climate information needs to be sent to its recipients in good time in order to help in the early adoption of coping/mitigating measures to ensure better crop/livestock production while conserving the environment at both the local and national levels.

Conclusion and Recommendations

This study reinforces the understanding by farmers that climate change is real as has been predicted by models. Though the farmers are aware of climate variability and climate change, their ability to adapt to the experienced shocks is low.

The findings imply that smallholders will face severe impacts of climate change if their capacity to cope is not enhanced. Reasons for low adaptation could be due to lack of capacity to employ necessary measures to cope with/mitigate the effects of climate change. It has been observed that where farmers received or had access to information on crop and livestock management or received weather and climate information, adoption of coping measures was higher. This could explain that information made available to farmers could only be on crop-related management rather than on livestock or land and water management. This explains why most adjustments were to crop variety/types, and change of planting dates. There is less capacity and fewer resources to employ irrigation schemes or tap into the potential for integrating crop and livestock and agroforestry systems to cope with/mitigate the effects of climate change and variability. The challenge is in how these (uncertain) changes should be considered in development policy design and investment operations to ensure that climate change does not compromise agriculture's role for development, but rather strengthen it. Government and development agencies have the opportunity to promote and guide adaptation of the agricultural systems to support the vulnerable members of farming communities in fragile environments to adapt to challenges of the growing effects of climate change. Packaging of appropriate information and technologies and upscaling them using suitable channels and in real time could build poor farmers' capacity to counteract this global phenomenon.

Acknowledgements The authors wish to show their appreciation for the input of community leaders and Ministry of Agriculture officers in various villages in the Gachoka division of Mbeere district and the Lare division of Nakuru district, without whose help we could not have conducted this study. We would also like to acknowledge the financial and moral support provided by the offices of FAO-Ethiopia and Rome, and the Director of the Kenya Agricultural Research Institute (KARI), Nairobi. Special appreciation goes to enumerators who conducted the interviews and the KARI researchers who supervised the enumerators, namely Mr Louis Gachimbi, Mr Edward Biketti, Mr Philip Mwangi, Mrs Jane Mwangi and Mrs Violet Gathara. The input of Ms Elizabeth Bryan and Dr Carla Ringler (IFPRI-Washington DC) during questionnaire design is acknowledged. We are also very grateful for the commitment of Mr Elias G. Thuranira and Ms Merciline Sikasa during data entry, cleaning and analysis.

References

Bryant RC, Smit B, Brklacich M, Johnston RT, Smithers J, Chiotti Q, Singh B (2000) Adaptation in Canadian agriculture to climatic variability and change. Clim Change 45:181–201

Funk C, Dettinger MD, Michaelsen JC, Verdin JP, Brown ME, Barlow M, Hoell A (2008) Warming of the Indian Ocean threatens eastern and southern African food security but could be mitigated by agricultural development. PNAS 105(32):11081–11086

Gbetibouo GA (2009) Understanding farmers' perceptions and adaptations to climate change and variability. The case of the Limpopo basin, South Africa. IFPRI Discussion Paper 00849, Feb 2009

Hageback J, Sundberg J, Ostwald M, Chen D, Yun X, Knutsson P (2005) Climate variability and land use change in danagou watershed China—examples of small scale farmers' adaptation to climate change. Clim Change 72:189–212

IPCC (2001) Climate variability 2001. In: McCarthy JJ, Canziani OF, Leary NA, Dokken DJ, White KS (eds) Impacts, adaptation, and vulnerability. Cambridge University Press, Cambridge

IPCC (2007) Climate change 2007: the physical science basis. Contribution of working group I to the fourth assessment report of the intergovernmental panel on climate change. Solomon S, Qin D, Manning M, Chen Z, Marquis M, Averyt KB, Tignor M and Miller HL. Cambridge University Press, Cambridge, United Kingdom

KNMI (2006) Climate change in Africa. Changes in extreme weather under global warming, Royal Netherlands Institute of Meteorology, Brochure. http://www.knmi.nl/africa_scenarios/

Maddison D (2006) The perception of and adaptation to climate change in Africa. CEEPA Discussion Paper No. 10. Centre for Environmental Economics and Policy in Africa, University of Pretoria, South Africa

McCarthy J, Canziani OF, Leary NA, Dokken DJ, White C (2001) Climate change 2001: impacts, adaptation, and vulnerability. Contribution of working group II to the third assessment report of the intergovernmental panel on climate change. Cambridge University Press, Cambridge

Mortimore M (1998) Roots in the African dust: sustaining the sub-Saharan drylands. Cambridge University Press, Cambridge

Mortimore M, Adams WM (2001) Farmer adaptation, change and 'crisis' in the sahel. Global Environ Change 11:49–57

Smithers J, Smit B (1997) Human adaptation to climatic variability and change. Global Environ Change 7(3):129–146

Thornton PK, Jones PG, Owiyo TM, Kruska RL, Herrero M, Kristjanson P, Notenbaert A, Bekele N and Omolo A with contributions from Orindi V, Otiende B, Ochieng A, Bhadwal S, Anantram K, Nair S, Kumar V and Kulkar U (2006) Mapping climate vulnerability and poverty in Africa. Report to the Department for International Development, ILRI, PO Box 30709, Nairobi 00100, Kenya, p 200

Tschakert P (2007) Views from the vulnerable: understanding climatic and other stressors in the sahel. Global Environ Change. doi: 10.1016/j.gloenveha,2006.11.008

UNEP/GoK United Nations Environment Programme (UNEP), Government of Kenya (GoK) (2000) Devastating drought in Kenya. Environmental impacts and responses. UNEP/GoK, Nairobi

Vedwan A, Rhoades RE (2001) Climate change in western Himalayas of India: a study of local perceptions and responses. Clim Res 19:109–117

Ziervogel G, Cartwright A, Tas A, Adejuwon J, Zermoglio F, Shale M, Smith B (2008) Climate change and adaptation in African agriculture. Report prepared for the Rockefeller foundation by the Stockholm Environment Institute

Author Biographies

Barrack Okoba, PhD is senior research scientist at the Kenya Agricultural Research Institute (KARI), Nairobi Kenya. He holds a PhD from Wageningen University, the Netherlands, in Production Ecology and Resource Conservation. His research interests cover soil and water management research issues and climate change and adaptation measures. He is the coordinator of the Soil and Water Management Programme at KARI and national coordinator of conservation agriculture research. He has published more than 20 papers in refereed journals and scientific papers. Currently, he is leading several partnership research projects.

A. Almeneh Dejene, PhD is a senior scientist and team leader of Environment and Climate Change Adaptation at the headquarters of FAO-Rome. His research interests include climate change adaptation research on smallholders in tropical regions of the world.

Meshack Malo, MSc is a natural resource officer with the FAO-Ethiopia regional office. He coordinates project implementation on environmental systems services in the region.

Chapter 11
Drought Monitoring in Food-Insecure Areas of Ethiopia by Using Satellite Technologies

Getachew Berhan, Tsegaye Tadesse, Solomon Atnafu and Shawndra Hill

Abstract The purpose of this paper is to develop a new concept and approach for extracting knowledge from satellite images for near real-time drought monitoring in areas experiencing food insecurity in order to mitigate climate change. The near real-time data downloaded from the Atlantic Bird satellite was used to produce the drought spatial distribution in the assessed areas. During the analysis, cloud-contaminated pixels were removed from each individual image by examining reflectance and temperatures. After completing the pre-processing of the satellite images, the NDVI values of the images were calculated, followed by the determination of the deviations associated with the normalised difference vegetation index (NDVI) values for the first decade (i.e. the first ten-day period) of October 2009. The actual drought condition data was obtained by comparing the NDVI from the MSG satellite for the first decade of October 2009 with the long-term mean NDVI from the NOAA satellite. Our results showed that approximately 40% of the observed areas exhibited negative deviation. This indicates drought conditions in 2009 in the East Africa and southern Asian regions. These results align with recorded rainfall data in 2009 for most parts of Ethiopia. In this study, the

G. Berhan (✉) · S. Atnafu
Addis Ababa University, PO Box 30726, Addis Ababa, Ethiopia
e-mail: getachewb1@yahoo.com

S. Atnafu
e-mail: satnafu@cs.aau.edu.et

T. Tadesse
National Drought Mitigation Center, University of Nebraska-Lincoln, Lincoln, NE, USA
e-mail: ttadesse2@unl.edu

S. Hill
University of Pennsylvania, Philadelphia, PA, USA
e-mail: shawndra@wharton.upenn.edu

W. Leal Filho (ed.), *Experiences of Climate Change Adaptation in Africa*,
Climate Change Management, DOI: 10.1007/978-3-642-22315-0_11,

possibility of using the near real-time spatio-temporal MSG data for drought monitoring in food-insecure areas of Ethiopia was tested and promising results were obtained. In 2009, there was drought in most parts of Ethiopia and Sudan. Our retrospective analysis of this study also confirms this fact. The output of this research is expected to assist decision-makers in taking timely and appropriate action in order to save millions of lives in drought-affected areas by using advanced satellite technology.

Keywords Drought monitoring · Geospatial information · Knowledge discovery · Satellite images

Introduction

Because of climate change and climate variability, drought has become a recurrent phenomenon. It is manifested in erratic and uncertain rainfall distribution in rainfall-dependent farming areas, especially in arid and semi-arid ecosystems. Frequent and severe drought has become one of the most important natural disasters in sub-Saharan Africa, resulting in serious economic, social, and environmental crises (Tadesse et al. 2008). Its effect is marked by creating uncertain agricultural economies (UNEP 2006).

Ethiopia is a country in the sub-Saharan region that has been affected by the drought. Millions of lives have been lost because of recurring droughts in the past several decades (UNEP 2006). These days, due to climatic change, drought occurs every two years in different parts of Ethiopia (UNEP 2006; NMSA 1996). In addition, over time, the drought recurrence cycle is shortening while the affected area is widening, impacting additional parts of the country that were once unaffected (NMSA 1996). In order to respond to the effects of drought, Ethiopia has been conducting drought assessment and monitoring missions.

In Ethiopia, drought assessment and monitoring efforts have been based on conventional methods that rely on the availability of meteorological data. The process of gathering meteorological data is very tedious and time consuming. Moreover, meteorological data and weather information dissemination is also a challenge. Consequently, millions of lives may be lost before the actual information is submitted to the appropriate decision-makers (UNEP 2006). The information that is produced in accordance with the conventional approach is usually highly uncertain for employing rescue missions. Therefore, producing reliable and timely information for decision-makers is of the utmost importance.

Traditionally, there are several operational indices in drought assessment and monitoring that are based on rainfall data. These indices are often not easily accessible, and not often tailored to be conveniently understood by decision-makers (Ji and Peters 2003). The common approach that is used to derive the necessary information is the application of climatic drought indices, such as the Palmer Drought Monitoring Index, which has been widely used by the US

Department of Agriculture (Jain et al. 2009). Another popular climatic drought index is the Standardized Precipitation Index (SPI) that was developed by McKee et al. (1993), which can identify data about emerging drought months for regional and global applications. Mishra and Desai (2005) have adopted the SPI for parts of India and, with that data, they compiled a drought severity area frequency curve. These drought severity and monitoring indices are based on point data measured at the different meteorological stations, which are located across a wide area. In remote areas without a dense network of stations, extrapolation of rainfall observation from nearby stations is commonly used, resulting in high uncertainty about its usefulness for real-time rescue missions.

At present, decision-makers in many countries use remote sensing for closing the gap and obtaining the desired information. Remote sensing data or data from satellite sensors can provide continuous data sets that can be used to detect the onset of a drought, its duration and magnitude (Thiruvengadachari and Gopalkrishana 1993). Remote sensing is far superior to conventional methods (Jain et al. 2009) for drought monitoring and early-warning applications. The challenge in applying remote sensing data in drought monitoring and in issuing early-warnings is that the various indices serving this purpose have to be validated and calibrated to the intended region and ecological conditions (Singh et al. 2003; Jain et al. 2009). So far, no significant efforts have been made to validate and calibrate remote sensing data in food-insecure areas within Ethiopia. Thus, the available information is unclear, uncertain, and difficult for decision-makers to access (FEWS NET 2009). In addition, even though drought has its own state and behaviour, there have been no past efforts to detect drought by its own properties as a spatial object (Rulinda et al. 2009).

Taking these challenges into account, the main objective of this research was to develop a new concept and approach for extracting knowledge from satellite imageries for near real-time drought monitoring in food-insecure areas for mitigating climate change. In line with this, the study intended to answer two research questions: (1) Is it possible to model and predict drought as a spatial object in food-insecure areas? and (2) What are the appropriate satellite imageries' temporal resolutions for modelling drought as a spatial object?

Advanced-technology satellite products with high temporal resolution (e.g. MSG data every 15 min) are cost effective and can serve to detect the onset of a drought and to predict its duration and magnitude. Such information can help decision-makers to take appropriate actions in a timely manner, reduce the impact of drought conditions, and mitigate drought's adverse effect on the environment. This effort is indicated to be one of the climatic change mitigation efforts for countries that have been affected by recurrent drought in the past (UNEP 2006).

An Overview of Drought Monitoring and Modelling

Drought is defined as "the naturally occurring phenomenon that exists when precipitation has been significantly below-normal recorded levels, causing serious hydrological imbalances that adversely affect land resource production systems"

(UNCCD 1999). Drought is also defined as a prolonged abnormally dry period when there is not enough water for users' normal needs, resulting in extensive damage to crops, and a loss of yields (Wilhite 2005). These definitions of drought are conceptual explanations that provide the basis for the operational meaning. The operational definition of drought focuses on identifying the beginning and end, spatial extent, and severity of the drought in a given region and it is based on scientific reasoning. The analysis is conducted by using hydro-meteorological information and is beneficial in developing drought policies, early-warning monitoring systems, mitigation strategies, and preparedness plans (Smakhtin and Hughes 2004). Generally, the most prominent types of drought are meteorological, agricultural and hydrological drought (Wilhite 2000; Obasi 1994) (Fig. 1a). Meteorological drought is usually defined according to the degree of dryness (i.e., in comparison to the "normal" or average amount of precipitation) and the duration of the dry period at a particular place and at a particular time. Hydrological drought is associated with the effects of periods of precipitation (including snowfall) shortfalls on the surface or subsurface water supply (i.e. stream flow, reservoir and lake levels, and groundwater). Although all droughts originate with a deficiency of precipitation, hydrologists are more concerned with how this deficiency plays out through the hydrological system. Hydrological drought is associated with the effect of low rainfall on the water levels of rivers, reservoirs, lakes and aquifers. Agricultural drought links various characteristics of meteorological and hydrological drought to agricultural impacts, focusing on precipitation shortages, differences between actual and potential evapotranspiration, soil water deficits, and reduced groundwater or reservoir levels (Wilhite 2000). Agricultural drought occurs when there is not enough water available for a particular crop to grow at a particular time. The focus of this research is on agricultural drought analysis and an early-warning system. The frequency of agricultural drought in Ethiopia is presented in Fig. 1b, c.

The process of monitoring agricultural (i.e. vegetative) drought usually requires remote sensing technologies and a large amount of temporal data in addition to traditional climate information. The major source of satellite data that is widely used is the NDVI (Rulinda et al. 2009). The NDVI is commonly calculated by using image data from polar orbiting satellites, which carry sensors that detect radiation in red and infrared wavelengths (Fensholt et al. 2006). The NDVI is used, in this case, by comparing the deviation of the current satellite observation from the historical average within a certain time period (i.e. window) of interest. In the analysis of droughts, their onset, duration and severity are often difficult to determine and the characteristics may vary significantly from one region to another (Rulinda et al. 2009). In rainfall-dependent agriculture production areas, seasonal rainfall variability is inevitably reflected in both highly variable production levels and in the risk-averse livelihoods of local farmers (Cooper et al. 2008). Africa has a long history of rainfall fluctuations of varying lengths and intensities (Nicholson 1994), mainly due to climatic change. Recent studies indicate varying behaviour of rainfall trends in Africa on different spatial and temporal scales. A recent study also demonstrated a decrease in rainfall in East

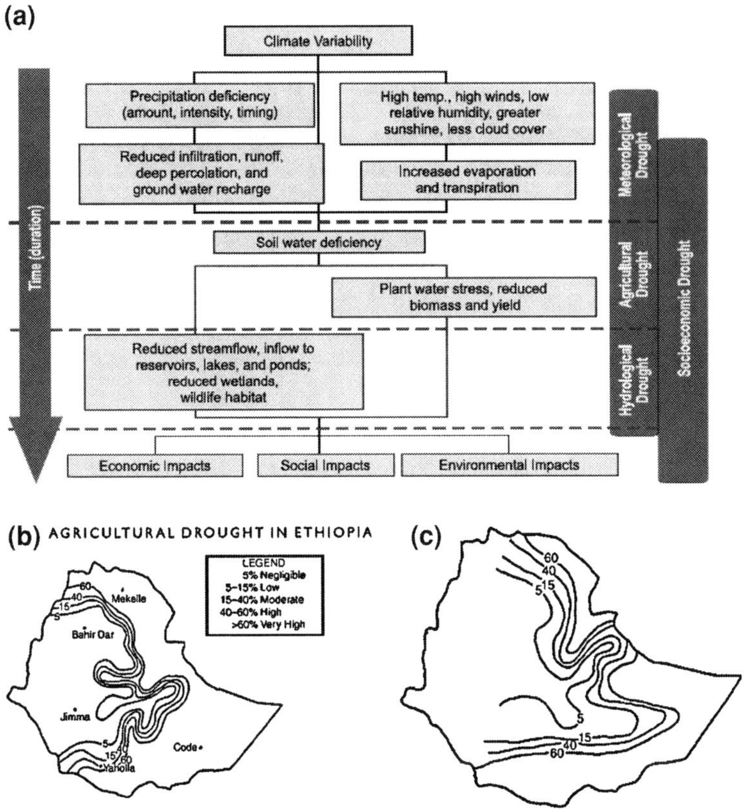

Fig. 1 Drought classification (adapted from UNISDR (2009) (**a**), frequency of agricultural droughts in Ethiopia during the first rainy seasons (February–May) (**b**), and second rainy seasons (July–September) (**c**) (adapted from NMSA1996)

Africa between 2003 and 2008 (Swenson and Wahr 2009), in which drought and famine situations were periodically reported (FEWS NET 2009). Drought has a particularly negative impact on agricultural production in the eastern African region, as most of its agriculture is dependent on rainfall (Thorton et al. 2009) rather than irrigation.

The conventional approach to drought monitoring and early-warning systems that uses ground-based data collection is tedious, time consuming and difficult (Prasad et al. 2007). In recent years, remote sensing data has been used for monitoring agro-climatic conditions, the state of the agricultural fields, vegetation cover, and for estimating crop yield in various countries. In particular, the advanced very high-resolution radiometer (AVHRR) NDVI information has been used in vegetation monitoring, crop yield assessment and forecasting (Hayes et al. 1982; Benedetti and Rossini 1993; Quarmby et al. 1993; Unganai and Kogan 1998; Kogan et al. 2003). The National Oceanic and Atmospheric Administration's (NOAA) AVHRR series satellite feedback provides a long-term record of NDVI data that can be used to predict crop yield (Prasad et al. 2007). Yield prediction is a component of the drought monitoring process, in that crop yield information is essential for determining the food assurance of a given region.

NDVI is presently used as a primary source of data in remote sensing-based drought monitoring. Rulinda et al. (2009) indicated that other parameters, such as soil moisture, rainfall (RF), and surface temperature, which can best explain drought, were not included in past analyses. The authors also noted that the development of drought monitoring and early-warning systems requires a priori knowledge of the characteristics of the study areas' vegetation and the notation of the time of the day for the acquisition of the satellite images. However, past research has not addressed these factors (Rulinda et al. 2009).

Drought Object Modelling

The concept of object identification and modelling has been an ongoing scientific effort for converting remotely sensed images into geographic phenomena (Stein et al. 2009). In this research, an object is noted in the context of object-oriented modelling (Woryboys et al. 1990). This is based on the basic principle in which an object has two characteristics: state and behaviour (Woryboys et al. 1990; Budd 2000). State is the attribute or information contained by an object and behaviour is the set of actions in which an object performs (Budd 2000). In identifying and modelling the drought object by using satellite images, state means the actual reflectance attributes (i.e. the digital numbers that are registered by the satellite sensors as pixel values or any index values, such as NDVI) and behaviour means that when drought object occurs on the ground, plants die or cease the process of photosynthesis (i.e. the red band of the spectrum is not used by the plant and is reflected back to the satellite sensor). As a consequence, yield is reduced in the long-term effects (Tucker 1979; UNISDR 2009).

There are two key questions to be asked in identifying the state and behaviour of any object: (1) What possible states (i.e. attributes) can a give object be in? and (2) What possible behaviour (i.e. actions) can this object perform when it happens (Budd 2000)? The geographic object that we are interested in in this research is agricultural drought—where it results in reduced biomass and yield (Wilhite 2005).

The concept of identifying and modelling drought as an object is new (Rulinda et al. 2009). Rulinda et al. (2009) further indicated that "a next step in drought modeling is an approach focusing on spatial object and this kind of object can be built from different temporal resolution images". In remote sensing, objects are identified and subsequently classified on the basis of pixel information, and the objects are subsequently tracked in time, during which their behaviour may be governed by external factors that must also be identified and quantified (Stein et al. 2009). This process is usually accomplished through image mining techniques.

Image mining is defined as the analysis of large sets of observational images to find suspected or unsuspected relationships and to summarise the data in novel ways that are both understandable and useful to stakeholders (Stein 2008). Object identification in remote sensing is usually conducted by converting raster pixel values to geographic objects. Usually the image is first segmented, providing approximately homogeneous segments, and then classified (Stein et al. 2009). Stein et al. (2009) further stated that various procedures for image segmentation are well documented, and include procedures based on mathematical morphology, on edge detection, and the identification of homogeneity in one band or in a set of bands. Classification practices include statistical routines, such as k-nearest neighbour classifiers, and increasingly fuzzy classification methods (Stein et al. 2009).

Materials and Methods

Theoretical Framework

One of the criteria for innovative academic research is that it must have a clearly defined theoretical framework, which helps to differentiate research from consultants' work (Gregor 2006). Defining the theoretical framework for a given study also helps to accumulate knowledge in a systematic manner. Such accumulated knowledge enlightens professional practice (Gregor 2006; Gregor and Jones 2007). Taking this fact into account, the theoretical framework for this research is "design science". Gregor (2006) indicated that "design theory" provides explicit prescriptions for constructing an artefact and mainly answers the question of how to do something.

Design science is a problem-solving process. In their problem-solving process, Hevner et al. (2004) presented seven guidelines with which to conduct design science research. These guidelines consider the design as an artefact, problem relevance, design evaluation, research contributions, research rigour, design as a search process, and finally, the communications of research. The study presented in this paper modified Hevner et al. (2004)'s seven guidelines into five steps: identification, modelling, tracking, prediction, and communication with stakeholders. The artefact for the process of knowledge discovery from satellite images is presented in Fig. 2.

The term artefact is used to describe something that is artificial or constructed by humans, versus something that occurs naturally (Simon 1996). In this research, the term artefact signifies the abstract representation of the design science research process and its final information delivery to decision-makers.

Study Area

The specific study sites that were selected for this research are presented in Fig. 3. These study sites are five of the most drought-affected areas of Ethiopia. These sites were selected for validating and calibrating the remote sensing data, where

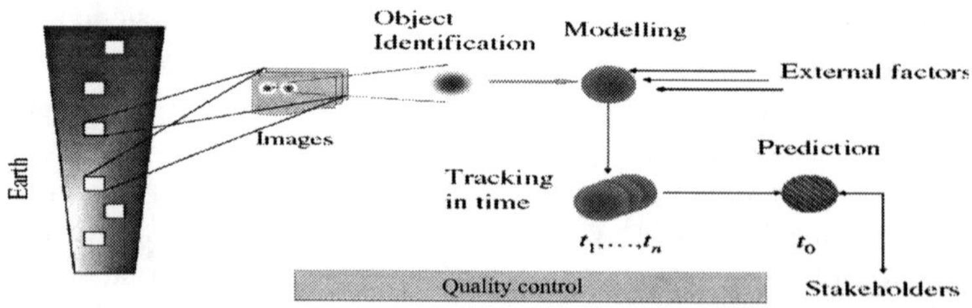

Fig. 2 An artefact for the process of knowledge discovery from satellite imageries (adopted from Stein et al. 2009)

the current research (part of a doctoral research project) was conducted. Currently, the first author is conducting preliminary surveys of the selected sites, including an assessment of the sites' socio-economic statuses. All of the selected sites contain meteorological stations that are administered by the National Meteorological Agency of Ethiopia. The sites were selected jointly with the Disaster Prevention and Preparedness and National Meteorological Agency of Ethiopia. The proposed sites include South Tigray (Alamata; WOALAM53 station), South Wello (WO-WERE94 station), North Shoa (SHANK014 station), East Shoa (SHDEBR42 station) and Somali Region (Jijiga HAJIJI11 station). These study sites are regularly affected by drought, and are safety net *weredas* (i.e. districts that are under continuous assistance of the Disaster Prevention and Preparedness Agency of Ethiopia). With respect to the preliminary results of this paper, drought assessment and analysis was conducted for the entire eastern African and southern Asian region.

Methodology

The satellite data and many of the biophysical variables were extracted using geographic information system (GIS) techniques. ILWIS 3.6 software was used in the analysis of the remote sensing imagery. The near real-time data downloaded from the Atlantic Bird satellite was used to produce the drought monitoring indices. During the analysis, cloud-contaminated pixels were removed from each individual image by examining the reflectance and temperatures. After completing the pre-processing of the satellite images, the NDVI values of the images were calculated using Eq. 1:

$$NDVI = \frac{\rho_{nir-}\rho_{red}}{\rho_{nir} + \rho_{red}} \tag{1}$$

where ρ_{red} (0.4–0.7 mm) and ρ_{nir}(0.75–1.1 mm) are reflectance in red and near infrared bands of the satellite imageries.

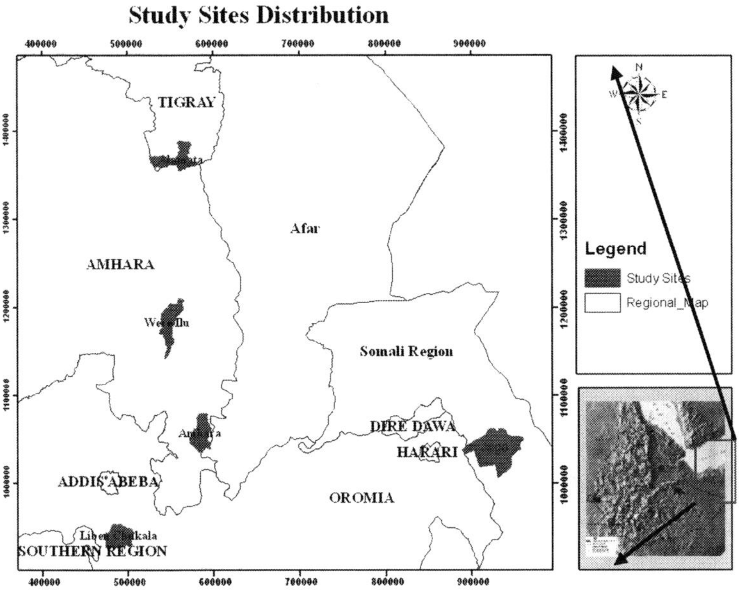

Fig. 3 Map showing the distributions of the five study sites in Ethiopia

NDVI is the most commonly used vegetation index. It has been shown to be related to vegetation vigour, percentage green cover, and biomass (Myneni and Asrar 1994; Anyamba and Tucker 2003; Tucker and Stenseth 2005). It is a non-linear function that varies between –1 and +1, and is undefined when both ρ_{red} and ρ_{nir} are zero. NDVI values for vegetated land areas generally range from approximately 0.1 to 0.7. Values greater than 0.5 indicate dense vegetation, whereas values lower than 0.1 indicate near zero vegetation, such as barren area, rock, sand, or snow (Tucker 1979).

In this research, the daily NDVI values were aggregated into decadal basis from MSG satellite data. In a year, there are 36 decades (one decade is equal to ten days). The decadal NDVI values were compared with the long-term mean NDVI value of the same decade from the NOAA AVHRR satellite data. The difference of these two data elements is called deviation of drought severity index, or the deviation of the NDVI (Dev_NDVI) (Tucker 1979). The Dev_NDVI was calculated using Eq. 2. When the Dev_NDVI is negative, it indicates below-normal vegetation conditions and might suggest a drought situation (Tucker 1979). We used Dev_NDVI to spatially locate the occurrence of drought.

$$Dev\_NDVI = NDVI\_i - NDVI\_Mean\_i \qquad (11.2)$$

where *NDVI_i* and *NDVI_Mean_i* are the actual ten-day composite NDVI and the long-term mean for the same decade NDVI values, respectively. *NDVI_i* was acquired from MSG and *NDVI_Mean_i* from NOAA.

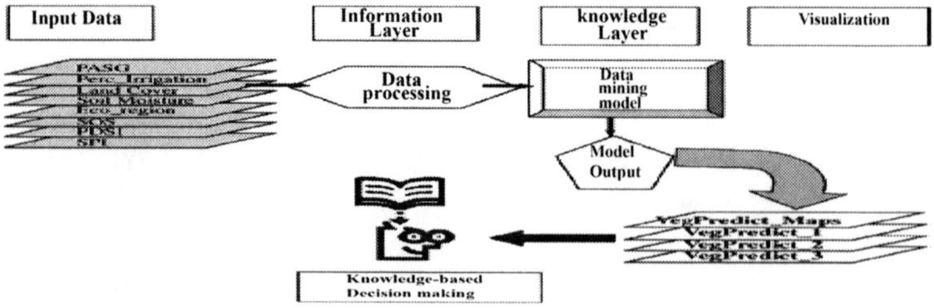

Fig. 4 Conceptual framework demonstrating the process used to produce information for decision-making (Tadesse et al. 2005)

The overall conceptual framework of the information produced from the various parameters mentioned above is integrated and the knowledge is delivered to the decision-maker in a user-friendly format. This can help decision-makers to formulate an action plan that is likely to save drought victims in food-insecure areas. The conceptual framework is depicted in Fig. 4.

The overall method for this research is presented in Fig. 5, which illustrates the major steps in knowledge discovery from satellite images for near real-time drought monitoring and early-warning systems.

Materials

For this study, satellite data from Meteosat Second Generation (MSG) and the National Oceanic and Atmospheric Administration (NOAA) AVHRR were used. MSG is the new European system of geostationary meteorological satellites together with the associated infrastructure. It was developed to succeed the highly successful series of original Meteosat satellites that has served the meteorological community for over two decades since they were first launched in 1977 (EUMETSAT 2005). The advanced spinning enhanced visible and infrared imager (SEVIRI) radiometer onboard the MSG series of geostationary satellites enables the Earth to be scanned in twelve spectral channels from visible to thermal infrared at 15 min intervals. Each of the twelve channels has one or more specific applications, either when used alone or in conjunction with data from other channels. From these twelve channels, in this research we used channels one and two for detecting vegetation condition. These two visible channels are well known from similar channels of the AVHRR instrument flown on NOAA satellites and can be used in combination to generate vegetation indices (such as NDVI) (EUMETSAT 2005).

NOAA is owned by the US government. The sensor onboard the NOAA mission that is relevant for earth observation is the advanced very high-resolution radiometer (AVHRR). NOAA and NASA have jointly produced long-term

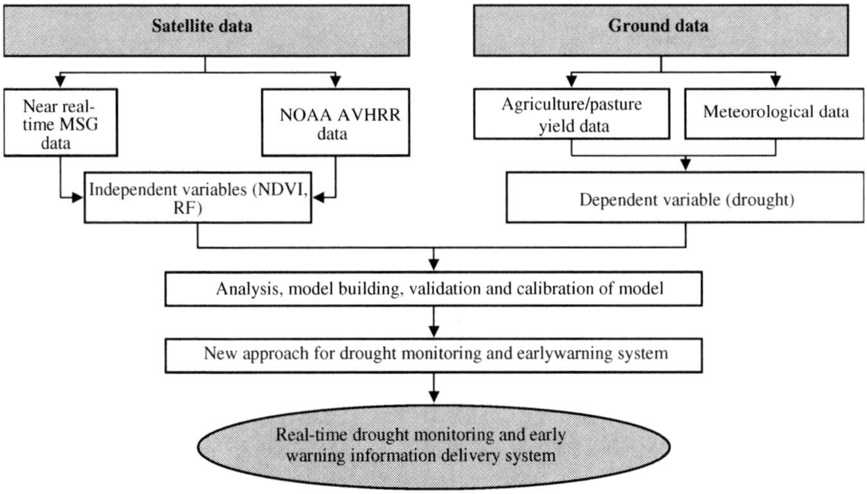

Fig. 5 The major steps in knowledge discovery from satellite images

AVHRR data sets that have been processed in a consistent manner for global change research. These data sets cover the period from July 1981 to the present. The data sets are ten-day composites of daily data (red, near infrared (NIR) and thermal wavelengths), mapped to a global equal area projection at 0.10 resolution (EUMETSAT 2005). There are three ten-day composites per month and the first is for days 1–10, the second is for days 11–20, and the third is for the remaining days. The data contains NDVI, a highly correlated parameter for surface vegetation, derived from the visible and near IR channel reflectance (EUMETSAT 2005; Holben 1986). This pathfinder data set has gone through many stages of calibration and correction (Smith et al. 1997).

GIS Data Processing and Preliminary Results

Relationships Between Precipitation and NDVI

To analyse the relationship between RF and NDVI values, data collected from 1982 to 2004 was used. Those years were selected because we found complete data sets for both RF and imageries during those time periods. The RF data was obtained from the National Meteorological Agency of Ethiopia and the NDVI values from http://earlywarning.cr.usgs.gov/adds/datathemephp.

To observe the relationships between RF and NDVI, Ethiopia was divided into 22 grids. Each grid was two degrees by two degrees (Fig. 6). The RF recorded by all stations inside the grids was averaged from 1982 to 2004 and an average point data was generated. The same procedure was followed for the NOAA AVHRR

Fig. 6 The map of Ethiopia with 2 × 2 degrees grids

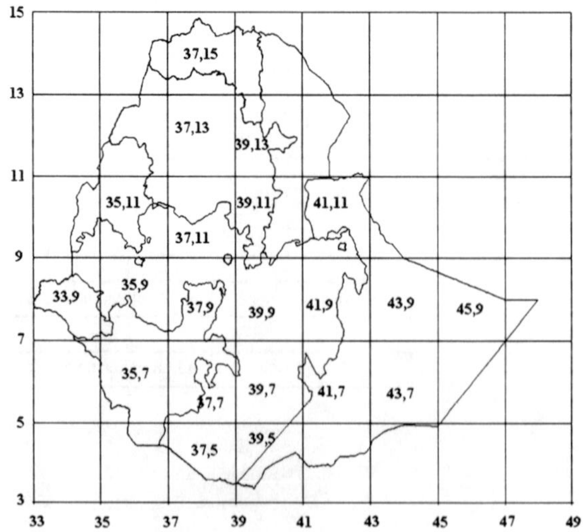

NDVI values of the two degree by two degree grids. The descriptive statistics for the six-month average values for both RF and NDVI is presented in Table 1. In addition, the scatter plots for the four-month period average values from June to September are presented in Fig. 7.

There is a strong relationship between the recorded RF data and the NDVI values obtained from each two degree by two degree grid. The highest R^2 value was observed for September, whereas the lowest value was recorded for July (Fig. 7). The strong relationship between September RF and NDVI values might be explained by the fact that during this month, if there is adequate RF, plants can have optimal photosynthesis, which can result in high NDVI values. The lowest relationship between July RF and NDVI value was unexpected and needs further research. In overall, it is convincing that we can use NDVI values to monitor drought (shortage of rainfall for monitoring agricultural drought).

NDVI and Deviation of NDVI for Spatially Locating Drought

In this section, we present the status of drought conditions in parts of eastern African and southern Asian countries in 2009 using the NDVI parameter. Northern Sudan and southern Asian countries were purposely included in the analysis window for controlling whether the result agrees with the ground reality or not. These areas have no natural vegetation and the assumption is that if the result is correct, there will be no vegetation condition deviation for these areas.

This analysis was also primarily conducted with the aim of testing the applicability of MSG data for spatio-temporal drought monitoring. The preliminary results were obtained by using October 2009 MSG data and the long-term average NDVI NOAA AVHRR data. The raw MSG data was acquired from the Ethiopian

Table 1 Descriptive statistics for average NDVI and average RF from 1982 to 2004

Items	June		July		August		September	
	NDVI	RF	NDVI	RF	NDVI	RF	NDVI	RF
Mean	0.35	84.73	0.33	141.97	0.36	147.03	0.39	104.35
Median	0.30	51.90	0.31	128.67	0.36	148.26	0.39	98.83
Standard deviation	0.14	89.80	0.15	126.73	0.16	124.19	0.17	76.92
Minimum	0.11	1.23	0.10	0.26	0.13	0.45	0.13	12.76
Maximum	0.66	311.71	0.66	359.76	0.68	345.69	0.70	315.76

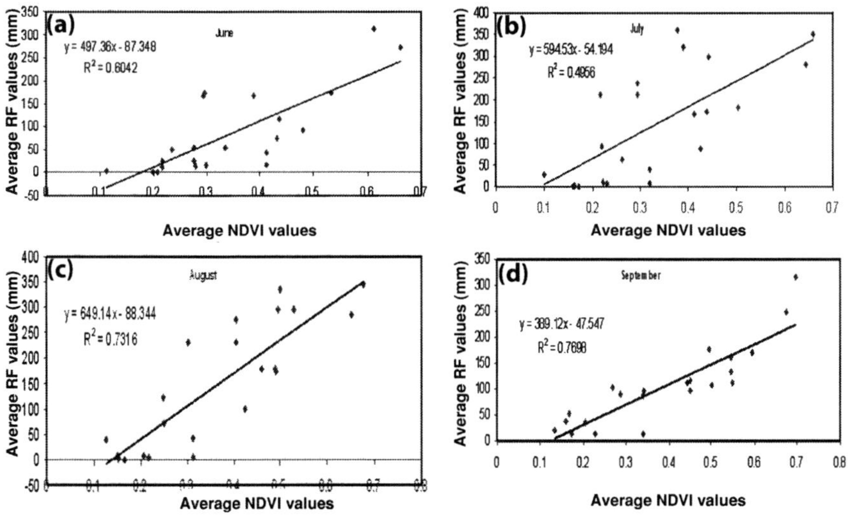

Fig. 7 Scatter plots showing the correspondence between the average NDVI and RF for 1982–2004. June (**a**), July (**b**), August (**c**), and September (**d**) are the major rainy and plant-growing months

Meteorological Agency in Addis Ababa. The long-term records of decadal NDVI data from NOAA were downloaded from http://earlywarning.cr.usgs.gov/adds/datathemephp and covered the first decade of October from 1982 to 2009. Using these two data sets, the Dev_NDVI was calculated.

The actual drought condition was determined by comparing the NDVI for the first decade of October 2009 with the long-term mean NDVI using NOAA satellite data. The data was analysed using ILWIS 3.6 software. The ten-day images of MSG (1–10 October 2009) were imported to ILWIS raster image format (Fig. 8) using the "Multiple times in one file" option. This means that we had all 10 bands stacked (maplist) together and ready for the NDVI calculation. After importing the three-band image data to ILWIS 3.6 raster format, a script was written for calculating the Dev_NDVI.

Our results show that approximately 40% of the area exhibited negative deviation (Fig. 9, 10). This indicates a prevalence of drought in 2009 in east African and southern Asian countries in the first decade of October 2009. These results

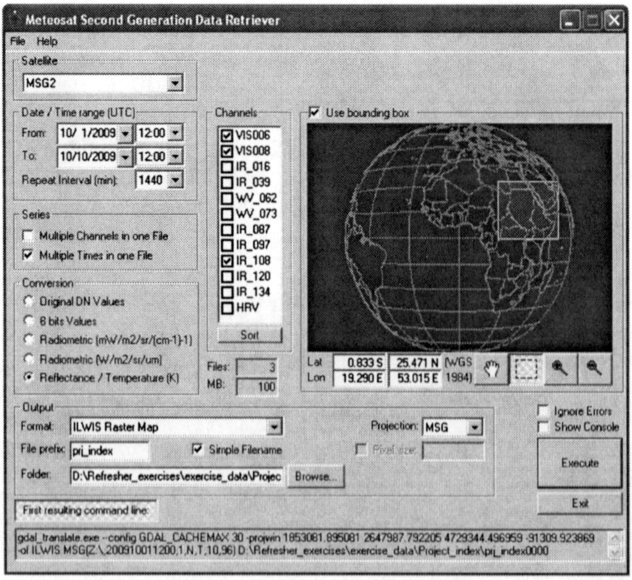

Fig. 8 MSG data retriever of ILWIS 3.6 for importing the raw data to raster image

Fig. 9 Dev_NDVI comparison of the change in vegetation

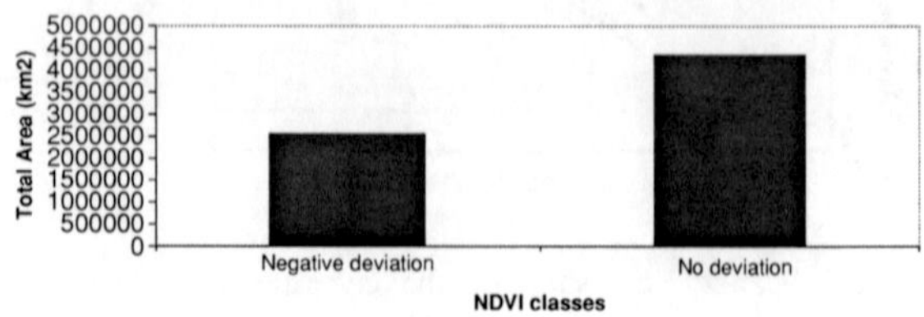

align with recorded RF in 2009 in most parts of Ethiopia. That is, the recorded rainfall amounts were below the overall average (FEWS NET 2009).

In Fig. 10, the grey areas are where there is either no change or positive deviation from the long-term average. The dark grey are areas with negative deviations, indicating the prevailing drought.

The analysis of the vegetation condition index (VCI) also indicates the occurrence of drought in the study area (Fig. 11). In Fig. 11, the areas in white are water bodies and/or areas without vegetation in the past. The VCI shows (in percentages) the vegetation condition of the actual decade NDVI compared to the long-term maximum and minimum of the corresponding decade. In principle, about 50% reflects a fair vegetation condition. Our geospatial analysis shows that approximately 37% of the total area had less than 40% VCI, indicating the occurrence of drought. Areas with below-normal vegetation cover were located in the central part of Sudan and northern and south-east Ethiopia. Only 18% of the area had optimal and above-normal vegetation conditions (Fig. 11). These areas are located in the central part of Sudan and the north-west corner of Ethiopia.

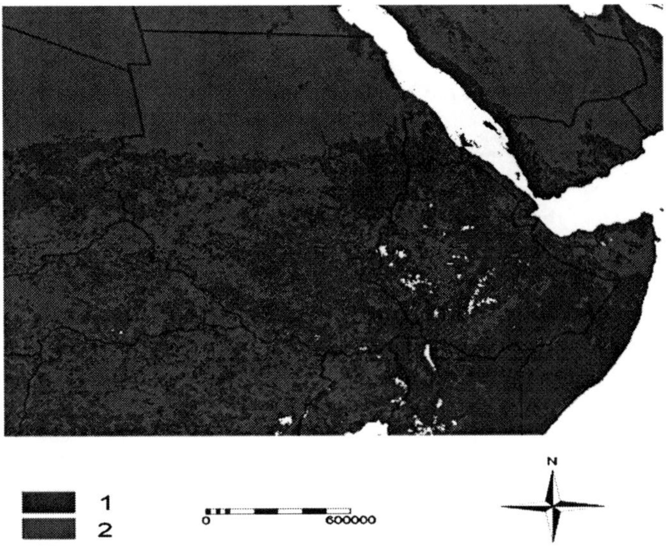

Fig. 10 Dev_NDVI spatial distribution

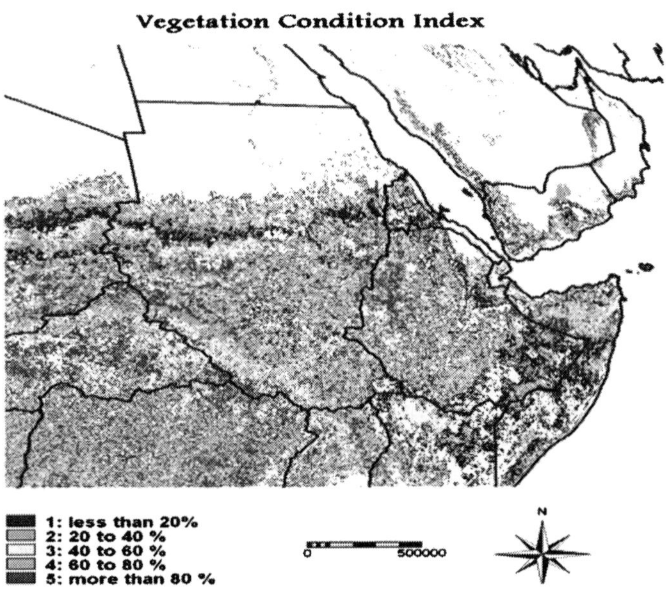

Fig. 11 Vegetation condition index (VCI) map for monitoring drought

Conclusions

The findings of this research are based on ongoing doctoral work. The preliminary results produced promising scientific outputs for implementing satellite data for drought monitoring. Currently, this research is in its early stages, although there is some convincing evidence that it is possible to model and predict drought conditions using real-time MSG data. The research question regarding identifying appropriate satellite imagery temporal resolutions for modelling drought as a spatial object is still pending and will be addressed in future research.

The preliminary results suggest that real-time spatio-temporal MSG data can be used for drought monitoring and early-warning systems in food-insecure areas. In 2009, there was drought in most parts of Ethiopia and Sudan due to the RF shortage during the crop-growing season from July to September. The results of our analysis confirm this fact.

Water and food shortage are long-term impacts of climatic change and are the major concern of the world community these days. Our results could help decision-makers to use advanced satellite technology for effective drought monitoring and early-warning systems in various regions. Combined with proper policies, these systems can help to prevent famine and starvation in food-insecure regions. In the past, satellite technologies have primarily been used in areas of meteorological applications. In this research, the main emphasis is on mining knowledge for drought hazard assessment and saving the lives of individuals who are affected by recurring droughts. The findings of this research can assist decision-makers in taking appropriate actions in time to save lives in drought-affected areas using advanced satellite technology.

References

Anyamba A, Tucker C (2003) Analysis of sahelian vegetation dynamics using NOAA-AVHRR NDVI data from 1981–2003. J Arid Environ 63:569–614

Benedetti R, Rossinni P (1993) On the use of NDVI profiles as a tool for agricultural statistics: the case study of wheat yield estimate and forecast in Emilia Romagna. Remote Sens Environ 45:311–326

Budd T (2000) Understanding object-oriented programming with Java, reading. Addison Wesley, Mass

Cooper P, Dimes J, Rao K, Shapiro B, Twomlow S (2008) Coping better with current climatic variability in the rain-fed farming systems of Sub-Saharan Africa: an essential first step in adapting to future climate change? Agric Ecosyst Environ 126(1-2):24–35

EUMETSAT (2005) Meteosat Second Generation In Orbit—In Use, ISBN 92-9110-071-4 MSG.02 Version 1, Am Kavalleriesand 31, 64295 Darmstadt, Germany, Web: http://www.eumetsat.int

Fensholt R, Sandholt I, Stisen S, Tucker C (2006) Vegetation monitoring with the geostationary meteosat second-generation SEVIRI sensor. Remote Sens Environ 101:212–229

FEWS NET (2009) United Nations' FEWS (famine early-warning systems) program, http://www.fews.net/Pages/default.aspx?l=en, verified on 8 December 2009

Gregor S (2006) The nature of theory in information systems. MIS Q 30(3):611–642

Gregor S, Jones D (2007) The anatomy of a design theory. J Assoc Inf Syst 85:312–335 Article 2

Hayes J, O' Rourke P, Terjung W, Todhunter P (1982) Yield: A numerical crop yield model of irrigated and rainfed agriculture. Publ Climatol 35(2)

Hevner A, March ST, Park J, Ram S (2004) Design science in information systems research. MIS Q 1(28):75–105

Holben BN (1986) Characteristics of maximum-value composite images from temporal data. Int J Remote Sens 7:1417–1434

Jain SK, Keshiri R, Goswami A, Sarkar A, Chaudhry A (2009) Identification of drought-vulnerable areas using NOAA AVHRR data. Int J Remote Sens 30:2653–2668

Ji L, Peters AJ (2003) Assessing vegetation response to drought in the northern Great Plains using vegetation and drought indices. Remote Sens Environ 87:85–98

Kogan F, Gitelson A, Zakarin E, Spivak L, Lebed L (2003) AVHRR-based spectral vegetation index for quantitative assessment of vegetation state and productivity:calibration and validation. Photogramm Eng Remote Sens 69:899–906

Mckee TB, Doesken NJ, Kleist J (1993) The relationship of drought frequency and duration to time scales. 8th Conference on applied climatology, Anaheim, CA, USA, 17–22 January, pp 179–148

Mishra AK, Desai VR (2005) Spatial and temporal drought analysis in Kansabati river basin, India. Int J River Basin Manag 3:31–41

Myneni R, Asrar G (1994) Atmospheric effects and spectral vegetation indices. Remote Sens Environ 47(3):390–402

Nicholson S (1994) Recent rainfall fluctuations in Africa and their relationship to past conditions over the continent. Holocene 4(2):121–134

NMSA (National Meteorological Services Agency) (1996) Assessment of drought in Ethiopia. Meteorological research reports series, no. 2, Addis Ababa, Ethiopia

Obasi GP (1994) WMO's role in international decade for natural disaster reduction. Bull Am Meteorol Soc 75(1):1655–1661

Prasad A, Singh R, Tare V, Kafatos M (2007) Use of vegetation index and meteorological parameters for the prediction of crop yield in India. Int J Remote Sens 28:5207–5235

Quarmby N, Milnes M, Hindle T, Silcos N (1993) The use of multi temporal NDVI measurements from AHRR data for crop yield estimation and prediction. Int J Remote Sens 14:199–210

Rulinda CM, Bijker W, Stein A (2009) Image mining for drought monitoring in eastern Africa using Meteosat SEVIRI data. Int J Appl Earth Observ Geoinform doi:10.1016/j.jag.2009.10.008

Simon HA (1996) The sciences of the artificial, 3rd edn. MIT Press, Cambridge, MA

Singh RP, Roy S, Kogan F (2003) Vegetation and temperature condition indices from NOAA AVHRR data for drought monitoring over India. Int J Remote Sens 24:4393–4402

Smakhtin V, Hughes D (2004) Review, Automated estimation and analysis of drought indices in South Asia. W.P. 83, Sri Lanka, International Water Management Institute

Smith PM, Kalluri S, Prince S, Defries R (1997) The NOAA NASA pathfinder AVHRR 8 km land data set. Photogramm Eng Remote Sens 63:12–31

Stein A (2008) Modern developments in image mining, science in China Series E. Technol Sci 51:13–25

Stein A, Hamm NAS, Ye Q (2009) Handling uncertainties in image mining for remote sensing studies. Int J Remote Sens 30:5365–5382

Swenson S, Wahr J (2009) Monitoring the water balance of lake Victoria, east Africa, from space. J Hydrol 370:123–176

Tadesse T, Brown JF, Hayes MJH (2005) A new approach for predicting drought-related vegetation stress: integrating satellite, climate, and biophysical data over the U.S. central plains. ISPRS J Photogramm Remote Sens 59(4):244–253

Tadesse T, Haile M, Senay G, Brian D, Cody W, Knutson L (2008) The need for integration of drought monitoring tools for proactive food security management in sub-saharan Africa. Nat Res Forum 32:265–279

Thiruvengadachari S, Gopalkrishana HR (1993) An integrated PC environment for the assessment of drought. Int J Remote Sens 14:3201–3208

Thorton P, Jones P, Alagarswamy G, Andresen J (2009) Spatial variation of crop yield response to climate change in East Africa. Glob Environ Change 19:54–65

Tucker C (1979) Red and photographic infrared linear combinations for monitoring vegetations. Remote Sens Environ 8:127–150

Tucker C, Stenseth NC (2005) Using the satellite-derived NDVI to assess ecological responses to environmental change. Trends Ecol Evol 20(9):503–510

UNCCD (United Nations Convention to Combat Desertification) (1999) Article 1. United Nations, Bonn, Germany

UNEP (United Nations Environment Programme) (2006) Climate Change and Variability in the Sahel Region: Impacts and Adaptation Strategies in the Agricultural Sector http://www.unep.org/Themes/Freshwater/Documents/pdf/ClimateChangeSahelCombine.pdf, verified on July 12, 2010

Ungani L, Kogan F (1998) Drought monitoring and corn yield estimation in southern Africa from AVHRR data. Remote Sens Environ 63:219–232

UNISDR (United Nations Secretariat of the International Strategy for Disaster Reduction) (2009) Drought risk reduction framework and practices. Contributing to the implementation of the hyogo framework for action. Geneva, Switzerland

Wilhite DA (2000) Drought as a natural hazard: Concepts and definition. In: Wilhite DA (ed) Drought: a global assessment, natural hazard and disasters series. Routledge Publishers, London, pp 3–18

Wilhite DA (2005) Drought and water crisis: science, technology and management issues. Taylor and Francis Group, London

Worboys M, Hilary M, Hearnshaw M, Maguire D (1990) Object-oriented data modeling for spatial databases. Int J Geogr Inf Syst 4:369–383

Author Biographies

Getachew Berhan received his MSc from the Geo-International Institute for Geo-Information Science and Earth Observation (ITC), the Netherlands, in 2006. Currently, Getachew is a PhD student under the IT PhD programme, Information Systems Track, at Addis Ababa University, Ethiopia.

Tsegaye Tadesse received his PhD in Agro-Meteorology from the University of Nebraska-Lincoln in 2002 and his MSc in Space Studies from the International Space University, Strasbourg, France, in 1998. Currently, Tsegaye is Assistant Research Professor/Climatologist at the National Drought Mitigation Center (NDMC), School of Natural Resources, University of Nebraska-Lincoln, USA.

Solomon Atnafu received his PhD in Computer Science from University of Lyon, France, in 2003 and his MSc in Computer Science from the Faculty of Informatics, Dresden University of Technology, Germany, in 1989. Currently, Solomon is Assistant Professor in the Department of Computer Science, Addis Ababa University, Ethiopia.

Shawndra Hill received her PhD in Information Systems from New York University in 2007 and her MPhil from New York University in 2003. Currently Shawndra is Assistant Professor of Operations and Information Management at Wharton University of Pennsylvania, USA.

Chapter 12
A Swot Analysis of Mitigation of Climate Change Through REDD

Emmanuel F. Nzunda and Tumaini G. Mahuve

Abstract Reducing emissions from deforestation and forest degradation (REDD) is currently given high priority as one of the means to mitigate the impacts of climate change. The purpose of this study is to examine the strengths, weaknesses, opportunities and threats that are related to reducing emissions from deforestation and forest degradation as a climate change mitigation strategy. REDD can contribute to funding sustainable forest management, carbon storage, biodiversity conservation and poverty reduction. However, the amount of mitigation that REDD can achieve is limited. Furthermore, REDD is limited by its funding mechanism and geographical and sectoral leakages. Success in REDD may result in unintended effects such as increased human/wildlife conflicts as a result of increase in abundance of animals due to better conserved habitat and more severe accidental fires due to accumulated fuel loads as a result of fire prevention. Poor governance may lead to inequality in the distribution of benefits and costs of REDD. In spite of these weaknesses of REDD, most developing countries are willing to participate in REDD. There is also a lot that can be learnt from previous conservation and development initiatives that can benefit REDD. However, the success of REDD will largely depend on political stability and the continued interest of the international community in climate change mitigation. This is the first article to explicitly analyse REDD in the light of its strengths, weaknesses, opportunities and threats. Existing literature has so far only discussed these aspects of REDD as part of general discussions on REDD, without explicitly naming them

E. F. Nzunda (✉) · T. G. Mahuve
Department of Forest Mensuration & Management, Sokoine University of Agriculture, PO Box 3013, Chuo Kikuu, Morogoro, Tanzania
e-mail: emmanuelnzunda@yahoo.com; nzundaef@gmail.com

T. G. Mahuve
e-mail: tumainimahuve@yahoo.com

W. Leal Filho (ed.), *Experiences of Climate Change Adaptation in Africa*, Climate Change Management, DOI: 10.1007/978-3-642-22315-0_12,

as strengths, weaknesses, opportunities and threats. Naming the aspects as such informs policy more directly.

Keywords Climate change · Governance · Leakage · Poverty alleviation · REDD · Sustainable forest management

Introduction

It is currently generally accepted that global climate change is caused mainly by anthropogenic activities (IPCC 2007). Since the beginning of the industrial revolution, atmospheric concentration of carbon dioxide (CO_2), the chief heat-trapping greenhouse gas (GHG), has risen 35%, from about 280 to 377 parts per million (ppm) (WRI 2005). This is because since the industrial revolution, the ability of humans to change land cover and produce chemicals from industrial processes has been increasing while the concern for preservation of nature has been decreasing until only recently. Atmospheric concentrations of methane (CH_4), the second leading GHG, have more than doubled over the past two centuries (WRI 2005). These and other GHG increases have led to a 0.6°C increase in the global average surface temperature since 1900 (WRI 2005). If current emissions trends are not altered, global temperatures are expected to rise a further 1.4–5.8°C by 2100 (IPCC 2007; WRI 2005). The effects of such temperature changes on agricultural production, water supply, forests and overall human development are uncertain, but are likely to be detrimental to a large portion of the world's population, particularly in developing countries (WRI 2005). To keep the global average temperature from rising more than 2°C above pre-industrial levels with a probability exceeding 85%, worldwide emissions from sources other than land-use change and forestry would need to peak around 2015 and subsequently decline by 40–45% by 2050 compared to 1990 levels (den-Elzen and Meinshausen 2005; WRI 2005). However, significant climate damages may still be associated with a 2°C increase in global temperatures (den-Elzen and Meinshausen 2005; WRI 2005). On the other hand, the global population is expected to increase by 40–100% and economic growth is projected to increase ten to 20-fold over this century (WRI 2005). Thus reducing emissions to levels that avoid dangerous human interference with the climate system, in the face of economic and population growth, will require substantial changes in energy use, including technological innovation plus advances in efficiency, conservation and alternative energy sources (WRI 2005). Although the emissions of many GHG need to be reduced, atmospheric carbon is receiving particular attention because of its greater proportion (Dudley 2010).

The main causes of increase in atmospheric carbon concentrations are industrial and transportation activities although land use, land-use change and forestry activities (LULUCF), especially deforestation and forest degradation, also contribute about 20% of CO_2 emissions (Dudley 2010; Mollicone et al. 2007).

Furthermore, the contribution of LULUCF to GHG emissions increases to about 25% when CH_4 and N_2O and other chemically reactive gases are included (Houghton 2005). It is important to note that the estimates of global emissions of CO_2 due to LULUCF vary depending on the methods used and thus may range from 1.6 ± 0.8 Gt C/year (12% of the world's emissions) to 2.4 Gt C/year (28%), a difference which is almost threefold (Corbera et al. 2009). On the other hand, the use of fossil energy sources contributes about 8.4 Gt C/year (Raupach et al. 2007, Tschakert et al. 2008). For the period 1970–2004, the rate of growth of LULUCF emissions was lower (40%) than that of the energy supply sector (145%), the transport sector (120%) and industry (65%) (IPCC 2007).

Among developing countries, Indonesia and Brazil contribute the largest proportion of emissions from LULUCF, at 34 and 18% respectively (WRI 2005). This amount is equivalent to 80% of the total reduction targets for Annex I countries in the Kyoto Protocol (Santilli et al. 2005). Some countries including Malaysia, Myanmar and the Democratic Republic of Congo are not major overall emitters of GHG but contribute significantly to emissions only from LULUCF (WRI 2005). National-level emissions figures have low reliability due to technical difficulties in estimation and the tendency of some nations to give figures that are less than what other sources show (WRI 2005). The uncertainties in estimates of emissions range from ±150% Mt CO_2/year for large fluxes to ±180% for estimates near zero (Corbera et al. 2009).

Three types of mitigation projects fall within the forestry sector, namely afforestation, reforestation and deforestation avoidance (Reyer et al. 2009; Watson et al. 2000). Afforestation entails conversion of historically non-forest land to forest with relatively free species selection, e.g. using non-native and fast-growing species. Reforestation involves conversion of recently non-forested land to forest, often with a conservation or landscape protection background, generally planting native species and focussing on restoration of natural ecosystems. Under deforestation avoidance, carbon-rich natural forests are protected from conversion to non-forest land. Reducing deforestation rates 50% by 2050 and then maintaining them at this level until 2100 would avoid the direct release of up to 50 GtC this century (equivalent to nearly six years of recent annual fossil fuel emissions, and up to 12% of the total reductions that must be achieved from all sources through 2100 to be consistent with stabilising atmospheric concentrations of CO_2 at 450 ppm (Gullison et al. 2007).

Despite the contribution of deforestation avoidance to mitigation, it was not included in the 2008–2012 commitment period of the Kyoto Protocol's Clean Development Mechanism due to a combination of sovereignty and methodological concerns (Gullison et al. 2007; Watson et al. 2000). After a number of forest-rich nations proposed the inclusion of avoided deforestation in the United Nations Framework Convention on Climate Change (UNFCCC) incentives for clean development, the UNFCCC launched a two-year initiative to assess technical and scientific issues and new policy approaches and positive incentives for reducing emissions from deforestation (RED) in developing countries (IPCC 2001). Reducing emissions from deforestation and forest degradation (REDD) was first considered as a possibility for climate change mitigation post-2012 by the UNFCCC (United Nations Convention on Climate Change) at its CoP 11 (11th Conference of Parties) in 2005 in Montreal, Canada (Olsen and Bishop 2009). Further discussions on REDD were held at CoP 12 in Nairobi, Kenya, in 2006 and an agreement to start REDD demonstration activities was reached by parties at CoP 13 in Bali, Indonesia, in 2007. Discussions on REDD further continued at CoP 14 in Poznan, Poland, in 2008 and it was expected that CoP 15 in Copenhagen, Denmark, in 2009 would come up with a complete package to guide the implementation of REDD but that was not the case. That notwithstanding, different non-Annex I countries, including Tanzania, Indonesia, Thailand, Vietnam, Cambodia and Laos are involved in REDD readiness initiatives, which aim to prepare the countries for the implementation of REDD (Phelps et al. 2010). To support REDD, a number of funding initiatives have been established including the World Bank-hosted Forest Carbon Partnership Facility and Forest Investment Programme, the UN-REDD programme, Norway's International Climate and Forest Initiative and Australia's International Forest Carbon Initiative (Corbera et al. 2009; Obersteiner et al. 2009).

Currently discussions on various aspects of REDD design and implementation are going on. In particular, the architectural aspects of REDD and ways that can make it more effective in terms of poverty reduction and climate amelioration are given a high priority (Angelsen 2009; Angelsen and Wertz-Kanounnikoff 2008; Peskett and Brockhaus 2009; Wertz-Kanounnikoff and Angelsen 2009). The aim of this article is to contribute to the discussions on REDD by highlighting its strengths and weaknesses; and the opportunities and threats that accompany it. First, the following chapter discusses the strengths of REDD in terms of its positive role in sustainable forest management, ecological conservation and poverty reduction. After that, the weaknesses of REDD are discussed by giving a comparison of what REDD can achieve against the total mitigation effort needed, a discussion of the problem of leakage, unintended positive effects, insufficiency, unreliability and complications of REDD funding and tendency to justify REDD on the basis that it is cheap. Then there is a discussion of the opportunities that favour REDD in terms of willingness of countries to participate and learning from similar past initiatives. Finally, the threats to REDD, namely poor governance, political instability and war and loss of interest in REDD by the international community are discussed.

Strengths of REDD

REDD Can Serve as a Source of Funds for Sustainable Forest Management

One of the main problems that makes sustainable forest management difficult in developing countries is shortage of funds. In turn, the shortage of money results in understaffing and inadequate resources for forest management. REDD can be used as a source of funds for sustainable forest management. The good thing about REDD is that it proposes to offer money not only for the administration of forests but also for financial compensation of the benefits that forest-dependent people would get through deforestation and forest degradation (Schwartzman et al. 2008).

Ecological Benefits of REDD

The increase in carbon sequestration and stock due to REDD will result in mitigation of climate change impacts. In fact, this is the reason for proposing REDD as a climate management strategy. In addition to this primary interest of REDD, there are other indirect benefits that can accrue from efficient REDD projects. These arise from the possibility of a positive relationship between REDD, biodiversity conservation, water catchment, soil conservation and poverty alleviation. By reducing deforestation and forest degradation, REDD can directly enhance conservation of plant diversity. In turn, the conserved forests can serve as habitats for animals and thus REDD can indirectly increase animal diversity as well. In return, conservation of biodiversity can result in better ecological functioning of the forests.

Deforestation and forest degradation reduce landscape water catchment ability. This results in excessive runoff, floods and soil erosion during the rainy season and drought during the dry season. Due to poor landscape water catchment, many rivers dry up completely or become seasonal. Furthermore, excessive soil erosion results in reduced soil productivity and siltation of water bodies, which reduces their depth. By enhancing landscape forest cover, REDD can contribute to water catchment and soil conservation and hence sustainable water quality and quantity and soil productivity that are needed for health and development.

REDD Can Contribute to Poverty Alleviation

REDD can contribute to poverty alleviation through sale of carbon credits, provision of employment and sale of other environmental services than carbon that are enhanced by implementation of REDD (Corbera et al. 2009). These benefits are especially relevant for a scenario whereby the alternative would be forest protection without the benefits that REDD can offer (Olsen and Bishop 2009). Furthermore, in comparison to most activities that forest land is converted to by the rural poor after deforestation such as pasture or annual cropping, preventing deforestation through REDD has higher returns (Chomitz et al. 2006; Corbera et al. 2009). However, in cases where deforestation is due to a more lucrative enterprise or the cost of REDD is high and where there is a possibility for rich investors or authoritarian government to grab the benefits of REDD, the potential for REDD to contribute to poverty alleviation becomes limited (Corbera et al. 2009).

Weaknesses of REDD

Ecological Versus Economic Rate of Operation

As a climate change mitigation strategy, REDD falls far short of what is needed. This is because REDD, even when 100% effective, will only contribute 20% of the needed solution to the problem of climate change. The remaining 80% has to be achieved through addressing carbon emissions due to other factors. Climate change is a problem caused by economic and technological factors that operate at a spatio-temporal rate that is much higher than the rate of the ecological processes that are the basis of effects of REDD on climate.

Data on the relative contribution to CO_2 emissions of fossil fuels versus deforestation and forest degradation for the period 1980–2008 shows that the contribution from fossil fuels has increased tremendously, while that from deforestation and forest degradation has actually dropped to almost 12% (van der Werf et al. 2009). This is due to the increasing rate of industrial development during the period, with new formerly less developed countries such as South Korea, China, India and Brazil joining the more developed countries in increased use of fossil fuels. By analogy, achieving the objectives of REDD alone is like scoring 20% in a class test (that has a maximum of 100%), which means an overall failure. The argument here is not that REDD is useless but that it has to be implemented alongside other strategies for effective climate change mitigation. Surprisingly, some authors argue as if REDD is an alternative to the other mitigation approaches while in fact it is a complement.

Spatial and Sectoral Leakages

When there is effective REDD in an area, forest-related livelihood needs of people in the area may need to be covered by another area. Unless these are completely replaced by their alternatives, people in the REDD project area will still need wood for fuel and shelter and other wood and non-wood forest products. Since the complete replacement of forests as a source of livelihood in developing countries (even in developed countries) is almost impossible, REDD will always be accompanied by some form of spatial leakage (Angelsen and Wertz-Kanounnikoff 2008; Olsen and Bishop 2009).

If a significant proportion of the population in the REDD project area obtained their income from deforestation and forest degradation before the REDD project, the shift in their location of activities may result in significant increase in deforestation and forest degradation in surrounding areas. This may especially be the case where the resulting REDD activities cannot absorb all the people that were earning their income through deforestation and forest degradation or where the income from REDD is not high enough to satisfy the people. Initial stages of the REDD project seem to attract the community as they obtain employment from afforestation and reforestation. However, over time, the employment opportunities are reduced as a result of reduction of operation tasks within the REDD project. The reduced employment opportunities will cause labour out-migration due to little incentive obtained from REDD (Olsen and Bishop 2009).

Local-level spatial leakage may be a more serious problem for project-based REDD initiatives than for national REDD initiatives although it is also possible to have international leakage. For example, a logging ban in one country has resulted a number of times in a logging boom in a nearby country. Since it is not likely to have international REDD programmes in the short term due to technical and financial limitations, the problem of leakage is highly likely to be a salient feature of REDD projects (Corbera et al. 2009). The problem of local leakage is more likely for projects covering small areas than those that cover large areas (Angelsen and Wertz-Kanounnikoff 2008).

Sectoral leakage can occur when REDD restricts agricultural expansion and thus forces agricultural intensification, which can result in increased emission of such gases as N_2O due to increased use of industrial fertilisers (Angelsen and Wertz-Kanounnikoff 2008; Obersteiner et al. 2009). Here, emissions of CO_2 are replaced by emissions of N_2O. For instance, REDD may necessitate replacement of shifting cultivation, such as is widely practised in miombo woodland areas, with settled agriculture. Without shifting cultivation, which allows fallow periods for soil to replenish its fertility, crop yields decline with time. Hence if agriculture is to continue, fertilisers must be used. This has occurred widely in southern highlands of Tanzania where, without the use of fertilisers, crop yield (especially maize) is negligible.

Unintended Positive Effects of REDD

REDD can result in unintended positive effects such as population increase due to reduced poverty and the resultant improvement of social services, increase in populations of wild animals due to better forest conservation and more severe accidental fires due to long fire intervals as a result of fire control. Although it is theoretically possible to include measures to control these effects in the REDD initiatives, practically there are some complications. For instance, human population is a complex issue that depends on national policy, societal tradition and culture, individual philosophy and disposition and availability of resources and social services. Human population can either be controlled intentionally by deciding not to have more than a certain number of children, especially where the rate of death has been controlled due to technological advancement, or unintentionally by the prevalence of an unfavourable death rate, which is common where technology has not liberated the human race. In most parts of the world for which REDD is intended (i.e. the developing countries), the second scenario, that of unintentional control of human population, is more prevalent than the first scenario. The increase in human population due to improvement of living conditions happened in Europe, North America and Asia, and has recently happened in Africa, particularly due to the reduction in death rate through vaccination and better health services and improvement in food availability due to better agriculture (compared to traditional practices). If REDD really succeeds in reducing poverty, it will most likely result in population increase, at least in the short term.

Better forest conservation usually results in increase in populations of wildlife. This leads to more human/wildlife conflicts in the form of increased crop and livestock raids and human injury and deaths. Unless REDD adopts measures to control this, such as fencing the protected areas, as is practised in South Africa, or culling the animals, these problems are going to tarnish the name of REDD. With current complications whereby there is an emphasis on animal rights, culling may be a difficult option to implement. On the other hand, fencing protected areas makes them more like zoos than wildernesses.

Fire control may present a real catch-22 situation for REDD, particularly in deciduous forests. On the one hand, fire severity may be reduced by using prescribed early burning, which prevents accumulation of fuel loads. On the other, early burning contributes to CO_2 emission and would thus not be acceptable as a strategy under a serious REDD initiative. Especially with the current increase in temperature, the dry season has become hotter than usual. This results in very dry fuel loads that may be very dangerous if left to accumulate. As a result, annual fires have invaded even areas that were not previously experiencing them. If REDD succeeds in protecting deciduous forests to the extent of making them change to more evergreen forests, where rainfall permits, then the problem of increased fire may be abated. But that may take time and thus the initial years of REDD initiatives in deciduous forests are likely going to be dogged by increased severity of accidental fires.

Insufficiency, Unreliability and Complications of REDD Funding

Based on experience with voluntary funding for climate change, it is going to be difficult to get sufficient voluntary funds for REDD (Corbera et al. 2009; Schwartzman et al. 2008; van der Werf et al. 2009). The minimum amount of funds estimated to be sufficient for global REDD is about US \$5 billion per year (Corbera et al. 2009; Grieg-Gran 2006). Between 1991 and 2005, the Global Environmental Facility was able to get only US \$1.75 billion for funding climate change initiatives (Corbera et al. 2009; Freestone 2009) and the replenishment fund for 2006–2010 amounts to only US \$3.13 billion (GEF 2010), which is far short of the minimum requirement for REDD (Corbera et al. 2009).

Until November 2009, there was only US \$54.2 million gathered by the UN-REDD Programme Fund, contributed only by two countries: Norway (US \$52.2 million) and Denmark (US \$2.0 million) (UNDP 2010). The fact that only two countries contributed to the UN-REDD Programme Fund suggests lack of commitment by other donor countries to the programme. This could be because there are other country-based REDD initiatives such as Australia's International Forest Carbon Initiative (which has committed some US $200 million to support REDD in Papua New Guinea, the Phillipines and Indonesia) and the Brazilian Amazon Fund (which has received contributions from Norway (US $110 million) and Germany (US $20 million) (Corbera et al. 2009). In addition to insufficiency, voluntary funding is also characterised by high unreliability due to budgeting procedures in donor countries, changes in priorities and policymaking or implementation delays (Corbera et al. 2009).

Market mechanisms can potentially provide sufficient funds for REDD needs (Corbera et al. 2009). In the long term, the carbon market can achieve an outlay of US $50–120 billion per year (Corbera et al. 2009; IEA 2005). However, based on experience from the CDM, market funding has complex modalities and methodologies that can result in the exclusion of a number of initiatives from funding (Corbera et al. 2009; Dudley 2010).

REDD can also be funded through a levy on the emissions reductions units (ERUs) issued or the assigned amounts units (AAUs) first traded in the carbon market, as well as a tax on carbon-intensive commodities and services in Annex I countries (Corbera et al. 2009). The amount of funds that can be obtained from ERUs in the period 2008–2010 is US $1.8 billion, equivalent to US $360 million per year) (Corbera et al. 2009). On the other hand, AAUs can create up to US $2.7 billion per year (Corbera et al. 2009, Haites 2004). When we add the funds from ERUs and AAUs, we get a total (US $3.1 billion) that is less than the minimum requirement of US $5 billion per year (Corbera et al. 2009, Grieg-Gran 2006). Hence this option alone cannot be sufficient to fund REDD.

Tendency to Justify REDD on the Basis of the Lower end of its Costs

Transaction, administration and opportunity costs of REDD are variable depending on the situation and the assumptions used in the calculations (Corbera et al. 2009; Olsen and Bishop 2009). The costs of REDD may depend on such technical factors as how timber harvesting and land-clearing costs are treated, what type of forest land is considered, how alternative land uses are modelled, which carbon density estimates are used and whether cost curves or point estimates for carbon abatement are calculated (Grieg-Gran 2006; Olsen and Bishop 2009). Furthermore, costs of REDD also depend on economic, social and geographical/physical factors such as primary commodity prices, the suitability of particular forest lands for different uses, soil and climate conditions that affect yields and hence returns to agriculture, scale of operation—small, medium, large—inputs and technology, distance from market and the quality of transport infrastructure (Grieg-Gran 2006; Olsen and Bishop 2009). For instance, in the Brazilian Amazon, the opportunity costs of

REDD range from below US \$0.01/tCO_2 to US \$8/tCO_2 depending on whether deforestation is carried out for logging, ranching, plantation farming or subsistence pasture and cropping (Chomitz et al. 2006; Corbera et al. 2009; Vera-Diaz and Schwartzman 2005).

To the developed countries, which are supposed to fund REDD, high costs of REDD means more burden for them. Thus a number of authors have argued for the adoption of REDD on the basis of its low cost compared to other options (Gullison et al. 2007; Schwartzman et al. 2008). On the other hand, to the developing countries, the assumption that REDD is cheap can have serious negative implications for their development and future. Unfortunately, there is more of a tendency to assume that REDD is a cheaper option than the converse and thus offer as low a value for REDD as possible. In a way, this is a legacy of the colonial relationship between the two blocks of the world. A rational approach would be to cost REDD as realistically as possible and adopt it because it is a complement to the other mitigation approaches and not because it is a mitigation option that is cheaper than the other options.

Opportunities for REDD

Willingness of Developing and Developed Countries to Participate in REDD

Compared to the reduction of emissions through adjustments of industrial processes, REDD has been favourably accepted by both Annex I and non-Annex I countries. Annex I countries are willing to fund implementation of REDD in non-Annex I countries, which are also willing to implement REDD. Despite the complications and a number of unresolved uncertainties concerning REDD, many developing countries are preparing themselves to implement REDD. For example, all 37 developing countries that were to be supported to develop national REDD programmes by the World Bank's Forest Carbon Partnership Facility were willing to participate in the programme. This is shown by the fact that all the 37 countries had developed REDD Readiness Plan Idea Note (r-PIN) by March 2009 (Corbera et al. 2009; World-Bank 2009). This is not surprising because discussions on the possibility of REDD were initiated in the first place in response to a request by several forest-rich (developing) countries to include avoided deforestation in the post-Kyoto climate mitigation policy (Gullison et al. 2007).

Learning from Similar Past Initiatives

There is a lot to learn from past initiatives involving conservation and development in developing countries using financial and technical support from developed countries, regardless of whether REDD is finally implemented through a market, fund or hybrid mechanism. For example, markets for coffee have been very unreliable and exploited the law of supply and demand to the effect that increased production did not help developing countries to reap more from the extra production, but ended up reducing the price of coffee on the global market. Due to experiences such as this, some authors have pointed out that it is possible in the end for carbon to flood the market to the extent that the carbon trade will not be profitable. This argument is based on the fact that as a general economic principle, the marginal return to investment in REDD decreases with increasing investment (Nepstad et al. 2007; Olsen and Bishop 2009).

Another example of the opportunities for REDD to learn from past phenomena is the fact that a good proportion of the formulation of the REDD payment strategy has been based on the principles of Payment for Ecosystem Services (PES), which was there before REDD. For instance, cost-benefit analyses of REDD (Grieg-Gran 2006; Olsen and Bishop 2009) have been carried out on the basis of PES principles.

Threats to REDD

Poor Governance

Poor governance is a threat to REDD in two ways. Firstly, poor governance is an important driver of deforestation and forest degradation and secondly, poor governance affects the design, development and implementation of REDD projects. Corruption, inequality in distribution of benefits, rights and responsibilities, lack of transparency, poor law enforcement and land-use conflicts all contribute to deforestation and forest degradation.

A recent example of poor governance as a driver of deforestation and forest degradation has been documented for southern Tanzania, where there was extensive logging (Milledge et al. 2007). The government put in place measures to control the logging but corruption was used to circumvent the measures. Manipulation of policy, laws and regulations governing the management of forest resources were facilitated by senior Tanzanian and foreign government officials in favour of foreign business interests. Rural communities were not empowered enough to fight with the illegal loggers and corrupt leaders, and thus their benefits were ignored in the timber trade. In general, there was low staffing for the forest sector in the whole country and there tended to be more staff where there were fewer forest resources to be managed. Thus southern Tanzania with a relatively large amount of forest resources was severely understaffed. Deforestation and forest degradation were further favoured by inter-sectoral and intra-sectoral conflicts of interest due to poor coordination of policy and planning. The forest sector itself has both the mandates to protect the forests and to destroy them—because it is also expected to raise revenues from timber royalties. The problem of poor governance is not limited to Tanzania—many developing countries have governance challenges that can hinder the implementation of REDD (Ebeling and Yasué 2008).

Political Instability and War

A major threat to the implementation of REDD is political instability, especially in the form of war. Where there is war, there is no order: everything is permitted. Although war can be a blessing in disguise or a curse to the forest, depending on the course of events that surrounds the war, and the situation before the outbreak of the war (Butler 2010), implementation of REDD as an intentional strategy to conserve forests would not include war as one of its acceptable tools. War may reduce deforestation and forest degradation by ousting governments that sell forest assets to service debt payments and finance weapons purchases, preventing forest-based foreign investment (e.g. in Liberia in the late 1990s) and by keeping forest users away from forests due to fear of landmines (e.g. in Vietnam, Cambodia, Laos, Mozambique, and Angola) and the fighters who may hide in the forests (Butler 2010).

War may act against REDD both directly through the destruction of forests to expose hiding enemies and improve accessibility and indirectly by making organised forest management impossible due to destruction of governance structures and absence of the rule of law. For instance, in the Democratic Republic of Congo (DRC)—formerly Zaire—in the late 1990s, war seriously undermined conservation institutions, destroying four world heritage sites; namely, Virunga National Park, Garamba National Park, Kahuzi-Biega National Park and the Okapi Wildlife Reserve (Butler 2010). Due to the absence of the rule of law, forest officials may fail to reinforce forest protection and sometimes even risk their lives. Such is the case in Thailand and Colombia, where forest officials are directly targeted by rebels with firearms (Butler 2010).

Loss of Interest in REDD by the International Community

It is possible for the international community to completely lose interest in REDD. This may happen due to a number of reasons, including the emergence of a more pressing issue than climate change, the decision to ignore the 20% contribution of REDD to climate change mitigation, assumption by developing countries that REDD is meant to deter their development, disappointment of the donor countries with the performance of donor-funded REDD and the collapse of the carbon market.

The international development community has a tendency to switch from one issue to the next, usually as if the issues were mutually exclusive. When an issue is fashionable, most of the funding is directed towards it and when its time passes, the funding for it drastically declines. How can we be assured that this will not happen to REDD?

REDD is being pursued as one of several components of a climate change mitigation portfolio. The best that REDD can do, assuming the present rates of deforestation and forest degradation, is to contribute 20% to the mitigation of climate change. Might people start to think that the 20% contribution to climate change due to REDD can be ignored?

The link between deforestation and forest degradation and development cannot be denied. Assuming that developing countries can reduce deforestation and forest degradation and hope to develop, using the carbon trade and donor funds from REDD is only one way of thinking. If developing countries think otherwise, they might not be interested in REDD because it is a way of keeping them underdeveloped. Why is it so difficult for developed countries to change their way of living in such a way as to favour climate? Why is it much cheaper to achieve climate change mitigation using REDD, which is connected to development of developing countries?

REDD depends on funds either from donors or from carbon market mechanisms. Continued willingness of donors to fund REDD will depend on satisfaction with what is being done by the money injected into the initiative. If performance is not satisfactory, donors can get disappointed and stop funding REDD. The only alternative would be carbon market mechanisms. However, even market mechanisms have their own likelihood and ways of failure.

Conclusions

Although REDD may contribute to sustainable forest management, ecological conservation and poverty reduction, the total mitigation of climate change that can result from successful REDD implementation is only a small fraction of what can be achieved through mitigation in the industrial sectors. Furthermore, implementation of REDD is probably going to be affected by problems of leakage, unintended positive effects, insufficiency, unreliability and complications of funding, poor governance, political instability and war. However, given the current willingness of developed and developing countries to participate in REDD, it is likely that REDD may be implemented with a reasonable level of success. In addition, success for REDD may be assured more by learning from similar past initiatives of payment for ecosystem services and conservation and development projects in developing countries.

Rather than thinking of REDD as an alternative to other climate change mitigation measures, REDD should be considered as a complement to them. Furthermore, when designing REDD schemes, the price of REDD assumed should not be the minimum because that will undermine development in developing countries and reduce their interest in REDD. With good design and implementation, REDD can create a win–win situation between developed and developing countries and result in leakage-free permanent reductions in deforestation and forest degradation.

Acknowledgments The publication of this chapter has been supported by the initiative of the Ecological Society for Eastern Africa (ESEA) and IJCCSM through presentation at ESEA's 3rd Scientific Conference and subsequent submission of the chapter to IJCCSM. We would like to express our gratitude to the ESEA secretariat and the editorial office of the IJCCSM are thanked for their auspices. Reviewers' comments on previous versions of this chapter are highly appreciated.

References

Angelsen A (2009) Introduction. In: Aw Angelsen B, Kanninen M, Sills E, Sunderlin WD, Wertz-Kanounnikoff S (eds) Realising REDD + : national strategy and policy options. CIFOR, Bogor, Indonesia, pp 1–12

Angelsen A, Wertz-Kanounnikoff S (2008) What are the key design issues for REDD and the criteria for assessing options? In: Angelsen A (ed) Moving ahead with REDD options and implications. CIFOR, Bogor, Indonesia, pp 11–22

Butler RA (2010) War. Tropical rainforests: imperiled riches—threatened rainforests. Available at http://rainforests.mongabay.com/0810.htm (accessed 20 April 2010)

Chomitz KMPB, de Luca G, Thomas TS, Wertz-Kanounnikoff S (2006) At loggerheads? Agricultural expansion, poverty reduction, and environment in the tropical forests. World Bank, Washington, DC

Corbera E, Estrada M, Brown K (2009) Reducing greenhouse gas emissions from deforestation and forest degradation in developing countries: revisiting the assumptions. Climatic change. pp 769–815

den-Elzen MG, Meinshausen M (2005) Meeting the EU 2°C climate target: global and regional emissions implications. Bilthoven: the Netherlands Environmental Assessment Agency

Dudley R (2010) A little REDD model to quickly compare possible baseline and policy scenarios for reducing emissions from deforestation and forest degradation. Mitig Adapt Strat Glob Ch 15:53–69

Ebeling J, Yasué M (2008) Generating carbon finance through avoided deforestation and its potential to create climatic, conservation and human development benefits. Philos Trans R Soc B 363:1917–1924

Freestone D (2009) The international climate change legal and institutional framework: an overview. In: Freestone D, Streck C (eds) Legal aspects of carbon trading: Kyoto, Copenhagen and beyond. Oxford University Press, London

GEF (2010) Replenishment fund; available at http://www.gefweb.org/replenishment/replenishment.html (accessed 24 April 2010)

Grieg-Gran M (2006) The cost of avoiding deforestation: report prepared for the Stern review of the economics of climate change. International Institute for Environment and Development, London

Gullison RE, Frumhoff PC, Canadell JG, Field CB, Nepstad DC, Hayhoe K, Avissar R, Curran LM, Friedlingstein P, Jones CD, Nobre C (2007) Tropical forests and climate policy. Science 316:985–986

Haites E (2004)Estimating the market potential for the clean development mechanism: review of models and lessons learned. PCFplus report, Washington DC; 19 June

Houghton RA (2005) Tropical deforestation as a source of greenhouse gas emissions. In: Moutinho P and B, S. S. (Eds) Tropical deforestation and climate change IPAM—Instituto de Pesquisa Ambiental da Amazônia; Environmental Defense; pp 13–22

IEA (2005) G8 climate change action plan and the investment framework the technology and financing context and issues. International Energy Agency, Montreal CoP11 UK/IEA Sideevent; 5 December

IPCC (2001) Climate change 2001: impacts, adaptation and vulnerability. Summary for policy makers. Intergovernmental panel on climate change. World Meteorological Organisation, Geneva

IPCC (2007) Climate change 2007:the physical science basis. Intergovernmental Panel on Climate Change, New York

Milledge SAH, Gelvas IK, Ahrends A (2007) Forestry, governance and national development: lessons learned from a logging boom in southern tanzania. An overview. TRAFFIC East/ Southern Africa/Tanzania Development Partners Group/Ministry of Natural Resources of Tourism, Dar es Salaam, Tanzania

Mollicone D, Freibauer A, Schulze ED, Braatz S, Grassi G, Federici S (2007) Elements for the expected mechanisms on 'reduced emissions from deforestation and degradation, REDD' under UNFCCC. Environmental Research Letters:045024

Nepstad DB, Soares-Filho B, Merry F, Moutinho P, Rodrigues HO, Bowman S, Schwartzman S, Almeida O, Rivero S (2007) The Costs and Benefits of Reducing Carbon Emissions from Deforestation and Forest Degradation in the Brazilian Amazon. Woods Hole Research Center, Falmouth, MA, USA

Obersteiner M, Huettner M, Kraxner F, McCallum I, Aoki K, Böttcher H, Fritz S, Gusti M, Havlik P, Kindermann G, Rametsteiner E, Reyers B (2009) On fair, effective and efficient REDD mechanism design. Carbon Balance Manage 4:11–111

Olsen N, Bishop J (2009) The financial costs of redd: evidence from brazil and indonesia: executive summary. In: Olsen N, Bishop J (eds) The financial costs of redd: evidence from brazil and indonesia. International Union for Conservation of Nature and Natural Resources (IUCN), Gland, Switzerland, pp 2–8

Peskett L, Brockhaus M (2009) When REDD + goes national: a review of realities, opportunities and challenges. In: Aw Angelsen B, Kanninen M, Sills E, Sunderlin WD, Wertz-Kanounnikoff S (eds) Realising REDD + : national strategy and policy options. CIFOR, Bogor, Indonesia

Phelps J, Guerrero MC, Dalabajan DA, Young B, Webb EL (2010) What makes a 'REDD' country? Global Environ Ch 20:322–332. doi:10.1016/j.gloenvcha.2010.01.002

Raupach MR, Marland G, Ciais P, Le-Quéré C, Canadell JG, Klepper G, Field CB (2007) Global and regional drivers of accelerating CO_2 emissions. Proc Natl Acad Sci USA 104: 10288–10293

Reyer C, Guericke M, Ibisch P (2009) Climate change mitigation via afforestation, reforestation and deforestation avoidance: and what about adaptation to environmental change? New Forests 38:15–34

Santilli M, Moutinho P, Schwartzman S, Nepstad D, Curran L, Nobre C (2005) Tropical deforestation and the Kyoto protocol: an editorial essay. Clim Change 71:267–276

Schwartzman S, Nepstad D, Mountinho P (2008) Getting REDD right: reducing emissions from deforestation and forest degradation (REDD) in the United nations framework convention on climate change (UNFCCC). The woods hole research centre/environmental defense/IPAM.10

Tschakert P, Huber-Sannwald E, Ojima DS, Raupach MR, Schienke E (2008) Holistic, adaptive management of the terrestrial carbon cycle at local and regional scales. Global Environ Change 18:128–141

UNDP (2010) UN-REDD programme fund. Available at http://www.undp.org/mdtf/un-redd/overview.shtml (accessed 24 April 2010)

van der Werf GR, Morton DC, DeFries RS, Olivier JGJ, Kasibhatla PS, Jackson RB, Collatz GJ (2009) Carbon dioxide emission from forest loss. Nat Geosci 2:737–738

Vera-Diaz MC, Schwartzman S (2005) Carbon offsets and land use in the Brazilian amazon. In: Moutinho P, Schwartzman S (Eds) Tropical deforestation and climate change IPAM and environmental defense, Belem, pp 35–44

Watson RT, Noble IR, Bolin B, Ravindranath NH, Verardo D, Dokken D (2000) Land use, land-use change, and forestry. A special report of the Intergovernmental Panel on Climate Change. Cambridge University Press, UK

Wertz-Kanounnikoff S, Angelsen A (2009) Global and national REDD+ architecture: linking institutions and actions. In: Angelsen AWB, Kanninen M, Sills E, Sunderlin WD, Wertz-Kanounnikoff S (eds) Realising REDD+ : national strategy and policy options. CIFOR, Bogor, Indonesia, pp 13–24

World Bank (2009) Forest carbon partnership facility. FY2009 annual report. World Bank, Washington, DC

WRI (2005) Navigating the numbers: greenhouse gas data and international climate policy. World Resources Institute, Washington, DC

Author Biographies

Emmanuel F. Nzunda holds a PhD in Ecological Science from the University of KwaZulu-Natal, South Africa, with an MSc and BSc in Forestry from Sokoine University of Agriculture, Morogoro, Tanzania. His PhD research was on "The role of resprouting and the persistence niche in forest structure and dynamics in coastal dune forest in KwaZulu-Natal province, South Africa". His MSc research involved the use of remote sensing and GIS in analysing landscape-level forest structural patterns in Udzungwa Mountains, Tanzania. His research interests include forest and landscape ecology, deforestation dynamics, forest governance, forests and climate change and ecosystem sustainability. He is competent in remote sensing and GIS, statistics, forest resource assessment, mapping and management planning, and geo-spatial modelling. He works as a lecturer, researcher and consultant in Forest Ecology and Management.

Tumaini G. Mahuve holds a BSc (Forestry) from Sokoine University of Agriculture, Tanzania. He is involved in research in forest ecology and management, particularly assisting in proposal development, fieldwork, data entry and analysis, literature search, report and publication writing.

Chapter 13
Environmental Mainstreaming in Development Policy and Planning in Sub-Saharan Africa: A Case Study from Kenya

Martin Ochieng Oulu and Emmanuel Kwesi Boon

Abstract The paper aims to map Kenya's national development planning framework, investigate any environmental (including climate change) mainstreaming (EM) strategies employed, and identify the prospects and challenges for EM furtherance. The country's heavy reliance on environment and natural resources for socio-economic development makes it particularly vulnerable to climate change. Effective EM will thus ensure better resource management and utilisation, improve livelihoods and enhance climate change adaptation. EM strategies are investigated through a review of published literature, government policy documents and statements, and key informant interviews with representatives of institutions and organisations involved in national development policy and planning. Effectiveness of normative, organisational and procedural EM strategies employed is evaluated from a horizontal environmental policy integration perspective. The country's political commitment and strategic vision, administrative culture and practices, assessment of and participation in planning, monitoring and learning from experience, and use of policy instruments to achieve EM are analysed. The participatory and integrated district development planning framework offers a unique opportunity for EM. However, absence of environmental representation in key development planning agencies and processes, no coherent national sustainable development strategy or office to coordinate EM, non-greening of public procurement, and non-diversification of policy instruments

M. O. Oulu (✉)
School of Tourism and Hospitality, Strathmore University,
PO Box 100828, 00101, Nairobi, Kenya
e-mail: ochiengmoulu@gmail.com

E. K. Boon
Human Ecology Department, Vrije Universiteit Brussel (VUB),
Laarbeeklaan 103, 1090 Brussels, Belgium
e-mail: eboon@vub.ac.be

W. Leal Filho (ed.), *Experiences of Climate Change Adaptation in Africa*,
Climate Change Management, DOI: 10.1007/978-3-642-22315-0_13,

are identified as major challenges. The paper is among the very few studies that have investigated cross-sectoral EM at a national level in Africa and will greatly contribute towards the achievement of Agenda 21s call for the integration of environment into national development policy and planning processes. The findings can be applied across sub-Saharan countries and beyond.

Keywords Development · Environmental mainstreaming · Climate change · Planning · Kenya

Introduction

Decoupling environmental degradation from socio-economic development is the defining challenge of sustainable development. Environmental mainstreaming (EM), technically known as environmental policy integration (EPI), offers great prospects for a sustainable future in Kenya and sub-Saharan Africa (SSA). EM is the continuous process of ensuring environmental issues are taken into account in all policymaking (EEA 2005). It involves a conscious inclusion of relevant environmental concerns into the decisions of institutions in charge of national, local and sectoral development policy, plans and action (Dalal-Clayton and Bass 2009). Essentially, environmental mainstreaming involves the incorporation of environmental objectives, including climate change mitigation and adaptation, into all policymaking stages in non-environmental policy sectors, with specific recognition of this goal as a guiding principle for the planning and execution of policy; accompanied by aggregation of presumed environmental (and climate change) consequences into an overall evaluation policy and a commitment to minimise contradictions between environmental and sectoral policies (Lafferty and Hovden 2003; Mickwitz et al. 2009).

The inability of conventional environmental policy to prevent pressure from being exerted on the environment by society is credited with the emergence of the EM concept (EEA 2005). Environmental concerns tend to be given insufficient weight in policy and political processes, with environmental policy having to react to negative impacts caused by socio-economic policies and practices rather than being integral in their very design. Environmental mainstreaming thus entails a fundamental recognition that the environmental sector alone is incapable of securing environmental objectives, hence the need for each sector to take on board environmental policy objectives (Lafferty and Hovden 2003). The rising importance of cross-cutting sustainability issues such as climate change, biodiversity loss, water scarcity and general environmental degradation will see a similar rise in the importance of EM over time (Ross and Dovers 2008). Sub-Saharan Africa's disproportionate reliance on environmental and natural resources for economic growth and livelihood support make them natural candidates for EM adoption.

The rationale for environmental mainstreaming can be distinguished into normative and organisational motives (Nilsson and Persson 2003). The normative viewpoint is based on concern for the integrity of the environment, i.e. that the environment is important, needs better protection and all sectors of society should help to achieve this objective. Moreover, due to the relative contribution of the environment to development and the potentially irreversible damage to life-support systems due to climate change, a higher priority should be given to environmental issues vis-à-vis traditional sector and economic objectives (Lafferty and Hovden 2003). The Government of Kenya affirms that the integration of climate information into government policies is crucial since climate is a major driving factor for most of the economic activities in the country (GoK 2010). The organisational rationale is concerned with the effectiveness and rationality of policymaking. Environmental mainstreaming allows environmental issues to be addressed in a more proactive and less ad hoc manner, it ensures coherence across different policy areas, enhances transparency and strengthens public participation (EEA 2005). The coming together of different policy actors increases the pool of knowledge and improves the chances of identifying previously unknown 'win–win' or more cost-effective opportunities (Nilsson and Persson 2003).

The EM impetus can be traced to long-standing debate at global and regional levels. Terming environmental mainstreaming the major institutional challenge of modern times, the Brundtland Report notes that the ability to choose sustainable policy paths requires simultaneous consideration of the ecological dimensions of policy and economic, agricultural, and energy ones (WCED 1987). Agenda 21 also weighed in by observing that prevailing decision-making systems in many countries often separate economic, social and environmental factors at the policy, planning and management levels (UNCED 1992). Reversing this trend requires adjustment or even fundamentally reshaping decision-making in line with country-specific conditions. Regionally, the New Partnership for African Development (NEPAD) Environmental Action Plan takes an integrated approach to the environment and development, while the African Union's Africa Convention on Conservation of Nature and Natural Resources (ACCNNR) calls on parties to integrate development and environmental concerns by treating both as integral parts of national and local development plans, while giving full consideration to ecological, social, economic and cultural factors in their development (UNEP 2006).

There is considerable experience with models and strategies for mainstreaming environment in development planning, but mainly in the developed world. This paper presents the results of a study that sought to review the state-of-practice of EM in Kenya from a developmental planning context. The paper's objectives are threefold: to map Kenya's national development planning framework, to investigate EM strategies employed, and to identify the challenges and prospects for EM adoption and furtherance in Kenya. The paper focuses on the macro or cross-sectoral level, i.e. the horizontal dimension of environmental policy integration. This is because unless the central government provides an appropriate national framework for EM, focus on any particular sector would have little impact (Lafferty and Hovden 2003).

Methodology

The study was conducted in Nairobi, Kenya, between July and September 2008. Nairobi is Kenya's capital city where most policy actors involved in national development planning in the country are located. The primary objective was to review the state-of-practice of EM in the country from the perspective of horizontal environmental policy integration (HEPI) in order to understand and enhance the progressive integration of environment and climate change into national development policy and planning processes. HEPI is the extent to which a government has developed a cross-sectoral strategy to achieve environmental mainstreaming (Lafferty and Hovden 2003). Kenya is one of the SSA countries whose socio-economic development is heavily dependent on natural and environmental resources. Hence, it is particularly vulnerable to climate change. Effective mainstreaming of environment and climate change into its development planning process will thus offer it the opportunity to better manage and utilise its natural resource base, improve livelihoods, and effectively mitigate and adapt to the adverse impacts of climate change. Since similar conditions generally prevail in neighbouring East African countries and many other SSA countries, the lessons learned will find wide application in these and similar countries worldwide.

The study focussed on developments between 1992 and 2007 for two primary reasons: the influence of the 1992 Rio Conference is expected to have created new framings of problems and issues at the national level (Nilsson and Persson 2003). The year 2007 saw the country's Economic Recovery Strategy for Wealth and Employment Creation (2003–2007) come to an end and its replacement by Kenya Vision 2030 (2008–2030). Moreover, Matland (1995) suggests that policy needs to be analysed in cycles of more than 10 years to allow for consideration of policy learning.

Secondary data sources included a review and analyses of existing literature, publications, government policy documents and statements related to EM. Primary data was collected through key informant interviews. Such interviews were conducted with 43 representatives of institutions and organisations active in the national policy cycle and development planning arena. This included representatives from government, environmental NGOs, business and industry, local political parties, civil society, international aid agencies and country representatives of the donor community, among others. Prior contact was made with potential key informants either via email or telephone where the purpose of the intended interview was explained to them. This ensured the interviewees were adequately prepared for the actual interview. Follow-up interviews were conducted to cross-check and/or confirm some of the information received.

To investigate the objectives and assumptions of the paper, information was gathered along the three broad classes of normative, organisational and procedural factors influencing successful EM implementation (Persson 2004). Existing EM strategies were evaluated using the European Environment Agency's criteria (EEA 2005). The evaluation encompassed an analysis of political commitment and strategic vision, administrative culture and practices, assessment and participation in planning, use of policy instruments to achieve EM, and monitoring and learning from experience. Conditions conducive to and/or barriers to environmental and climate change mainstreaming were identified and analysed.

Results and Discussion

There is no constitutional or legal provision for environmental or climate change mainstreaming in Kenya. The country's development planning framework is clearly separated into socio-economic (herein referred to as development planning) and environmental planning processes, with few inter-linkages.

Development Planning Framework

Kenya's development planning framework is largely concerned with socio-economic planning and, apart from the traditional sectoral planning, includes the two planning frameworks described in the next sub-sections.

The District-Focus Planning Framework

In operation since 1983, the district-focus planning framework is a participatory planning process outlined in the District Focus for Rural Development Strategy (GoK 1995). At its core is the District Development Committee (DDC), responsible for rural development planning, coordination, implementation and oversight. Headed by a District Commissioner (DC), members of the DDC include a district development officer, district-level central government representatives including environmental officers, local NGOs and community-based organisations (CBOs). The DDC is widely representative of district-level stakeholders. The DDC is mandated to prepare district development plans (DDPs) every 5 years through a consultative process. The DDPs are eventually consolidated into a national development plan. The process begins with a situation analysis of the particular district after review and interpretation of national policies and guidelines, culminating in the preparation of a district fact sheet and a district profile. The district profile incorporates information on settlement structure, physiographic and natural conditions, population profiles and projections, and a summary of various sectors. The physiographic and natural conditions include analysis of prevailing environmental conditions, major ecosystems, ecosystem services and trends in their use, climate and weather reviews and occurrence of climatic shocks such as droughts, floods or landslides affecting the district. Such environmental assessments are intended to feed into and guide the preparation of the DDPs, providing an integrated planning framework and a strong basis for environmental mainstreaming. A draft District Development Planning Handbook incorporates guidelines on how to mainstream environment into the district development planning process. The district-focus planning framework is however, under threat from the NESC-led framework (see below) that seems to have taken over most of its development planning functions, yet is not as participatory.

The National Economic and Social Council-led Planning Framework

The National Economic and Social Council (NESC) is an autonomous advisory body that was established in 2004. Chaired by the President or, in his absence, the Minister for Finance, its functions include developing and recommending socio-economic policies to the Cabinet; appraising government programmes and activities; gathering, analysing and compiling information on economic developments and trends and submitting it to the Cabinet with appropriate recommendations (GoK 2004). It also serves as a forum for government interaction with the private sector and labour unions, although its link with other planning frameworks remains unclear.

Members of the Council include heads of several sectoral ministries, representatives of the private sector, professional societies, labour unions, farmer associations, academia, private healthcare providers and nominated international experts. Poignantly missing is any organised environmental representation in the NESC. Although the Minister for Environment was incorporated in 2009, neither environmental NGOs nor representatives of the National Environmental Management Authority (NEMA) are members. Paradoxically, NEMA is the country's premier environmental agency charged with ensuring the integration of environmental considerations into development policies, plans, programmes and projects. Under the chairmanship of the President, the NESC is one of the most powerful planning organs in the country, perhaps only second to the Cabinet. Although passed off as an 'advisory' council, its preparation of Kenya Vision 2030, the country's long-term development plan (2008–2030) currently under implementation, is testimony to its central position in Kenya's development planning. Indeed, Vision 2030 has been criticised for not adequately taking into account the impact of climate change, a situation that environmental representation in the NESC would have solved. The absence of environmental representation in the NESC essentially means environmental and climate change concerns are systematically missing from the highest levels of decision-making in the country, making it increasingly difficult to mainstream them at lower decision-making levels, while EM should be implemented at all stages of the decision-making chain (Fang et al. 2006).

The choice of the name NESC betrays its socio-economic focus. Put to one of the officials, the response was that economic and social councils are a common feature the world over and do not imply a lesser view of the environment. True as this may be, the lack of any organised environmental representation in the NESC given its high position in Kenya's development planning is worrisome. If this is not an indication of according the environment and climate change issues a low status in the country's development planning, it certainly illustrates the separation of socio-economic and environmental factors at the policy and planning levels. Evidently, the NESC is one of the best entry points to mainstreaming environment and climate change into development planning in Kenya. Not only is it strategically placed and at a high-level to ensure political commitment, its mandate already partially deals with sustainable development. However, transforming the NESC into a National Council for Sustainable Development (NCSD) would better serve to integrate economic, social and environmental (including climate change) dimensions of sustainable development as well as policy and action at the different government levels. NCSDs assist governments by providing clear guidance on policy tools, regulations and indicators of sustainable development as well as monitoring and evaluating progress (ECA 2005).

Environmental Planning Framework

Kenya's environmental planning has a long history but a short past. Save for the recent publication of the National Climate Change Response Strategy (NCCRS) in April 2010, Kenya currently has no policies or laws dealing directly or exclusively with climate change (GoK 2010). Consequently, the institutions in place to govern climate change are still inadequate, if existing at all. Climate change governance in the country is therefore best analysed within the broader environmental framework and by reviewing the proposals in the NCCRS. Despite the formation of the National Environment Secretariat (NES) in 1974 as the lead agency to coordinate and oversee environmental activities, a coherent and elaborate national environmental planning framework only emerged in the late 1990s with the enactment of the Environmental Management and Coordination Act, 1999 (EMCA). To date, the country does not have a comprehensive national environment policy, although one is currently under preparation.

Coming into force in 2000, the EMCA is Kenya's premier environmental legislation. The Act established the National Environment Management Agency (NEMA) as the principal instrument for implementing all environmental policies. Importantly, NEMA is responsible for promoting the integration of environmental considerations into development policies, plans, programmes and projects (GoK 2000). The Ministry of Environment and Natural Resources together with the National Environment Council (NEC) oversee general environmental policy formulation and direction. Made up of representatives from all sectoral ministries, NEC membership exposes other ministries to the issues and challenges facing the environment and impresses upon them the relationship between environment, climate change and development and the significance of their own sectoral ministries' mandates for environmental protection. The High Court and the National Environment Tribunal (NET) arbitrate on legal aspects related to the environment.

Vertically, the EMCA establishes district environment committees (DECs) and provincial environment committees (PECs) modelled alongside the district-focus development planning framework. Under the chairmanship of the District Commissioner (DC) and with the District Environment Officer (DEO) as its secretary, the DEC is responsible for environmental management at the district level. Conveniently, the same set of actors forming the DDC under the district-focus development framework are also members of the DEC, with the DC heading both the DDC and the DEC. This offers a *sui generis* framework and opportunity for mainstreaming environmental and climate change issues into the district-focus development planning process.

The DEC is mandated to prepare a district environmental action plan (DEAP) every 5 years. Once prepared, the DEAPs are forwarded to the respective PECs where they are amalgamated into a provincial environment action plan (PEAP). The PEAPs are subsequently used to prepare a national environment action plan (NEAP). Interestingly, the NEAP committee is chaired by the Permanent Secretary in the ministry responsible for economic planning and development and not that of the environment. This offers another opportunity for environment-development integration.

Certain provisions in EMCA oblige all environmental action plans (district, provincial and national) to propose guidelines for mainstreaming environmental protection standards into development planning and management, and to recommend appropriate legal and fiscal incentives to encourage mainstreaming of environmental requirements into development planning and operational processes (GoK 2000). Moreover, one of the strategic objectives of the Ministry of Environment's Strategic Plan (2006–2010) is to mainstream environmental and climate change concerns into overall planning and implementation of programmes and projects. Planned activities under this objective include strengthening of district, provincial and national environmental action plans and enhancing their use for environmental mainstreaming (GoK 2006a). If implemented, these provisions would positively influence environmental and climate change mainstreaming in the country.

Kenya's Environmental Mainstreaming Strategies

Normative Strategies: Political Commitment and Strategic Vision

Normative factors are the values, norms and traditions that set the general parameters for policymaking, hence determine the significance of EM in the policymaking system (Persson 2004). Among these, political will is considered a *sine qua non* for the attainment of sustainable development by making environmental mainstreaming credible and an active aspiration rather than just a principle on paper (Lafferty and Knudsen 2007). A national sustainable development strategy (NSDS) is a strong indicator of a government's view of environmental and climate change mainstreaming by providing a framework for ensuring an effective translation of commitment into concrete action (Lafferty and Hovden 2003). Unfortunately, Kenya does not have an NSDS, an indictment on the country's commitment to EM. Most of the interviewees during the field survey were remotely aware of what an NSDS is, its objective or significance. They believed that a diligent implementation of the EMCA is sufficient. Although regulations are essential, NSDSs are an important mechanism for general environmental mainstreaming and climate change integration (Cline-Cole and O' Keefe 2006).

The absence of an NSDS and absence of environmental representation in the NESC illustrates a policy paradigm and tradition of separation of socio-economic and environmental factors in the country's policymaking system, with a strong socio-economic focus. The government itself admits that successive development plans have emphasised economic growth and environmental conservation and protection without considering the links between them (GoK 1999).

Although still fledgling, use of knowledge and science in decision-making is evident in the annual preparation of state of environment (SoE) reports and routine incorporation of professional groups, academia, civil society and environmental NGOs in development planning. The EMCA requires presentation of the SoE before Parliament for adoption, increasing the chances of its recommendations being widely implemented. Unfortunately, the study could not confirm that any of the four SoEs published has so far been presented to Parliament. The government admits that climate information has also not been adequately factored into most of the sectors of the economy, including government development policies and plans (GoK 2010).

Organisational Strategies: Institutional Arrangements and Practices

The way governments function, particularly their 'compartmentalisation', has been identified as a key factor inhibiting EM. The administrative challenge is to do away with the culture of fragmentation by making departments less single-minded and more responsive to environmental and climate change issues (EEA 2005). Kenya's institutional EM arrangement is fragmented. There is no designated high-level cross-sectoral office, unit or department charged with coordinating and guiding EM or sustainable development in general. The significance of such an office has been demonstrated across the border in Tanzania where the successful greening of the country's Poverty Reduction Strategy Paper (PRSP) was coordinated by a non-sectoral and high-level Vice-President's Office (VPO). This office housed both the Poverty Eradication Division and the Division of Environment (Assey et al. 2007). The VPO proved to be an effective location for coordinating the environmental mainstreaming work of the 'greened' PRSP. Similarly, horizontal coordination of sustainable development in the United Kingdom (UK) is under the high-level Prime Minister's Office (Russel 2007).

Kenya's environmental agency NEMA is charged with promoting and not coordinating environmental mainstreaming in Kenya. NEMA has in turn created an internal department—the Environmental Planning and Research Coordination Department—to spearhead this function. However, as Lafferty and Hovden (2003) point out, a separate sectoral environmental outfit will rarely, if ever, have the authority necessary to impose environmental objectives into the decision-making premises of other sectoral ministries and authorities. This reinforces the logic of anchoring the responsibility of coordinating or promoting EM in an overarching authority structure. Indeed, citing the Tanzanian case, (Assey et al. 2007) report that had the process been led by a single environmental organisation, it would have run the risk of being seen as an environmental 'special pleading' or 'territory building', hence not fully integrated into the PRSP's deliberations. NEMA's environmental and climate change mainstreaming efforts run such a risk.

Institutional fragmentation can however, be overcome by facilitating increased communication without necessarily changing organisational mandates or hierarchical relationships (EEA 2005). NEMA's mandate of promoting environmental mainstreaming falls within this ambit. Environmental officers are routinely seconded to line ministries, sector departments and agencies. At a general governmental level, inter-ministerial committees and multi-disciplinary taskforces are often used where necessary, bringing together different ministries, experts and stakeholders. Most are, however, temporary, ad hoc and neither transparent nor accountable other than to the appointing authority, hence their impact on EM is difficult to assess (Bird and Kirira 2009).

Procedural Strategies: Assessment and Consultation

Ex ante impact assessments such as strategic environmental assessments (SEA) are procedural strategies of a continuous nature and a way of ensuring that the likely impacts on the environment are identified beforehand, thus allowing for mitigation or adaptive measures to be put in place. SEA is provided for under Section 6 of Kenya's Environmental (Impact Assessment and Audit) Regulations, 2003. All proposals for public policy, plans and programmes must be subjected to an SEA. The 2004 State of Environment Report identifies 15 land-related policies and plans that should have undergone SEA. The study however, found no policy or plan that was subjected to SEA, a situation attributed to a lack of national SEA guidelines. NEMA is reportedly developing draft guidelines for the eventual application of SEA in the country.

Although limited, procedures for ensuring use of environmental information in decision-making are evident in several areas. One example is the already mentioned annual preparation, issuance and presentation to Parliament of the state of the environment (SoE) report by NEMA. The set-up of a natural resource accounting system has been in the pipeline for many years and is reportedly under serious consideration in the forthcoming environmental policy. In addition, the NEMA Strategic Plan (2005–2010) proposes to develop a manual and issue regular guidelines for incorporating environmental concerns into development activities (NEMA 2005). The Ministry for Planning and National Development has also developed a draft District Development Planning Handbook (2008) with guidelines on how to mainstream environment and climate change into the district-focus development planning process.

Consultation and participation of non-governmental actors in decision-making has improved since the early 1990s, taking the form of inter-ministerial, stakeholder committees and taskforces. They are however, not formalised or institutionalised and in many cases remain the prerogative of the sectoral ministries involved. There is also a need to build capacity of NGOs on the environment-development nexus, potential impacts of climate change and on various tools for environmental and climate change mainstreaming at the different decision-making levels.

Use of Policy Instruments for Environmental Mainstreaming

Government regulation as a policy instrument, particularly in the form of the Environmental Management and Coordination Act 1999 (EMCA), dominates. The existence of comprehensive legal standards is an important driver for EM and an important ingredient in the policy mix (EEA 2005). The regulations vary greatly in nature, from outlining environmental planning procedures, setting minimum environmental standards to economic and fiscal incentives. Environmental impact assessment (EIA) is now mandatory for certain types of projects as outlined in the EMCA and the Environmental (Impact Assessment and Audit) Regulations, 2003. However, progress reports of the government's own Monitoring and Evaluation Directorate (GoK 2006b, 2007) show that no public projects have undergone EIA. It therefore appears that the government itself is operating outside the requirements of the EIA statute, thus undermining its intent.

The Standards and Enforcement Review Committee is charged with setting environmental quality standards and ensuring compliance. The slow pace of developing the standards is a major setback, indicating a lack of experience with the impact and challenges of their implementation. This may explain the confidence placed in the EMCA to achieving general environmental protection and EM in particular by the planning fraternity in Kenya. The bubble will soon burst.

Economic instruments are a way of integrating environmental concerns into market conditions (Herodes et al. 2007) by putting into practice the "polluter pays" principle (Roberts 2004). Section 57 of EMCA titled 'Fiscal Incentives' proposes the use of economic instruments for environmental management. Although few of these measures are routinely applied through the annual budget, use of economic instruments remains largely insignificant. This is partly due to a lack of guidelines since the stalling of a draft bill titled The Use of Economic Instruments in Environmental Conservation and Management Regulations of January 2004. The economic instruments in use are also poorly designed and haphazardly employed, ending up as mere revenue collection mechanisms.

Green public procurement represents an opportunity to mainstream environmental concerns into public administration and into all sectors that supply the public sector with goods and services (Herodes et al. 2007). However, Kenya's Public Procurement and Disposal Act, 2005 (GoK 2005) does not have any provisions for promoting environmentally conscious public procurement. The Act's purpose makes no mention whatsoever of any environmental objective despite the fact that central government procurement stood at 6.13% of GDP in 1999/2000 and 10.87% of GDP in 2000/2001 (OECD 2003). Having come into operation in January 2007, 7 years into the life of NEMA, the Procurement Act was a major lost opportunity for EM in the country and seriously needs greening.

There is no legal requirement for businesses or industry to incorporate environmental management systems (EMS) into their management or production systems in Kenya. However, mainly due to pressure from foreign markets, EMS is being steadily embraced and implemented, particularly ISO 14001, by many businesses.

Conclusion and Recommendations

The post-Rio Conference era was a watershed for environmental management in Kenya despite continued separation of environment and development at the policy and planning levels. While the need for EM is expressly acknowledged in many development policies and strategies, the absence of environmental representation in the all-important NESC is a major setback. The NESC should be transformed into a National Council for Sustainable Development (NCSD) for better integration of environmental objectives, including climate change, into the country's development policy and planning processes.

Without a designated high-level strategic office or unit to coordinate environmental and climate change mainstreaming efforts, many activities in this area will remain fragmented and limited. The residence of EM responsibility with NEMA, a sectoral environmental agency with little political clout, is rather ineffective and insufficient. A high-level strategic office, unit or department, preferably under the Office of the President (OP) or Prime Minister's Office, should oversee the EM coordination mandate. This should be followed by the development of a comprehensive national sustainable development strategy (NSDS) to help concretise action towards EM achievement.

The dictum of policy implementation involves a mix of policy instruments. The heavy reliance on command and control measures needs to give way to experimentation with newer policy instruments such as economic instruments, environmental management systems and voluntary agreements. Greening public procurement, particularly the Public Procurement and Disposal Act (2005), is an obvious starting point. Ultimately, mainstreaming environment and climate change into Kenya's and other developing countries' development planning processes will require strong political commitment, organisational and procedural restructuring and a willingness to experiment with new ideas, approaches and policy instruments.

References

Assey P, Bass S, Cheche B, Howlett D, Jambiya G, Kikula I, Likwelile S, Manyama A, Mugurusi E, Muheto R, Rutasitara L (2007) Environment at the heart of Tanzania's development: lessons from Tanzania's national strategy for growth and reduction of poverty (MKUKUTA). International Institute for Environment and Development (IIED), London

Bird N, Kirira N (2009) Government institutions, public expenditure and the role of development partners: meeting Kenya's environmental challenges. Overseas Development Institute (ODI), London

Cline-Cole R, O'Keefe P (2006) Mainstreaming the African environment? Rev Afr Political Economy 33(109):377–389

Dalal-Clayton B, Bass S (2009) The challenges of environmental mainstreaming. IIED, London

ECA (Economic Commission for Africa) (2005) National councils for sustainable development in Africa: a review of institutions and their functioning. ECA, Addis Ababa

EEA (European Environment Agency) (2005) Environmental policy integration in Europe: state of play and an evaluation framework. EEA, Copenhagen

Fang QH, Hong HS, Zhang LP, Chen WQ, Xue XZ (2006) Methods for integrating environment into decision-making in China. Aquat Ecosyst Health Manage 9(1):73–78

GoK (1999) Sessional paper no. 6 of 1999 on environment and development. Ministry of Environmental Conservation, Nairobi

GoK (2000) Environmental management and coordination act, 1999. Kenya gazette supplement no. 3, Government Printer, Nairobi pp 43–175

GoK (2004) The national economic and social council. Kenya gazette notice no. 7699, vol. CVI-No. 82, Office of the President, Nairobi pp 2107–2108

GoK (2005) The public procurement and disposal act, 2005. Kenya gazette supplement no. 77, Government Printer, Nairobi pp 45–133

GoK (2006a) Ministry of environment and natural resources (MENR) Strategic Plan 2006–2010. MENR, Nairobi

GoK (2006b) Annual progress report 2004/2005 economic recovery strategy. Ministry of Planning and National Development, Nairobi

GoK (2007) Third annual progress report 2005/2006: the economic recovery strategy for wealth and employment creation 2003–2007. Ministry of Planning and National Development, Nairobi

GoK (2010) National climate change response strategy. GoK, Nairobi

GoK (Government of Kenya) (1995) District focus for rural development. Office of the President, Nairobi

Herodes M, Adelle C, Pallemaerts M (2007) Environmental policy integration and modes of governance–a literature review. Ecological Institute for International and European Environmental Policy, Berlin

Lafferty WM, Hovden E (2003) Environmental policy integration: towards an analytic framework. Environ Politics 12(3):1–22

Lafferty WM, Knudsen J (2007) The issue of 'balance' and trade-offs in environmental policy integration: How will we know EPI when we see it?. Ecological Institute for International and European Environmental Policy, Berlin

Matland RE (1995) Synthesizing the implementation literature: the ambiquity-conflict model of policy implementation. J Administration Res Theor 5(2):145–174

Mickwitz P, Aix F, Beck S, Carss D, Ferrand N, Gorg C, Jensen A, Kivimaa P, Kuhlicke C, Kuindersma W, Manez M, Melanen M, Monni S, Pedersen AB, Reinert H, Van Séverin Bommel (2009) Climate policy integration, coherence and governance. Partnership for European Environmental Research, Helsinki

NEMA (2005) NEMA strategic plan 2005–2010. NEMA, Nairobi

Nilsson M, Persson Å (2003) Framework for analysing environmental policy integration. J Environ Policy Planning 5(4):333–359

OECD (2003) Public procurement: lessons from Kenya, Tanzania and Uganda; Available at: http://masetto.sourceoecd.org/vl=2414088/cl=20/nw=1/rpsv/cgi-bin/wppdf?file=5lgsjhvj7843.pdf (accessed 9 April 2008)

Persson Å (2004) Environmental policy integration: an introduction. Stockholm Environment Institute, Stockholm

Roberts J (2004) Environmental policy. Routledge, London

Ross A, Dovers S (2008) Making the harder yards: environmental policy integration in Australia. Aust J Pub Administration 67(3):245–260

Russel D (2007) The United Kingdom's sustainable development strategies: leading the way or flattering to deceive? Eur Env 17:189–200

UNCED (1992) Earth summit'92. Regency Press, London

UNEP (2006) Africa environment outlook 2: our environment, our wealth; Available at: http://www.unep.org/dewa/africa/docs/en/aeo-2/chapters/aeo-2_ch01_THE_HUMAN_DIMENSION.pdf (accessed 24 February 2008)

WCED (1987) Our common future. Oxford University Press, Oxford

Author Biographies

Martin Ochieng Oulu is a lecturer in Wildlife Tourism and Environmental Management at the School of Tourism and Hospitality of Strathmore University, Nairobi, Kenya. He is also in private practice as an environmental consultant. He holds a Master's degree in Human Ecology from the Vrije Universiteit Brussels (VUB), Belgium, and is a recipient of the European Union Marie Curie Fellowship in Sustainability Evaluation. His teaching and research interests are multidisciplinary and include governance and policy-relevant issues at the confluence of environment and development.

Emmanuel Kwesi Boon is a professor of Human Ecology and Environmental Management at the Human Ecology Department of the Vrije Universiteit Brussels (VUB), Belgium. He is also the founder and Chairman of the International Centre for Enterprise and Sustainable Development (ICED), an NGO based in Accra and Wa in Ghana.

Chapter 14
Net Primary Productivity Response to Climate Change in the Mount Kenya Ecosystem

Ochieng Aggrey Adimo, Njoroge John Bosco and Wamocho Leonard

Abstract This chapter evaluates methods to estimate terrestrial ecosystem carbon assimilation and to monitor it over time at both local and national scales. This study presents simulated models that rely on vegetation indices and cover from Landsat 7 ETM sensors and vegetation stress scalars of temperature and moisture generated from the Thornthwaite water balance model. Temperature and moisture are the main climatic variables that affect vegetation productivity in the context of climate change at the landscape level. The temperature stress scalar in this study was derived from the seasonal optimal temperature for plant production from 28 years of climate data, while water stress was estimated from monthly water deficits, based on comparison of observed moisture supply (precipitation) to potential evapotranspiration (PET) demand derived using the Thornthwaite and Mather method (1957). The Carnegie-Ames-Stanford approach (CASA) model was used to estimate monthly patterns of carbon fixation using 30 M resolution satellite remote sensing images of the surface vegetation characteristics and driven by spatially interpolated climate data. IPCC climate general circulation models (GCM) scenarios were downscaled using a statistical downscaling model (SDSM) to generate locally useful data for impact analysis. Different vegetation classes show significant levels of sensitivity to temperature and moisture variability depicted by the varied net primary productivity (NPP) levels. Indigenous vegetation is well adapted and generally exhibited high NPP around Mount Kenya.

O. A. Adimo (✉) · N. J. Bosco · W. Leonard
Department of Horticulture, Jomo Kenyatta University,
P. O. Box 62000, 00200 Nairobi, Kenya
e-mail: radimoochg@yahoo.com

N. J. Bosco
e-mail: njbosco@hotmail.com

W. Leonard
e-mail: wamocho@yahoo.com

W. Leal Filho (ed.), *Experiences of Climate Change Adaptation in Africa*, Climate Change Management, DOI: 10.1007/978-3-642-22315-0_14,

This method can be useful in monitoring and evaluating ecosystem function of carbon sequestration and in facilitating optimisation of reducing carbon emissions from deforestation and degradation (REDD) through vegetation selection in forestry programmes.

Keywords Production efficiency model · Carbon sequestration · Downscaling · Ecosystem function · Mount Kenya

Introduction

The capacity of a land area to sequester carbon from the atmosphere is an important issue for policymakers and land managers. Policies and markets are emerging rapidly to compensate for reducing carbon emissions from deforestation and degradation (REDD), presenting an enormous opportunity to fund forest conservation at a local and national level, especially in Africa where conservation efforts frequently lack funding at the local level. Seizing this opportunity requires robust methods to estimate forest and other land cover carbon, and to monitor it over time at both local and national scales. Approaches for measuring and monitoring land cover carbon flux are therefore very important in areas with few or no eddy covariance towers, such as Kenya and other African countries. It is important to aid small-scale farmers to benefit from the carbon market by having a tool that can be used to estimate the amount of carbon their land-use activities (such as tree planting) contribute in terms of carbon dioxide sequestration, so that they can be adequately compensated.

Regional to global-scale relationships between carbon uptake and climatic gradients have been explored, but capturing landscape-scale patterns has been a challenge (Potter et al. 2007). A number of studies have however addressed sub-landscape level variation in forest production through remote sensing and ecological process models (Potter et al. 1993; Melillo et al. 1993). One such ecological process that has been used is net primary productivity (NPP) (McCallum et al. 2009). NPP is not only the net flux of carbon from the atmosphere into green plants per unit of time, but also a rate process, i.e. the amount of vegetable matter produced (net primary production) per day, week or year. NPP is a fundamental ecological variable, not only because it measures the energy input to the biosphere and terrestrial carbon dioxide assimilation but also because of its significance in indicating the condition of the land surface area and status of a wide range of ecological processes.

Carbon dioxide is usually removed from the atmosphere by means of photosynthesis by plants. Upon entering the terrestrial ecosystem it is termed gross primary productivity (GPP), with the difference between carbon gain via GPP and carbon loss through plant respiration defined as NPP (McCallum et al. 2009). NPP is difficult to measure (in situ) over large areas owing to spatial variability of environmental conditions and limitations in the accuracy of allometric equations (Melillo et al. 1993).

Despite this, a variety of methods have been developed to estimate carbon fluxes, including flux towers (e.g. Friend et al. 2007), carbon accounting (e.g. Shvidenko and Nilsson 2003), global vegetation models (e.g. Beer et al. 2006), atmospheric measurements (e.g. Stephens et al. 2007) and satellite-based techniques that rely on production efficiency models (PEM) (e.g. Running et al. 2004).

PEM have been developed to monitor primary production, taking advantage of available satellite data. PEMs are based on the theory of light use efficiency (LUE) which states that a relatively constant relationship exists between photosynthetic carbon uptake and radiation receipt at the canopy level (Potter et al. 2003a). PEMs usually require inputs of meteorological data (i.e. radiation, temperature and others) and the satellite-derived fraction of absorbed photosynthetically available radiation (FAPAR). The main PEM models include: the Carnegie-Ames-Stanford approach (CASA); the global production efficiency model (GLOPEM); terrestrial uptake and release of carbon (TURC); C-Fix; moderate-resolution spectroradiometer (MODIS)/MOD17; and biosphere model-integrating eco-physiological and mechanistic approaches using satellite data (BEAMS) (McCallum et al. 2009).

Temperature and moisture are the main climatic variables that affect vegetation productivity in the context of climate change at the landscape level.

This study presents simulated models that rely on vegetation indices and cover from Landsat 7 ETM sensors and vegetation stress scalars of temperature and moisture deficits generated from the Thornthwaite water balance model. The objectives were, firstly to assess the functional attributes of the vegetation to carbon assimilation via the CASA model and then, to quantify carbon sequestration potential for different landscape vegetation cover.

Methodology

Study Area

The study was conducted in the Mount Kenya ecosystems. The area is about 6,143 km^2, between coordinates 37° 0′ E to 37° 50′ E, and 0° 10′ N to 0° 30′ S.

Mount Kenya is an equatorial mountain and rises to 5,199 m.a.s.l. It is a Pleistocene composite dormant volcano and a major water tower that serves lowland inhabitants up to the Indian Ocean. It was designated a protected national park in 1949. The mountain has a rich biodiversity and is the only glacial remnant in the country, serving the whole country with hydropower. Its vegetation zonation is one of the global wonders and its preservation is critical (Ojany 2008). The vegetation groups include bamboo, indigenous forest vegetation, moorland,

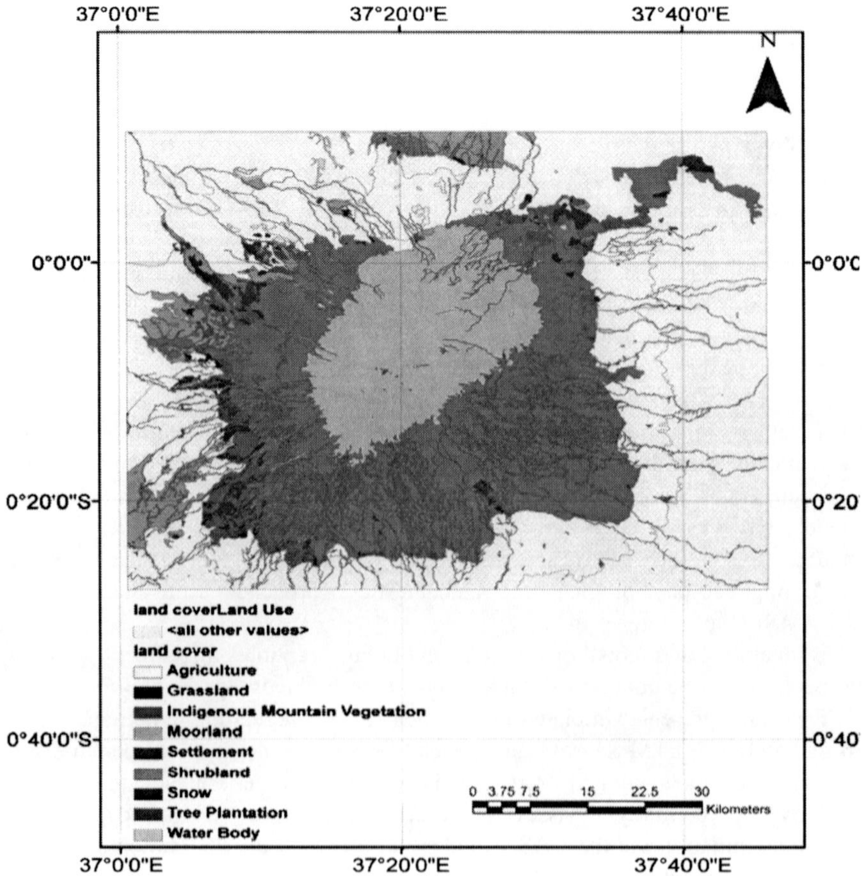

Fig. 1 Study area land cover use

cyprus, eucalyptus, grevillia, closed grassland, closed shrubland, pine, meru oak, cedar, camphor, small tea farms, open grassland and open shrubland (Fig. 1).

CASA Model

The CASA is a numerical model of monthly fluxes of water, carbon and nitrogen in terrestrial ecosystems. Estimates of terrestrial NPP fluxes depend on inputs of global satellite observations for land surface properties and on gridded model drivers from interpolated weather station records (Potter et al. 2005). LUE is set uniformly at 0.39 g C MJ-1 PAR, a value that derives from calibration of predicted annual NPP of previous more than 1,900 field estimates (McCallum et al. 2009). This model calibration has been assessed globally (Potter et al. 2003b).

Temperature stress is computed with reference to derivation of optimal temperatures for plant production. CASA includes a water stress scalar estimated from monthly water deficits, based on a comparison of moisture supply to potential evapotranspiration demand (Potter et al. 2003b). This is the only model that does not separately calculate GPP. Instead, it models NPP directly, thus avoiding autotrophic respiration calculation.

NPP is estimated as a product of time-varying surface solar irradiance (Sr), enhanced vegetation index (EVI) from Landsat satellite data, and LUE term (e_{max}) that is modified by time-varying stress scalar terms for temperature (T) and moisture (W) effects (Eq. 1) (McCallum et al. 2009).

$$\mathrm{NPP} = \mathrm{S_r EVI}\ \mathrm{e_{max}TW} \tag{1}$$

The T stress scalar is computed from the observed mean monthly air temperature (°C) with reference to derivation of optimal seasonal temperature for plant production. W stress scalar is estimated from monthly water deficits, based on a comparison of moisture supply (precipitation) to potential evapotranspiration (PET) demands using the Thornthwaite and Mather method (McCabe and Markstrom 2007).

For this study, a monthly water-balance model that is driven by a graphical user interface (GUI) developed by the US Geological Survey was used. The model is referred to as the Thornthwaite water-balance program. It was used as a method to examine the various components of the hydrological cycle (for example, precipitation and evapotranspiration). Such models have been used to estimate the global water balance (Mather 1969; Legates and Mather 1992; Legates and McCabe 2005) and to develop climate classifications (Thornthwaite 1948).

EVI was computed using Eq. 2 (Liu and Huete 1995).

$$\mathrm{EVI} = G\frac{\rho_{\mathrm{NIR}} - \rho_{red}}{\rho_{\mathrm{NIR}} + C_1\rho_{red} - C_2\rho_{blue} + L} \tag{2}$$

Where L is the canopy background adjustment that addresses non-linear, differential NIR and red radiant transfer through a canopy, and C1, C2 are the coefficients of the aerosol resistance term, which uses the blue band to correct for aerosol influences in the red band. The coefficients adopted in the MODIS-EVI algorithm are: L = 1, C1 = 6, C2 = 7.5, and G (gain factor) = 2.5 (Huete et al. 1999).For Landsat 7 satellite data used in this study, Eq. 3 was used.

$$\mathrm{EVI} = 2.5 * (\mathrm{B4} - \mathrm{B3}) / (\mathrm{B4} + 6.0 * \mathrm{B3} - 7.5\mathrm{B1} + 1) \tag{3}$$

where B4 is band 4 NIR, B3 is band 3 *red,* and B1 is band 1 *blue* in Landsat 7 Enhanced Thematic Mapper (ETM) imagery. EVI was modelled using ERDAS IMAGINE software.

EVI is an optimised index designed to enhance the vegetation signal with improved sensitivity in high-biomass regions and improved vegetation monitoring through a decoupling of the canopy background signal and a reduction in atmosphere influences (Huete et al. 1999). EVI was developed to optimise the

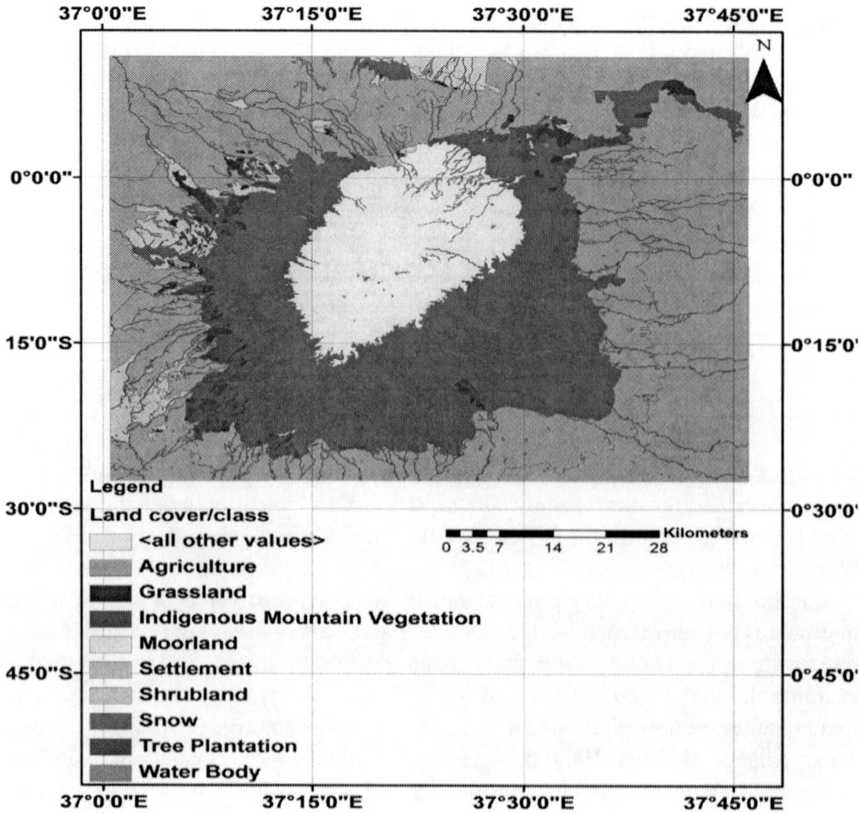

Fig. 2 Land cover classes

greenness signal, or area averaged photosynthetic capacity. Enhanced vegetation index has been found useful in estimating absorbed photosynthetically active radiation (PAR) related to chlorophyll in vegetated canopies and has been shown to be highly correlated with processes that depend on absorbed light (Zhang et al. 2005; Xiao et al. 2004; Rahman et al. 2005).

Data Used

Landsat imagery that was used to generate the EVI for the CASA model was collected on 25 February 1987 and 14 February 2002 respectively. February is usually a dry season and in most cases annual crops are not in the fields, so this avails an opportunity to capture cloud-free forest vegetation, tree plantations, and trees on farms that are critical for carbon sequestration estimation. Satellite data was also used to specify the predominant land cover class (Fig. 2).

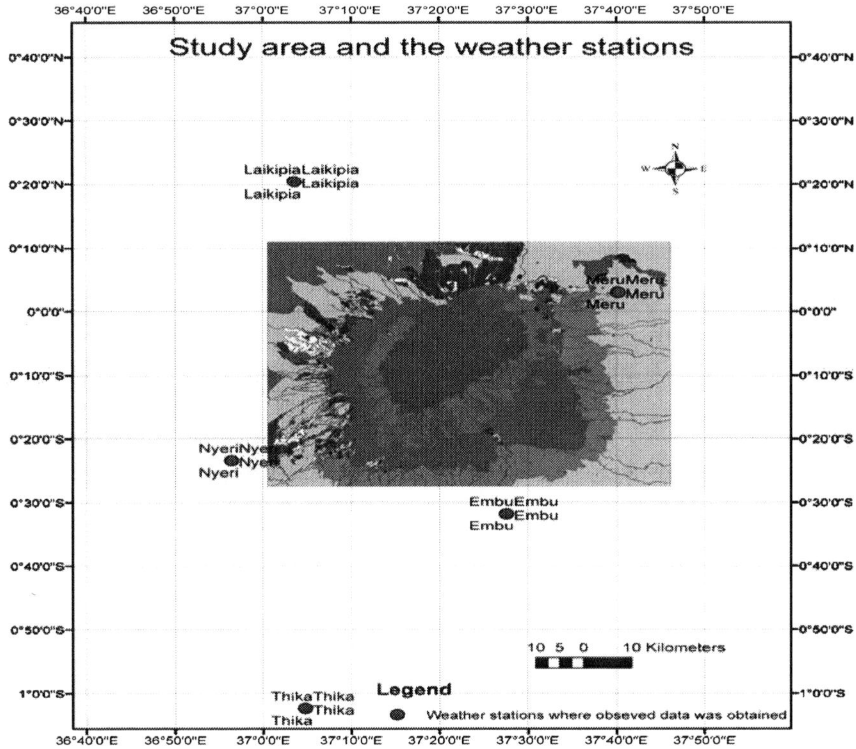

Fig. 3 Study area and the surrounding weather stations

Climate data that included daily precipitation, daily sunshine hours, daily solar irradiance and daily mean air temperature was collected from Kenya Meteorological Department, which provided data for the last 28 years (1980–2007) from seven weather stations around the mountain. These data was entered into the Thornthwaite water balance model (McCabe and Markstrom 2007). Water deficit, potential evapotranspiration, soil moisture and precipitation were estimated from this model and then interpolated to a 0.1 degree grid via the Spline method in ArcGIS 9.3. The temperature stress scalar in this study was derived from the seasonal optimal temperature for plant production from the 28 years of climate data obtained from Kenya Meteorological Department, while water stress was estimated from monthly water deficits based on comparison of observed moisture supply (precipitation) to potential evapotranspiration (PET) demand derived using the Thornthwaite and Mather method (1957). The CASA model was used to estimate monthly patterns of carbon fixation using 30 M resolution satellite remote sensing of the surface vegetation characteristics and driven by spatially interpolated climate data.

Climate data from stations around Mount Kenya included the following: Embu, Nyeri, Laikipia, Thika, Katumani and Meru stations (Fig. 3).

General Circulation Model Downscaling

General circulation model (GCM) simulations from Hadley circulation models' (HadCM3) scenario A2 (1961–2099) were used. The A2 storyline and scenario was adapted for this study. The A2 storyline and scenario family describes a very heterogeneous world. The underlying theme is self-reliance and preservation of local identities. Fertility patterns across regions converge very slowly, which results in continuously increasing global population. Economic development is primarily regionally oriented and per capita economic growth and technological changes are more fragmented and slower than in other storylines (IPCC 2000). GCMs are restricted in their usefulness for local impact studies by their coarse spatial resolution (typically in the order of 50,000 km^2) and inability to resolve important sub-grid scale features such as clouds and topography.

The statistical downscaling model (SDSM) was used and methods by Wilby et al. (2002) followed for downscaling. Generation of synthetic series of daily weather data at a local site based on empirical relationships between local-scale predictions (daily temperature and precipitation) and large-scale predictions (atmospheric variables) was conducted.

It involved identification of large-scale predictors (X) that could control the local parameters' predictands (Y) and then to find a statistical relationship between X and Y, and to validate the relationship with independent data. Finally Y is generated using values of X from HadCM3 A2 1961–2099 data. For this study, 20 ensembles were run for each parameter.

The simulated data was then grouped into five categories at an interval of 30 years, which included 1961–1991, 1992–2022, 2023–2053, 2054–2084 and the remaining 2085–2099.

Results and Discussions

GCM Downscaling and Validation

The 20 ensembles had very high correlation and therefore for this study only one was picked and used for impact analysis. Only three parameters that were used as a time-variable scalar in the model were downscaled, including maximum temperature and minimum temperature used for getting optimum temperature for plant growth, precipitation and radiation of the respective weather stations.

Observed precipitation, maximum and minimum temperature was used to validate the simulated and downscaled data. The simulated data between 1961–1991 and 1992–2022 captured the period 1980–2007, which is the period the observed data was recorded in various weather stations. The simulated data between these periods did not vary much from the observed ones (Figs. 4, 5 and 6).

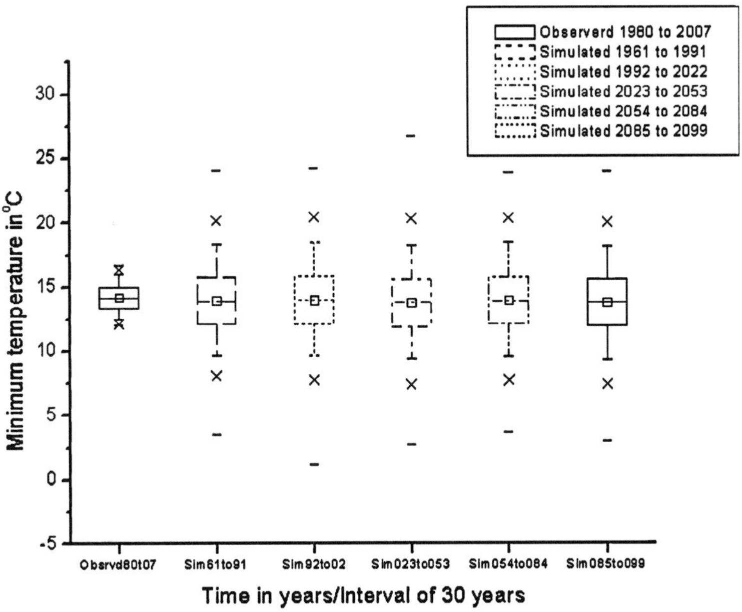

Fig. 4 Comparison of observed and simulated Embu minimum temperature

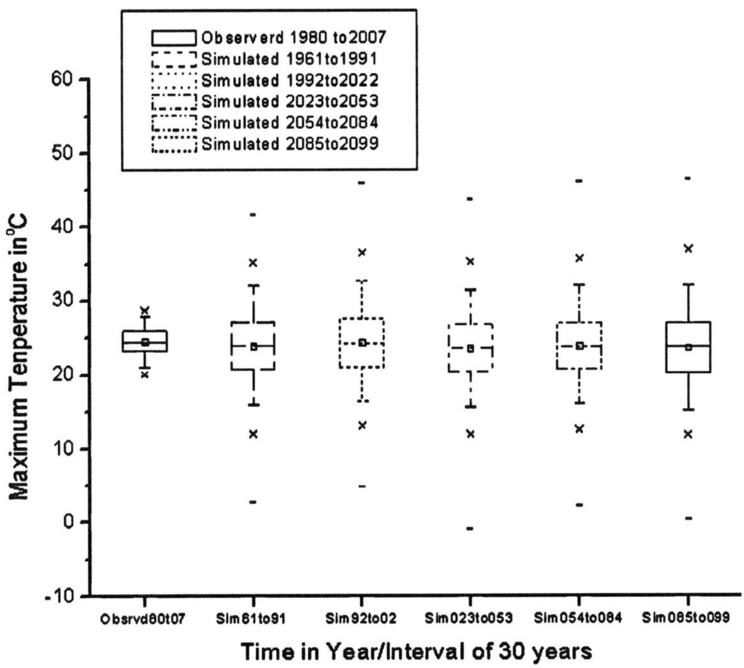

Fig. 5 Comparison of observed and simulated Embu maximum temperature

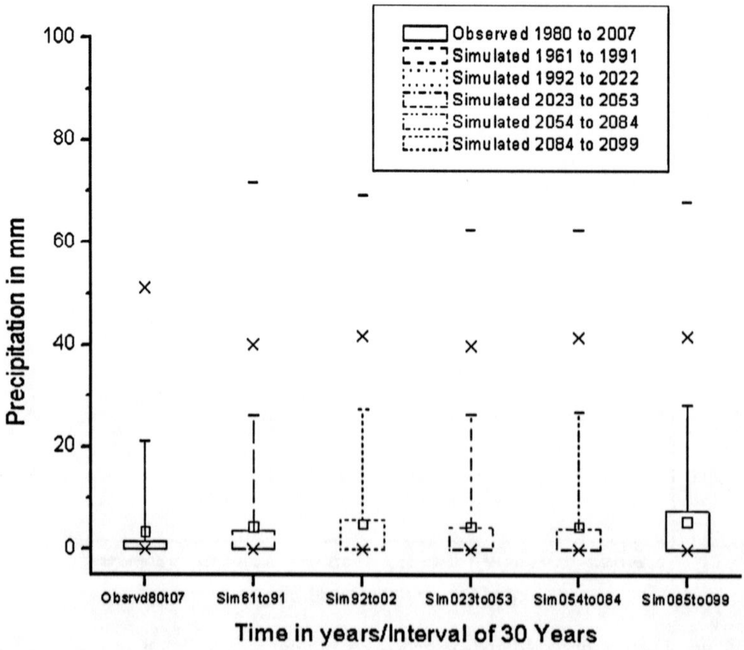

Fig. 6 Comparison of observed and simulated Embu precipitation

Table 1 Wilcoxon scores (rank sums) for maximum temperature classified by 30 year time interval

Timeline (years)	N	Sum of scores	Expected under H0	SD under H0	Mean score
1961–1991	65,700	9,671,745,296	9,862,916,850	19,635,208.6	147,210.7
1992–2022	65,700	9,778,193,837	9,862,916,850	19,635,208.6	148,831.0
2023–2053	65,700	22,027.76579	9,862,916,850	19,635,208.6	156,505.6
2054–2054	65,700	9,365,321,005	9,862,916,850	19,635,208.6	142,546.7
2085–2099	37,440	5,974,502,677	5,620,511,520	15,690,073.4	159,575.3

The median was about the same but the upper and lower quartile ranges were different in one of the stations, Embu.

Climatic Scalar Variability

The simulated data was divided into five timelines at an interval of 30 years (Table 1). Non-parametric analysis of variance (Kruskal–Wallis test) showed significant statistical variation at 5% level of significance in the timelines and the variables (Tables 1, 2).

Table 2 Scalar parameter variability between 5 and 30 years timelines/intervals (1961–2099)

Variable	Kruskal–Wallis test		
	Chi-square	DF	Pr > Chi quare
Maximum temperature	1,392.3893	4	<0.0001
Minimum temperature	227.0028	4	<0.0001
Precipitation	636.6887	4	<0.0001
Radiation	754.1101	4	<0.0001

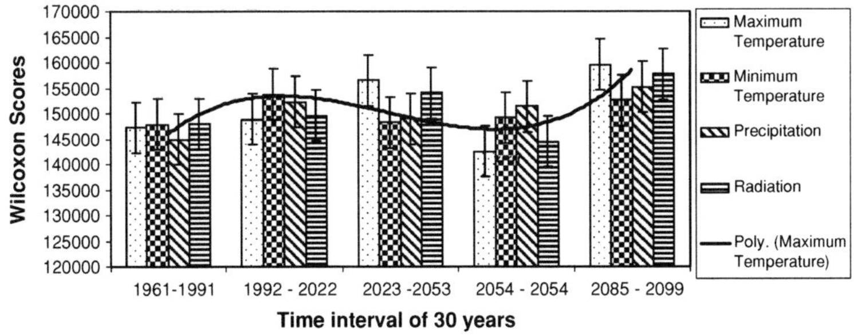

Fig. 7 Simulated maximum temperature trend 1961–2099 compared with other scalar variables

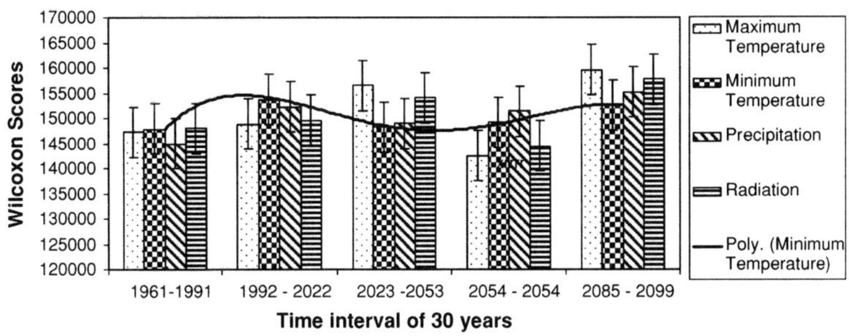

Fig. 8 Simulated minimum temperature trend 1961–2099 compared with other scalar variables

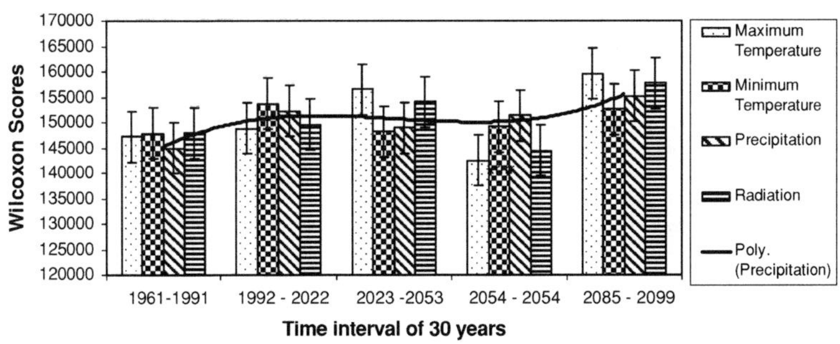

Fig. 9 Simulated precipitation trend 1961–2099 compared with other scalar variables

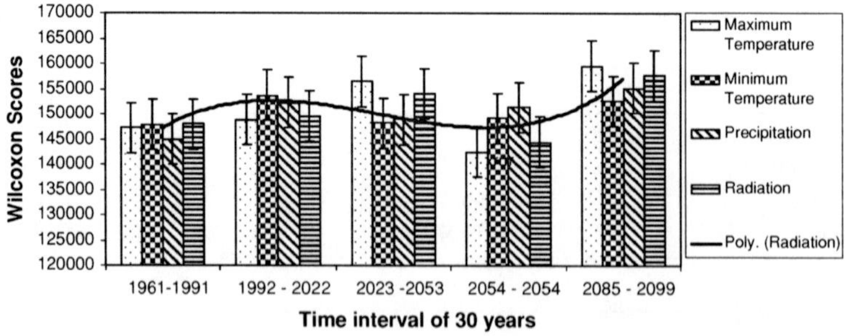

Fig. 10 Simulated radiation trend 1961–2099 compared with other scalar variables

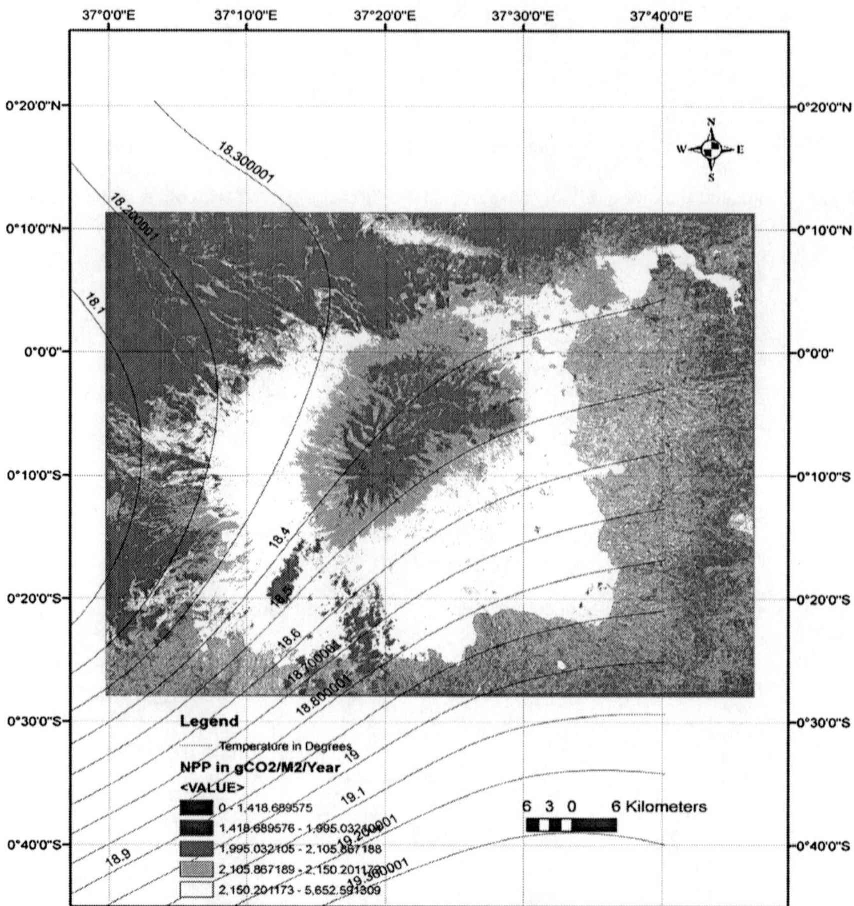

Fig. 11 Net primary production variation with temperature regime in 1987

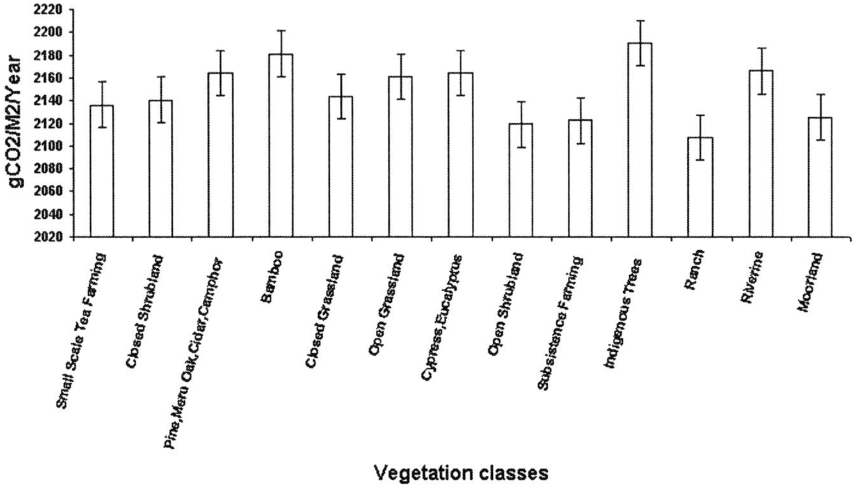

Fig. 12 1987 NPP estimates of different vegetation classes

Table 3 Summary statistics of vegetation classes, 2002 NPP

Vegetation class	Area (m^2)	Min. NPP	Max. NPP	Range	Means of NPP $gCO_2/m^2/year$	STD
Small-scale tea farming	657,250,000	3,259	5,475	2,217	4,171	255
Closed shrubland	93,948,000	3,192	5,409	2,217	3,951	311
Pine, meru, oak, cidar (OK so? Cedar?), camphor	88,204,400	3,458	5,542	2,084	4,327	360
Bamboo	503,308,000	2,704	5,630	2,926	4,561	206
Closed grassland	32,294,200	3,502	5,475	1,973	4,049	380
Open grassland	95,637,800	3,369	5,431	2,062	4,258	359
Cypress, eucalyptus	45,099,000	3,525	5,542	2,017	4,365	396
Open shrubland	135,292,000	3,126	5,320	2,195	3,830	292
Subsistence farming	1,811,400,000	2,106	5,254	3,148	3,915	219
Indigenous trees	905,532,000	3,081	5,653	2,571	4,632	320
Riverine	329,357,000	2,483	5,542	3,059	4,348	435
Moorland	679,649,000	0	5,586	5,586	3,751	349

The chi-square test results in Table 2 show that the changes in maximum temperature, minimum temperature, precipitation and radiation was not random but clearly indicated variability/change among the 30-year timelines/interval classes that were used.

The time-variable scalars depicted a similar trend, which indicates a general increase during the 1992–2022 period, a decline in the period 2023–2053 before starting to rise again towards the end of the century (2085–2099) (Figs. 7, 8, 9, 10).

Estimated CASA Model NPP

The predicted NPP from the model show variations among vegetation classes (Fig. 11).

Overall, the highest estimated NPP was from indigenous mountain vegetation, followed by tree plantations for both 1987 and 2002 (Fig. 12 and Table 3).

Conclusions

This study used the IPCC general circulation model (GCM), scenario A2 storyline, for climate simulation. From the results, the selected and downscaled A2 parameters for this study area (minimum temperature, maximum temperature, precipitation and radiation) indicate there will be an increasing trend in the levels of temperature, precipitation and radiation around the Mount Kenya region before declining and rising again towards the end of the century (Fig. 7). This trend to some extent ties in with some local observation. Mangoes, which were conventionally grown on the lower slopes due to low temperatures up the slopes, are now finding their way up the slopes of Mt Kenya, meaning temperatures have been increasing over the years. IPCC scenario A2 generally describes a very heterogeneous world, and the underlying theme being self-reliance and preservation of local identities. NPP assessed here took into consideration heterogeneity in vegetation species' response to climate variability over time.

NPP estimated from 1987 imagery falls within the first 30 years (1961–1991) timeline, and 2002 NPP falls in the subsequent timeline (1992–2022). From the analysis of variance (Tables 1, 2) there is evidence that these two timelines are significantly different in terms of the NPP climatic scalars and therefore the variation observed in the vegetation classes in terms of NPP can be attributed to the climatic change or variability observed/simulated. Estimated CASA mean annual NPP prediction during this period showed a consistent trend with the climatic variable scalars (Figs. 8, 9, 10, 12 and Table 3). In 2002 CASA-estimated NPP was higher than 1987 NPP due to increased temperatures, precipitation and radiation. There was variability among the broad vegetation classes; overall the indigenous forest vegetation had the highest carbon dioxide fixation both in 1987 and 2002 (Fig. 12 and Table 3). All the vegetation classes showed an increase in carbon dioxide fixation with increased scalar. The variation observed in open shrubland and closed shrubland where the former has less NPP may be an indication of harvesting of biomass by browsers and humans for wood fuel and building material. In contrast to open grassland and closed grassland, open grassland is subjected to frequent defoliation and thus rejuvenated and is more photosynthetically active than closed grass areas, which may be overgrown and drying around the time the satellite image was captured, resulting in low NPP values.

NPP is a fundamental ecological variable, not only because it measures the energy input to the biosphere and terrestrial carbon dioxide assimilation but also because of its significance in indicating the condition of the land surface area and status of a wide range of ecological processes. From these results therefore it is evident that the Mount Kenya ecosystem regulatory service function of carbon sequestration will vary with climate change.

Policies and markets that are emerging to compensate for reducing carbon emissions from deforestation and degradation (REDD) can use this approach as a tool for monitoring and evaluating carbon flux in areas with few or no eddy covariance towers, such as Kenya and other African countries. This can aid small-scale farmers and landowners to benefit from the carbon market by having a tool that can be used to measure the amount of carbon their land-use activities such as tree planting contribute in terms of carbon dioxide sequestration so that they can be adequately compensated. The basic approach can involve considering the dominant vegetation cover on a particular land unit, which is then used to calculate the possible rate based on the broad vegetation class annual mean NPP estimated from the model. Carbon units' validation under a clean development mechanism for forestry or reforestation projects in such areas as Kenya can also use these annual estimates as a foundation for encouraging carbon trade based on the estimated potential of each land area.

References

Beer C, Lucht W, Schmullius C, Shvidenko A (2006) Small net carbon dioxide uptake by Russian forests during 1981–1999. Geophys Res Lett 33:L15403

Friend AD, Arneth A, Kiang NY, Lomas M, Ogee J, Roedenbeck C, Running SW, Santaren JD, Sitch S, Viovy N (2007) FLUXNET and modelling the global carbon cycle. Global Change Biol 13:610–633

Huete A, Justice C, Van Leeuwen W (1999) MODIS vegetation index (MOD 13)-algorithm. Theoretical basis document; Available online: http://modis.gsfc.nasa.gov/data/atbd/atbd_mod13.pdf Accessed 10 Nov 2009

IPCC (2000) A special report of IPCC Working Group III. Published for the intergovernmental panel on climate change. 92-9169-113-5

Legates DR, Mather JR (1992) An evaluation of the average annual global water balance. Geogr Rev 82:253–267

Legates DR, McCabe GJ (2005) A re-evaluation of the average annual global water balance. Phys Geogr 26:467–479

Liu HQ, Huete AR (1995) A feedback based modification of the NDV I to minimize canopy background and atmospheric noise. IEEE Trans Geosci Remote Sens 33:457–465

Mather JR (1969) The average annual water balance of the world. Symposium on water balance in North America, Series No. 7, Proceedings: Banff, Alberta, Canada, American Water Resources Association, pp 29–40

McCabe GJ, Markstrom SL (2007) A monthly water-balance model driven by a graphical user interface. US geological survey open-file report 2007–1088, p 6

McCallum I, Wagner W, Schmullius C, Shvindenko A, Obersteiner M, Firtz S, Nilson S (2009) Satellite-based terrestrial production efficiency modelling. Carbon Balance Manage 4:8

Melillo JM, McGuire AD, Kicklighter DW, Moore B III, Vorosmarty CJ, Schloss AL (1993) Global climate change and terrestrial net primary production. Nature 363:234–240

Ojany FF (2008) Mount Kenya biosphere reserve. Int J Environ Sustainable Dev 7(2):2004

Potter CS, Randerson JT, Field CB, Matson PA, Vitousek PM, Mooney HA, Klooster SA (1993) Terrestrial ecosystem production: a process model based on global satellite and surface data. Global Biogeochem Cycles 7:811–841

Potter C, Klooster S, Genovese V, Myneni R (2003a) Satellite data helps predict terrestrial carbon sinks. Eos Trans Am Geophy Union 84:502–508

Potter C, Klooster S, Myneni R, Genovese V, Tan PN, Kumar V (2003b) Continental-scale comparisons of terrestrial carbon sinks estimated from satellite data and ecosystem modeling 1982–1998. Global Planet Change 39:201–213

Potter C, Klooster S, Tan P, Steinbach M, Kumar V, Genovese V (2005) Variability in terrestrial carbon sinks over two decades: Part 2 Eurasia. Global Planet Change 49:177–186

Potter CS, Gross P, Genovese V, Smith ML (2007) Net primary productivity of forest stand in New Hampshire estimated from Landsat and MODIS satellite data. Carbon Manage 2:9

Rahman AF, Sims DA, Cordova VD, El-Masri BZ (2005) Potential of MODIS EVI and surface temperature for directly estimating per-pixel ecosystem C fluxes. Geophys Res Lett 32: L19404

Running SW, Nemani RR, Heinsch FA, Zhao M, Reeves M, Hashimoto H (2004) A continuous satellite-derived measure of global terrestrial primary production. Bioscience 47:545–560

Shvidenko A, Nilsson S (2003) A synthesis of the impact of Russian forests on the global carbon budget for 1961–1998. Tellus Ser B Chem Phys Meteorol 55:391–415

Stephens BB, Gurney KR, Tans PP, Sweeney C, Peters W, Bruhwiler L, Ciais P, Ramonet M, Bousquet P, Nakazawa T (2007) Weak northern and strong tropical land carbon uptake from vertical profiles of atmospheric CO_2. Science 316:1732–1735

Thornthwaite CW (1948) An approach toward a rational classification of climate. Geogr Rev 38:55–94

Thornthwaite CW, Mather JR (1957) Instructions and tables for computing potential evapotranspiration and the water balance. Publications in Climatology, vol 10, no 3. C W Thornthwaite & Associates, Centerton

Wilby RL, Dawson CW, Barrow EM (2002) SDSM–a decision support tool for the assessment of regional climate change impacts. Environ Model Softw 17:145–157

Xiao X, Zhang Q, Braswell B, Urbanski S, Boles S, Wofsy S, Moore B III, Ojima D (2004) Modeling gross primary production of temperate deciduous broadleaf forest using satellite images and climate data. Remote Sens Environ 91:256–270

Zhang Q, Xiao X, Braswell B, Linder E, Baret F, Moore B III (2005) Estimating light absorption by chlorophyll, leaf and canopy in a deciduous broadleaf forest using MODIS data and a radiative transfer model. Remote Sens Environ 99:357–371

Author Biographies

Ochieng Aggrey Adimo has a Master of Science in Landscape Plants and is currently pursuing a PhD degree in Landscape Planning and Conservation at Jomo Kenyatta University of Agriculture and Technology, Kenya. He works at the same institution as a tutorial fellow. His research interest is climate change modelling, mitigation and adaptation.

Dr. Njoroge John Bosco is a senior lecturer in the area of Landscape Planning and Conservation at Jomo Kenyatta University, Kenya. His research interests include landscape ecology and planning.

Prof. Wamocho Leonard is a professor of Horticulture. His research interests include using mycorhiza to enhance plant performance in water-stressed environments.

Chapter 15
Adapting Agriculture to Climate Change by Developing Promising Strategies Using Analogue Locations in Eastern and Southern Africa: Introducing the Calesa Project

Walter Leal Filho and Franziska Mannke

Abstract The impacts of climate change on agriculture in Africa are significant and call for concrete measures that allow a better understanding of these problems, as well as the identification of the means to address them. One of the means to tackle them is by using analogue locations, i.e. locations that have the climatic characteristics *today* that are expected *tomorrow*. This paper introduces the project "Developing promising strategies using analogue locations in Eastern and Southern Africa" (CALESA). Using a combination of model-based *ex ante* analyses and iterative field-based research on station and in farmers' fields, the project will test potential agricultural adaptation strategies for rain-fed agriculture in the semi-arid and dry sub-humid tropics. This will be achieved through choosing four currently important crop production zones (two in Kenya and two in Zimbabwe) and then identifying corresponding 'spatially analogue locations' for each production zone, providing experiences which may be replicable elsewhere.

Keywords Rain-fed agriculture · Adaptation · Temperature · Analogue locations

Introduction

Rain-fed agriculture will remain vital for food security in sub-Saharan Africa (SSA). Nearly 90% of staple food production will continue to come from rain-fed smallholder farming systems. However, it is here that some of the world's poorest

W. Leal Filho (✉) · F. Mannke
Faculty of Life Sciences, Hamburg University of Applied Sciences,
Lohbruegger Kirchstraße 65, 21033 Hamburg, Germany
e-mail: walter.leal@haw-hamburg.de

W. Leal Filho (ed.), *Experiences of Climate Change Adaptation in Africa*,
Climate Change Management, DOI: 10.1007/978-3-642-22315-0_15,

and most vulnerable rural communities live. However, in spite of its vital food security role for SSA, rain-fed agriculture has stagnated.

Whilst all general circulation models (GCMs) agree that it will become warmer across sub-Saharan Africa, the degree of warming predicted is quite variable. The Fourth Assessment Report of the IPCC suggests that the median temperature rise in eastern and southern Africa will be 3–4°C by the end of the twenty first century (Christensen et al. 2007), with a greater temperature rise in June, July and August (median rise 3.4°C) than from September to February (3.1°C) in eastern Africa and a greater temperature rise in September, October and November (3.7°C) than in December to May (3.1°C) in southern Africa. Indeed, evidence of changes in climate extremes, in particular with regard to temperature, is already emerging in southern and West Africa (New et al. 2006). However, with regard to the percentage changes in rainfall amounts, the uncertainty is considerably greater.

Nevertheless, there appears to be a consensus regarding the predicted trend of wetting in eastern Africa and of drying in the winter rainfall regions of southern Africa. Christensen et al. (2007) predicted a median increase in rainfall of 7% (2–11% for the 25th and 75th percentiles) for eastern Africa, while annual rainfall is predicted to decrease by 4% (from −9 to +2% for the 25th and 75th percentiles) for southern Africa by the end of the twenty first century. More detail regarding climate change scenarios and their implications for adaptation of small-scale farming in eastern and southern Africa can be found in Turner (2008).

Given the constraint of both current climate-induced production risk and the predicted change in nature of that risk in the future, it is now evident that a two-pronged approach to adaptation to climate change is required (Burton and van Aalst 2004; DFID 2005; Washington et al. 2006; Cooper et al. 2008; ICRISAT 2008).

Added to the constraints imposed by poor supportive policies, extreme poverty and often a degrading resource base, is the inherent climate-induced production risk associated with the current season-to-season variability of rainfall. This situation is likely to be made worse by global warming and its predicted impacts on seasonal rainfall amounts and distribution patterns, which threaten to exacerbate the climate-induced risk problems already faced by rain-fed farmers. Such communities, already struggling to cope effectively with the impacts of current rainfall variability, will face a daunting task in adapting to future climate change. However, they will need to adapt their farming practice to these new climate risks as they evolve. It is believed that a two-pronged approach is required.

In the *short to medium* term, it is essential to help poor and vulnerable farmers to build up their livelihood resilience (and hence adaptive capacity) through coping better with *current* climate-induced risk as a prerequisite to adapting to *future* climate change. Secondly, it is accepted that in the *medium to longer* term and as climate change begins to have noticeable impacts, farmers will have to progressively adapt their farming practices to a new set of climate-induced risks and opportunities. It is on the second '*medium to longer term*' aspect of adaptation that this proposal will place its emphasis, namely adaptation to progressive climate change with a special emphasis on increasing temperature.

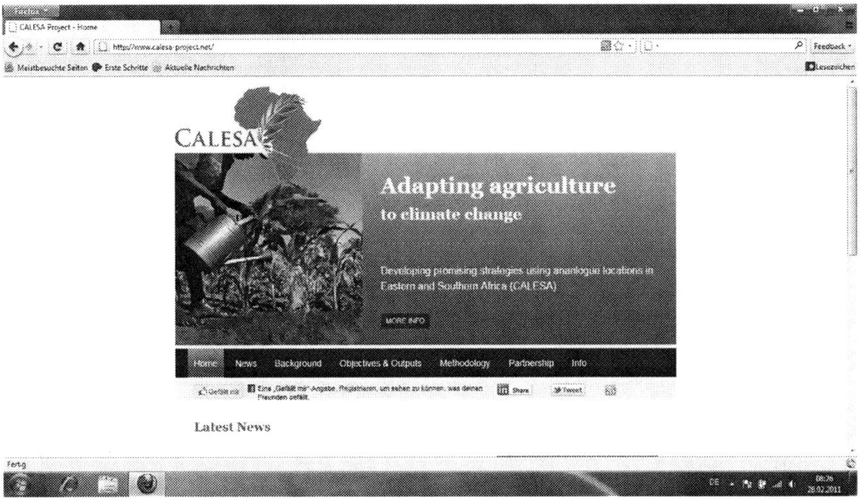

Fig. 1 CALESA website (http://www.calesa-project.net/)

The project "Developing promising strategies using analogue locations in Eastern and Southern Africa" (CALESA) has been initiated based on the need to address the above issues.

The project is coordinated by the International Crop Research Institute for the Semi-Arid Tropics (ICRISAT), funded by the German Agency for International Cooperation (Deutsche Gesellschaft für Internationale Zusammenarbeit (GIZ)) on behalf of the German Ministry of Cooperation and Development (Bundesministerium für wirtschaftliche Zusammenarbeit und Entwicklung (BMZ)).

The cooperation partners are:

- Kenya Meteorological Dept. (KMD), Kenya.
- Kenya Agricultural Research Institute (KARI), Kenya
- Midlands State University, (MSU), Zimbabwe
- Zimbabwe Meteorological Department (ZMD), Zimbabwe
- Hamburg University of Applied Sciences, Faculty of Life Sciences, Germany

The project is also supported by the International Climate Change Information Programme (ICCIP), which assists with the dissemination elements.

The aim of CALESA, which runs until the end of 2013, is to improve the ability of rain-fed farmers in the semi-arid tropics of Africa to adapt to progressive climate change through crop, soil and water management innovations and appropriate crop genotype choices. Figure 1 shows the project's website.

CALESA combines simulations and field assessments, with an analysis of the views and perceptions of the concerned stakeholders, paying a special attention to gender issues.

Scope of the Project and Methodology

Previous projects dealing with climate change in farming have provided the following lessons:

(a) Over decades, farmers have developed farming strategies that are designed to cope with the natural season-to-season variability of rainfall that they experience. However, these coping strategies are more often than not designed to mitigate the impact of the poorer seasons and fail to exploit the opportunities of the better seasons.
(b) In particular, agricultural diversity, livestock management and gender aspects are important aspects of climate risk management in dry environments.
(c) Farmers often relate changes in their rain-fed farm productivity to changes in rainfall patterns and attribute this, usually incorrectly, to climate change.
(d) In general, farmers are able to accurately recollect the characteristics of past seasons for up to five years. Such short-term trends of rainfall patterns (when viewed in the context of long-term climate data of 40+ years) are often shown to be a characteristic of the natural season-to-season variability at any given location and are not due to climate change *per se*.
(e) In most cases, the impacts of other drivers of change that directly impinge on the natural resource base are *currently* more important than climate change in being responsible for trends or declines in farm productivity.
(f) There are wide ranges of weather-driven climate risk assessment tools that provide valuable insights on climate-induced production risk as well as allowing the evaluation of the potential of a range of crop, pasture, and soil and water management innovations to mitigate such risks. Such tools can also evaluate the implications of a range of climate change scenarios.

The project CALESA will draw on these lessons. Using a combination of model-based *ex ante* analyses and iterative field-based research on stations and in farmers' fields, the project will test potential agricultural adaptation strategies for rain-fed agriculture in the semi-arid and dry sub-humid tropics. This will be achieved through choosing four currently important crop production zones (two in Kenya and two in Zimbabwe) and then identifying corresponding 'spatially analogue locations' for each production zone, providing eight study locations in all.

The project defines 'analogue locations' as those locations that have the climatic characteristics *today* that are expected *tomorrow* in our four chosen production zones. In defining the locations, special attention will be given to adaptation to temperature increases. Altitudinal effects on mean air temperature will facilitate this. Given the potential of 'analogue locations' to provide a solid basis for such research across sub-Saharan Africa, special attention will also be given to the continuous documentation and dissemination of project activities and achievements through the Web, newsletters and dissemination events. A strong element of participatory research with farmers within the project locations will ensure that the project activities and outputs remain relevant to their needs and expectations.

The project will also evaluate agricultural adaptation strategies to climate change through the use of 'analogue locations', with special reference to predicted temperature increases. A clear relationship between altitude, air temperatures and crop-growth and yield facilitates the identification of appropriate temperature-analogue locations in eastern and southern Africa. The work will be undertaken in close collaboration with the NMS and NARS in Kenya and Zimbabwe and with the Hamburg University of Applied Sciences, and through the training of two African PhD students. Across these two countries, the project will identify analogue locations for four food production areas that currently experience (1) cool/dry, (2) cool/wet, (3) warm/dry and (4) warm/wet growing conditions. The following areas of research will be addressed:

1. Develop the criteria for the selection of analogue locations and identify four paired analogue locations for four currently important food production areas across Kenya and Zimbabwe using CLIMEX software (www.climatemodel.com/climex.htm).
2. Access long-term daily climatic data for those paired locations (40 years +) and undertake detailed climate risk and climate trend analyses though the use of the statistical package In-Stat. (www.graphpad.com/instat/instat.htm).
3. Through participatory research with farming communities, fully characterise the four paired sets of locations with regard to crops, soils, climate, current farming practices, the roles of men and women, crop diversity, livestock management, farmers' perceptions of current climate-induced risk and climate change, and possible adaptation strategies. Added depth will be given to this through two PhD students who will focus on the gender aspects.
4. Through detailed on-station agronomic and physiological research, field calibrate the weather-driven Agricultural Production Systems Simulator (APSIM) for important locally grown food crops at the four sets of paired locations (http://www.apsim.info/apsim/Documentation/.).
5. Use GIS, downscaled GCM predictions and other innovative tools for incorporating and extrapolating spatial and temporal effects of climate change, including gender and environmental impacts.
6. Through a combination of field research on stations and in farmers' fields and simulation-based research over the three-year period, iteratively test the potential of improved soil, water and crop management strategies together with contrasting crop genotypes to mitigate the impacts of increased temperature.

Special attention will be given to informing and communicating the project outputs through the use of the Web, logo and poster development, newsletters and dissemination events.

Stakeholders and Expected Outputs

Stakeholders in this project include farmers and farmer groups within the chosen production areas and their analogue locations in Zimbabwe and Kenya; staff of the NARS and NMSs in those two countries who will receive hands-on training in climate risk analyses, participatory interaction with farmers and the approaches associated with the use of analogue locations.

In addition, two postgraduate students (from Zimbabwe and Kenya) will work under the supervision of the Hamburg University of Applied Sciences, who will gain extensive experience of evaluating the gender-related aspects of agricultural climate adaptation strategies through the use of analogue locations.

The expected outputs from the project are:

1. Four important crop-growing areas in Kenya and Zimbabwe which comprise (1) cool/dry, (2) cool/wet, (3) warm/dry and (4) warm/wet growing conditions and their temperature-analogue locations, identified and fully characterised.
2. Through the combined use of long-term daily climate data, crop-growth simulation models and participatory surveys with farmers, the implications of both current and future (climate change) production risk at the study locations identified and quantified.
3. Through iterative field research both on stations and in farmers' fields over a two-year period, potential crop, soil and water management and crop genotype adaptation options evaluated and adaptation strategies formulated for the target locations.
4. Through the wide promotion of the project, dissemination of its activities, results and hands-on capacity building, the strengthened institutional capacity (both in understanding climate change impacts and developing effective adaptation responses) will be ensured.

In terms of the use of research results, a clear strategy will be used. Firstly, the information gained through testing the use of analogue locations to evaluate adaptation strategies with farmers is potentially of great value to NARS and NMS, not only in Kenya and Zimbabwe but in all countries in SSA where rain-fed agriculture is important. Secondly, the outputs of such research will also be of value to decision-makers and policymakers in SSA as well as the wider global community. Finally, the results will be very useful to smallholder rain-fed farmers, who will make use of the results of climate change adaptation research to ensure their future livelihoods under the influence of climate change.

Conclusions

CALESA is essentially a research-for development project, which integrates climate risk analyses, crop-growth simulation modelling, field-based research both on-station and in farmers' fields with participatory research with farmers to assess their perceptions of current and future climate risk and their preferred climate change adaptation strategies. It comprises research-oriented activities for knowledge and technology creation and development-oriented activities for information sharing and capacity building. Bearing in mind that rain-fed agriculture is vital for food security and yet stagnating in sub-Saharan Africa and that current and future climate-induced risk poses an added constraint to the adoption of innovation, the CALESA project will provide a concrete contribution to evaluating the impacts of climate change and how it can be addressed vis-à-vis agricultural development that bears in mind the needs of the poor and vulnerable.

References

Burton I, van Aalst M (2004) Look before you leap: a risk management approach for incorporating climate change adaptation into world bank operations. World Bank Monograph, Washington DEV/GEN/37 E

Christensen JH, Hewitson B, Busuioc A, Chen A, Gao X, Held I, Jones R, Kolli RK, Kwon WT, Laprise R, Magaña Rueda V, Mearns L, Menéndez CG, Räisänen J, Rinke A, Sarr A, Whetton P (2007) Regional climate projections. In: Solomon S, Qin D, Manning M, Chen Z, Marquis M, Averyt KB, Tignor M, Miller HL (eds) Climate change 2007: the physical science basis. Contribution of working group I to the fourth assessment report of the intergovernmental panel on climate change. Cambridge University Press, Cambridge

Cooper PJM, Dimes J, Rao KPC, Shapiro B, Shiferaw B, Twomlow S (2008) Coping better with current climatic variability in the rainfed farming systems of sub-saharan Africa: an essential first step in adapting to future climate change? Agric Ecosyst Environ 126(1–2):24–35

DFID (2005) Climate proofing Africa: climate and Africa's development challenge. Department for International Development, London

ICRISAT (2008) Adaptation to climate change in the semi-arid tropics. http://www.icrisat.org/aes-climatechange-sat.htm

New M, Hewitson B, Stephenson D, Tsiga A, Kruger A, Manhique A, Gomez B, Coelho C, Masisi D, Kululanga E, Mbambalala E, Adesina F, Saleh H, Kanyanga J, Adosi J, Bulane L, Fortunata L, Mdoka M, Lajoie R (2006) Evidence of trends in daily climatic extremes over southern and west Africa. J Geophys Res 111(D14102):1–11. doi:10.1029/2005JD006289

Turner N C (2008) Adaptation to climate change: research needs for smallholder farmers growing sorghum and pigeonpea in eastern and southern Africa. Report commissioned by ICRISAT-ESA, Nairobi, Kenya

Washington R, Harrison M, Conway D, Black E, Challinor AJ, Grimes D, Jones R, Morse A, Kay G, Todd M (2006) African climate change: taking the shorter route. Bull Am Meterol Soc 87(10):1355–1366. doi:10.1175/BAMS-87-10-1355

Author Biographies

Walter Leal Filho work at the "Applications of Life Sciences" Research and Transfer Centre at the Hamburg University of Applied Sciences, in Hamburg, Germany.

Franziska Mannke work at the "Applications of Life Sciences" Research and Transfer Centre at the Hamburg University of Applied Sciences, in Hamburg, Germany.

Chapter 16
The Unitar Climate Change Programme: Innovative Adaptation Initiatives in Africa

Mamadou Moussa Diakhite, Audrey Avanzi and Sharon Oseku

Abstract This chapter describes UNITAR Climate Change Programme and outlines the initiatives it has been undertaken to foster the cause of climate change adaptation in Africa.

Keywords UNITAR · Africa · Adaptation · Innovation

Introduction

The United Nations Institute for Training and Research (UNITAR) was created in 1965 with the mission to deliver innovative learning products, conduct research and provide training to diplomats from the newly-emerged independent States. Since its inception, UNITAR has developed unique expertise, knowledge and sustainable partnerships worldwide. The institute meets the growing needs for capacity building development training in the fields of Environment; Peace, Security and Diplomacy; and Governance.

As climate change became increasingly recognized as one of the greatest challenges to the survival of mankind and capacity building identified as a key element to solving this continuing phenomenon, the UNITAR launched in 1994 its

M. M. Diakhite · A. Avanzi (✉) · S. Oseku
Climate Change Programme (CCP), United Nations Institute for Training and Research (UNITAR), Palais des Nations, 1211 Geneva, Switzerland
e-mail: audrey.avanzi@unitar.org

M. M. Diakhite
e-mail: mamadou.diakhite@unitar.org

W. Leal Filho (ed.), *Experiences of Climate Change Adaptation in Africa*, Climate Change Management, DOI: 10.1007/978-3-642-22315-0_16,

Climate Change Programme (CCP). Within this framework, the UNITAR/CCP seeks, among other objectives, to enhance the capacity of developing countries to increase their participation in the United Nations Framework Convention on Climate Change (UNFCCC) process. The UNITAR/CCP works towards the achievement of excellence in capacity building for climate change, carrying out the following activities: knowledge sharing and building of expertise through fellowship and internship programmes, online and face-to-face training courses, customized training workshops, as well as project design and implementation. Through regional centers of excellence based mainly in developing countries, the CCP has developed a unique network and strong partnerships for project implementation worldwide.

It is now well recognized that human societies face the challenge to adjust economic and social systems to meet the adverse effects of climate change while fostering opportunities generated by this phenomenon, a process mostly known as "adaptation to climate change".[1] Vulnerable areas of the world, suffering from extreme climatic changes and high geo-physical alterations, are also mostly affected by climate change and variability; the African continent is one of the most challenged. As underlined by the Intergovernmental Panel on Climate Change (IPCC) report on climate change in Africa,[2] adaptation strategies have already been put in place by African farmers and other vulnerable populations, but these options are often still limited and further capacity development is required for sufficient and effective adaptation. The CCP contributes to the efforts to address this challenge through adaptation and policy projects in Africa, such as the Advancing Capacity to support Climate Change Adaptation (ACCCA), the Climate Change Capacity Development (C3D) and support to the preparation and implementation of National Adaptation Programmes of Action projects (NAPAs), as well as to develop partnerships with other UN, multi- and bilateral development agencies.

---

[1] IPCC definition of *Adaptation to climate Change*: "Adjustment in natural or human systems to a new or changing environment. Adaptation to climate change refers to adjustment in natural or human systems in response to actual or expected climatic stimuli or their effects, which moderates harm or exploits beneficial opportunities. Various types of adaptation can be distinguished, including anticipatory and reactive adaptation, private and public adaptation, and autonomous and planned adaptation." http://www.ipcc.ch/ipccreports/tar/wg3/454.htm.

[2] *Africa. Climate Change 2007: Impacts, Adaptation and Vulnerability. Contribution of Working Group II to the Fourth Assessment Report of the Intergovernmental Panel on Climate Change*: "Human or societal adaptive capacity, identified as being low for Africa in the Third Assessment Report, is now better understood and this understanding is supported by several case studies of both current and future adaptation options. However, such advances in the science of adaptation to climate change and variability, including both contextual and outcome vulnerabilities to climate variability and climate change, show that these adaptations may be insufficient to cope with future changes of climate."

Current and Recent Activities

From 2006 to 2010, the programme developed the Advancing Capacity to support Climate Change Adaptation (ACCCA) project, building up nineteen pilot actions in Asia and Africa.[3] Implemented in several countries with a focus for Africa in Burkina-Faso, Cameroon, Ethiopia, Ghana, Kenya, Malawi, Mali, Niger, Nigeria, South Africa, Tanzania and Tunisia, the project benefited from lessons learned about effectively communicating climate risk information that is relevant to decision-makers and local communities by addressing climate risks and adaptation in an integrated and multidisciplinary way. It underscored the importance of fully engaging stakeholders and the long-term benefits of partnering with institutions of the scientific and policy communities for understanding and managing climate risks. One of the main specificities of the initiative consisted in the integration of not only local scientific communities but also local policy-makers into risk-communication projects, acknowledging that "climate change has pronounced impacts on Africa's resources, which has also affected local decisions on the use and management of these resources".[4] ACCCA comprehensive methodology includes raising awareness, enhancing human capacities, establishing sustainable partnerships, implementing funding schemes and recommendations for adaptation and identification of remaining critical knowledge gaps to build future training courses. Adaptation projects in Africa focused in the sectors of fisheries (Nigeria, Tanzania), livelihoods (Nigeria, Ghana, Tanzania, and Cameroon), food security (Ghana, Ethiopia), responsive water management (Mali, Burkina-Faso), dryland agriculture (Kenya, Tunisia/Niger, Cameroon, and Ethiopia), disaster management (Malawi, Kenya, Ghana, South Africa) and human health (Kenya, Ghana, Burkina-Faso, South Africa).

Options built with stakeholders through ACCCA projects ended up being an added-value as assessed by local beneficiaries. For instance, in the dryland regions of Taita and Makueni, Southern Kenya, potential strategies to deal with water shortage were discussed with local communities, who acknowledged the need for conservation of water. The final common decision was to "implement water reservoirs to conserve water, grow drought resistant crops, and sink bore holes and water recycling".[5] In Nsanje and Salima districts, Malawi, rural vulnerable communities are heavily hit by droughts and floods. The ACCCA project "Audiovisual tools for community-based adaptation: bridging the Meteorological Service and the Red Cross's work" aimed at strengthening capacity at a local level.[6] Main outputs were communication tools, videos, posters and other

---

[3] ACCCA (2010. p. 30) http://www.acccaproject.org/accca/files/ACCCA_Brochure_19 pilotactions.pdf.

[4] ENDA (2010). *Synthesis report of five pilot actions* (p.2).

[5] ACCCA (2009, p. 30).

[6] *Malawi: farmers become filmmakers* (2009). http://www.youtube.com/watch?v=WxScWzI3Rrg.

materials to support their adaptation at the community level through a participatory process. Six adaptation options were identified: the diversification of crops, using storm drains and elephant grass to contain running water, using ducks instead of chickens as they swim during floods, store food, using flood alert systems in case of a storm and develop irrigation farming.[7] Ideas were inspired by local communities and other ACCCA project strategies. In this example, the participatory video enabled communities to get involved in the identification of solutions.

ACCCA five pilot actions in Burkina-Faso, Cameroon, Ethiopia, Kenya and South-Africa, implemented from August 2007 to July 2010, revealed a relatively increasing trend of temperature along with factors increasing vulnerability in pilot sites that restrained partners' action, namely "poor feeder roads, lack of access to markets, poor pricing of farm products, lack of storage facilities, poor housing structures and lack of farming equipments, poor housing, poor sanitation, tenure insecurity, lack of water reserves, dependence on surface water, week political control and water management, inappropriate social behavior and lack of reliable climate information". On the benefits side, local adaptive measures were supported and succeeded through the presence of positive factors, namely "extension services, livestock ownership, gender of the head of household being male, access to climate change information and perceived change in temperature".[8] Before launching adaptation and risk-communication projects in the region, such supporting and limiting factors need to be identified through appropriate socio-economic studies.

Outputs of ACCCA

As highlighted in the above examples, ACCCA partners implemented innovative and necessary solutions designed for specific communities, with a focus on creating communication materials, gathering data and baselines for future studies, taking adaptation actions, enhancing capacities and networking.[9] Outputs of ACCCA adaptation projects in Africa include but are not limited to: audiovisual tools, capacity building at several stakeholders' levels (household, community, private sector, government), data collection, communications to decision-makers, comparative adaptation analysis, constitution of self-help livelihood groups, exploration of new adaptive strategies, posters dissemination, hydrological and meteorological analysis, indigenous mechanism studies, Integrated Vector Management implementation, participatory assessments, risk communication products, socio-economic analysis, training on agriculture and other relevant

[7] ACCCA (2009, p. 30).

[8] ENDA (2010). *Synthesis report of five pilot actions* (p. 2, 39).

[9] Ibid. p. 39.

fields, vegetation cover analysis, vulnerability assessments in two Sahara and Sahel Observatories (OSS), workshops, etc. Above all, ACCCA implemented creative learning methodologies, such as learning by doing, to empower learners from vulnerable communities in the learning process.

Several lessons learned on developing adaptation projects could be extracted from this experience[10]: (1) That understanding expectations of all stakeholders for collective learning, and especially with local scientific and policy-making communities, is a real challenge because of cultural differences, different levels of partners' expertise and competences, as well as various spatial and temporal scales, leading sometimes to a misunderstanding of the importance of climate change impacts on a medium or long-term basis (2) "In light of the time frame, resources and expertise available within each of the pilot actions", communication tools must be customized; ACCCA partners developed creative learning sessions as mentioned above (3) Interactive lectures and knowledge sharing during training workshops require concrete involvement and actions of concerned stakeholders taken afterwards.[11] (4) Local communities need to formulate their own adaptive measures for those to be efficient and sustainable; the inclusion of local and national institutions also facilitates the process. (5) Communication on climate change must be undertaken in a simple and vernacular language with very clear messages combined with the highest accuracy levels. (6) Meteorological stations and departments in these countries lack accurate information on climatic historical data, a serious barrier to take into consideration in risk communication projects.[12]

The overall successful implementation of ACCCA nineteen pilot actions, including fourteen projects in Africa,[13] will lead to continuous activities in the upcoming years, for instance with the integration of project experiences and outcomes into the AfricaAdapt Knowledge online platform. They will also form the basis of the AfricaAdapt Knowledge Exchange Fair and be an entry point for the Adaptation Academy project of the Climate System Analysis Group (CSAG) in Cape Town, South Africa.

In 2003, the CCP launched the Climate Change Capacity Development (C3D) project to answer capacity needs for climate change in Least Developed Countries and other developing countries through innovative training and capacity-building partnerships, supporting seven regional centers of excellence based in Africa and Asia. The C3D project, followed in 2009 by the C3D+ initiative (Capacity Development for Adaptation to Climate Change & GHG Mitigation in Non-Annex I Countries), focused on education and training, national and regional support for adaptation and mitigation, research action and case studies, educational tool development and workshops, and capacity-building for authorities and institutional

---

[10] Zermoglio and Devisscher (2009, p. 28).

[11] Ibid. p. 29.

[12] ENDA (2010). *Synthesis report of five pilot actions* (p.3).

[13] More detailed information on ACCCA are available on the website http://www.acccaproject.org/accca/.

members. One of C3D's specific objectives is to promote gender-equality, as for example in supporting women's access to land, micro-finance and education programmes.

Running until 2011, this initiative includes six centers of excellence, of which three are located in Africa: ENDA TM (Environnement et Développement du Tiers-Monde) for West and Central Africa, and CSAG (Climate Systems Analysis Group) and ERC (Energy Research Center) for Southern and Eastern Africa. Each of these institutions have specific expertise, which together serve as a group resource, delivering targeted training and capacity development at the national and regional levels.[14] They have built sustainable partnerships over the years while supporting a wide range of projects within their respective regions and have shared and disseminated the knowledge acquired and tools developed as widely as possible.

C3D centers of excellence translate research into actions that local communities can apply to enhance their capacity and knowledge to address adaptation challenges.[15] Six thematic guidelines have built the C3D methodology: climate change context framing, stakeholder analysis, adaptation to climate change decision screening, climate change risk communication, and energy, poverty and climate change mitigation. During field visits in Banjul, Gambia and in the Maradi district, South of Niger, ENDA-TM developed a new conceptual thinking, encouraging people-led initiatives. These communities needed to cope with food insecurity and to adapt their agricultural practices due to climate variability. ENDA-TM helped them to design solutions through its unique community network, the Indigenous Knowledge Bank, gathering knowledge, adaptation strategies and ENDA communities' best practices within a single database. Among main tools developed by partners such as IISD and SEI, CRISTAL, an adaptation screening tool, aimed at highlighting potential risks attached to adopting specific climate adaptation actions by vulnerable communities. The tool was created in 2009, along with a version of CRISTAL targeting specifically at the forest sector, in partnership with CIFOR. One of the main training successes of the C3D project in Africa is CSAG winter schools, held every year at the University of Cape Town and focusing on climate change analysis with a scientific view. Among main outputs, C3D has developed: climate change training modules, such as the vulnerability and adaptation module, targeting regional training workshops, with a focus on Africa for Mali, Mauritania, Niger, South Africa, Senegal, Sri Lanka and Zambia, education and training through C3D partners, fellowships and researchers, and the equipment of centers of excellence with informatics and training materials.

C3D partners in Africa have learned from C3D experience that (1) Communication tools must be customized for partner countries and developed through specific interactive sessions to communicate climate science, (2) Working with C3D, the centers have strengthened their own capacity to deliver training to

---

[14] More detailed information on C3D+ are available on the website http://www.c3d-unitar.org/.

[15] UNITAR (2010, March,. p. 5).

national ministries, NGOs and academic stakeholders and (3) Above all, partners gained experience from building projects in a collaborative way; working together resulted in being a strong asset and partners weaved a useful network through C3D collaboration. Even though distance made their work difficult, partners enhanced their ties and gained a network: they are now well-recognized within their region. Centers finally acknowledged the need for each partner to be specialized in a thematic area for better efficiency.

As C3D activities were running on a long-term basis, it was decided to extend the project to a C3D+ initiative to pursue and finalize ongoing projects. In 2011, the partners' plan focuses on further developing CRISTAL and WeAdapt Google Earth Application tools, as well as on implementing workshops to disseminate these new learning methodologies.[16]

While assisting in the National Adaptation Programme of Action (NAPAs) project,[17] aimed at providing support to Least Developed Countries (LDCs) in the preparation and implementation of their adaptation strategies, the CCP addressed adaptation issues in Africa at a country level. The purpose of NAPAs is to "identify the urgent and immediate needs of a country to adapt to the present threats from climate change".[18] The Climate Change Programme has worked towards NAPA preparation since 2003, a work developed as a joint effort with the UNFCCC secretariat and other international development partners, such as the Global Environmental Fund (GEF) and its implementation agencies (IAs). To date, 45 country NAPAs were submitted to the UNFCCC Secretariat and the implementation phase has been launched by the above mentioned group of partners.

Along with projects detailed above, CCP has developed strong partnerships with the United Nations Framework Convention on Climate Change (UNFCCC) and the United Nations Economic Commission for Africa (UNECA) on adaptation in Africa. As an example, in November 2009, UNECA staff received a four-days training workshop on the impacts of climate change and the on-going negotiations for a post-Kyoto deal.

Based on CCP activities, some key factors for successful implementation of adaptation related activities in Africa could be: building projects and activities on research dedicated to the targeted communities and specific purpose served, providing partners with training and education on adaptation, implementing local training and keeping in mind lessons learned from past projects and working within a network of partners to facilitate projects implementation.

The purpose of the ACCCA initiative was to bring together stakeholders and scientific communities of the developing world to jointly execute pilot projects on adaptation in Asia and Africa. The purpose of C3D+ was to deliver top-skill training and workshops to address capacity-building needs in developing

[16] UNITAR (2010).

[17] More detailed information on NAPAs are available on the website www.napa-pana.org/.

[18] UNFCCC website: http://unfccc.int/cooperation_support/least_developed_countries_portal/submitted_napas/items/4585.php.

countries. Assisting in NAPAs preparation and providing training to NAPA teams reinforced countries' capacity and knowledge on issues related to climate change and also implemented effective adaptation strategies.

Conclusions

As underlined throughout this paper, three key-factors ensured the success of adaptation initiatives in Africa with the UNITAR Climate Change Programme: (1) pilot actions, case studies and projects undertaken relied on empowering local communities on the ground and considering their specific socio-economic context; not only on practical research and results (climate change data collection, assessments, cost effective methodologies, materials, etc.). CCP integrative methodologies include bottom-up, learning by doing, learning by observing and social learning approaches, participatory processes and assessments, case study highlights or local farming strategies, (2) The disseminated results and analysis should not only inform local authorities, scientific communities and policy makers at the final stage of the project but also include these stakeholders at main levels of the project. With that view, the Climate Change Programme relies on two dedicated centers of excellence and several partners' focal points in Africa and (3) These projects have advanced stakeholders' experience and knowledge acquired, as well as adaptation strategies and measures at local and national levels. For these reasons, adaptation projects should always build on existing approaches and constitute a ground basis for future projects. Going beyond current training and educational approaches, being integrative and continuously offering innovative methodologies for future developments are key ingredients for successful adaptation projects in Africa.

References

ACCCA (2009) Advancing capacity to support climate change adaptation. Final project report, p 30

ACCCA (2010) Description of nineteen pilot actions in Asia and Africa. p 30 http://www.acccaproject.org/accca/files/ACCCA_Brochure_19pilotactions.pdf

Adapting to climate variability and change (2010, 21 min) Video produced by the agricultural information resource centre. Prof. Agnes W. Mwang'ombe, University of Nairobi

Malawi: farmers become filmmakers (2009) Video. http://www.youtube.com/watch?v=WxScWzI3Rrg

UNITAR (2010) Strengthening capacity in developing countries for training purposes on climate change with the C3D+ partners. Final narrative report, February 2008 to July 2009

UNITAR (March 2010) C3D+ brochure

Zermoglio F, Devisscher T (2009) Africa synthesis report: lessons learned on climate change science and risk communication in the ACCCA project. CSAG § SEI, p 54

Home Pages, Videos and Resources

www.ipcc.ch
www.unitar.org
www.c3d-unitar.org/
www.napa-pana.org/
www.acccaproject.org/accca/

Author Biographies

Mamadou Moussa Diakhite work at the Climate Change Programme (CCP), at the United Nations Institute for Training and Research (UNITAR).

Audrey Avanzi work at the Climate Change Programme (CCP), at the United Nations Institute for Training and Research (UNITAR).

Sharon Oseku work at the Climate Change Programme (CCP), at the United Nations Institute for Training and Research (UNITAR).

Chapter 17
A Grassroots Initiative to Disseminate Solar Energy Technologies in Ethiopia: Implications to Climate Change Education

Aklilu Dalelo

Abstract As a tropical country, Ethiopia has an enormous potential to develop and use environment-friendly sources of energy like solar power. Despite such huge potential to tap electricity form the sun (not to mention the 45,000 MW potential from water), access to electricity is only about 35%. Solar energy technologies, if used widely, could alleviate the problem of rural electrification. This, however, demands a sustained effort to raise awareness about the technologies and develop skills required to make use of them. To this end, a school-based pilot project was initiated in 2004 by a Faith-based Organization in Ethiopia. The principal aim of the project was to bring about a positive change on the environment by using primary schools as centers of such change. The major activities meant to achieve the aim of the pilot project included, among others, assessing the place of energy and environment in the existing curricula; and dissemination of solar energy know-how and technologies. This paper reports the findings of a comprehensive study conducted to examine the extent to which the aforementioned objectives have been achieved. The findings indicate that the existing textbooks offer adequate opportunity to enable schools to address issues related to energy and environment. A closer look at content of the textbooks and the way the issues have been presented reveals, however, that most of the topics related to energy and environmental management have been presented neither comprehensively nor systematically. On the other hand, the solar energy technologies installed in the school compounds attracted the attention of individuals and

A. Dalelo (✉)
Alexander Von Humboldt Research Fellow at the Institute of Environmental and Sustainability Communication Faculty of Sustainability Sciences, University of Luneburg, Luneburg, Germany
e-mail: akliludw@gmail.com; dalelowa@leuphana.de

A. Dalelo
Addis Ababa University, Addis Ababa, Ethiopia

W. Leal Filho (ed.), *Experiences of Climate Change Adaptation in Africa*, Climate Change Management, DOI: 10.1007/978-3-642-22315-0_17,

community based organizations. More importantly, the solar home systems are found to have been operating (in nine of the eleven schools) without a problem, years after the official end of the project. Based on this and other evidences on the ground, the paper concludes that schools in Ethiopia have a potential not only to serve as centers of dissemination of knowledge about alternative energy technologies but also as centers where skills are developed to seek sustainable solutions to key environmental problems like climate change.

Keywords Solar energy · Energy technology · Climate change education · Ethiopia

Introduction

Energy use is inextricably bound with the development of modern economy. (Sokona 2002, p. 39) rightly underscores that energy "is a critical and irreplaceable element of life and development, in the sense that it plays a key role in the establishment of economic, social and environmental conditions". It is also argued that societies are "characterized by their level of, ability for, access to, and use of energy" (Shakya 2000, p. 145). Energy still maintains its pivotal position in the era of information technology that marks the twenty first century. Those who command the energy related know-how and technology are in a position to compete in the world development race if not in a position to command it. The last two centuries witnessed an upsurge in scientific and technological inventions that helped human beings to convert energy from naturally available forms to those that perform various tasks (Jadhav 1997).

Despite the availability of natural sources of energy like solar radiation, running water and wind, access to electricity in most countries of Africa remains a luxury reserved to a small proportion of the population. Affordability stands, perhaps, at the heart of factors hindering rural electrification in these countries. The other problem is pattern of population settlement. The latter problem, i.e., the difficulty to connect the dispersed villages in the rural areas using the central grid system, will remain a challenge even if the financial problem is adequately addressed. It is obvious that electricity from the grid is cheaper when compared to, for instance, that generated by photovoltaic (PV) system (Anderson et al. 1999). The latter, on the other hand, is known to be more appropriate for rural situations where grid connection is often problematic due to poor infrastructure and dispersed pattern of most of the rural settlements. It must also be underlined that comparison of cost between the grid and stand-alone systems of electricity supply is totally irrelevant for those pockets of remote settlements that can barely be connected to the grid. This justifies, beyond any doubt, the need for use of alternative sources of energy at present and in the years to come.

Ethiopian is believed to have a huge potential to develop and utilize renewable sources of energy. These include hydroelectric power, solar power and geothermal energy (to a lesser extent). The country's hydropower generation capacity is, for instance, estimated at 45,000 MW. However, the utilization of this potential is limited to 2,000 MW which is less than 4.5% of the existing potential (Ghibe III Office 2010). Despite the availability of such a potential to generate electricity, access to electricity is only about 35%. Most people (65%!) still live in the darkness. What is more, the average per-capita energy consumption is about 36 kWh (Ghibe III Office 2010). On the other hand, consumption of petroleum products in rural parts of the country is limited mostly to the use of kerosene for lighting in wick and lantern lamps.

More recently, energy issue in general and problems related to rural electrification in particular are beginning to get better attention in Ethiopia, along with the ongoing efforts aimed at poverty eradiation. Over the last decade or two, a number of proclamations, which focus on environmental protection in general and energy use and management in particular, have been issued. The following are, for instance, some of the guiding principles set for the country's energy resources development and management (EPA 1997, pp. 83–84).

- Adopting an inter-sectoral process of planning and development which integrates energy development with energy conservation, environmental protection and sustainable utilization of renewable resources;
- Supplying energy to remote and isolated areas using decentralized energy supply systems in order to minimize investment costs;
- Making institutional, pricing and regulatory arrangements so as to ensure the optimal and efficient development and utilization of energy resources;
- Increasing reliance on energy efficient technologies, sustainable use of renewable resources, and the development of indigenous energy resources;
- Promoting the development of renewable energy sources and reducing the use of fossil energy sources both for ensuring sustainability and for protecting the environment; and
- Encouraging the private sector and providing with the necessary incentives to participate in the development of the country's energy resources.

This list of principles shows, among other things, that most of the issues necessary for a successful development and dissemination of alternative energy technologies have been duly recognized at policy level.

The Problem and Rationale

As noted earlier, energy is a driving force for socio-economic development. One should also underline here the complex relationship between use of energy and environmental pollution and degradation. A recent report by the Intergovernmental Panel on Climate Change (IPCC) reconfirmed that human activities (mainly through use of non-renewable sources of energy) make the greatest contribution to global warming. The report specifically asserts that "the global increases in carbon dioxide concentrations are due primarily to fossil fuels use and land-use change" (IPCC 2007, p. 2). It is widely recognized that although climate change impacts will affect all countries, the poor will be disproportionately affected. With regard to the impacts of climate change on Africa, a recent report by the Economic Commission for Africa (ECA) and African Union Commission (AUC) shows that climate change "is already eroding decades of hard-won development gains" (ECA and AUC 2010, p. 1).

In developing countries like those in Africa, there is a unique problem with regard to use of energy resources. On the one hand, there is only limited capacity to make use of energy resources thereby leading to underdevelopment. On the other, some types of energy resources have been overused thereby causing massive environmental degradation. (Kazoora and Wandera 2003, p. 107) also express concern about the delicate balance between energy use and economic development in Africa: "The crucial dilemma for Africa is how to reconcile development goals and the elimination of poverty which will require increased use of energy and row materials". The use of alternative sources of energy, sources that enhance development without generating pollution, appears to be one of the appropriate responses to the dilemma. The role schools and institutions of higher education play in disseminating the know-how and technologies related to less polluting sources of energy is worth emphasizing.

Rationale Behind the Grassroots Initiative

Rural schools and community environmental education programs are often believed to represent an important strategy for mitigating environmental problems in countries of the Third World where the majority of the population live in rural regions and where most of the environmental problems prevail there (Ham and Castillo 1990). (Schleicher 1989) presents some concrete points to justify his argument that the development of a future-orientated sense of environmental responsibility depends largely on school teaching. Schools, in his opinion, are so important (1) because comprehension of ecological interdependence requires more systematic insight than the media or the public opinion can offer; (2) because a reflection of cultural and scientific paradigms can hardly be achieved by administrative means; and (3) because attitudinal changes towards nature and natural resources are psychologically easiest with young. Furthermore, schools can, by working hand in hand with extension services, encourage an experimental approach and a willingness to accept change (Malassis 1976). A relatively more recent study that evaluated the effectiveness of an environment education project carried out in Denmark concludes that the "schools (and their pupils) do actually have possibilities for acting as a catalyst for environmental changes in the local community" (Jensen 2002, p. 328).

Background to the Grassroots Initiative

An empirical study was conducted in southern Ethiopia to investigate, among others, students' awareness of and attitude towards the problem of natural resource degradation (Dalelo 2001). More than 1,100 students took part in the study from junior and senior secondary schools of the then Kembata-Alaba-Tembaro zone; and Awassa Teacher Training Institute (TTI). The study revealed a remarkable deficiency in awareness about issues related to natural resource use and management. Only one-fourth of the students were, for instance, able to mention three correct responses about the causes, consequences and solutions to the most serious environmental problem in the country, namely, land degradation. Students' awareness about some key issues like the impact of population growth on resource use and management was also found to be too low to be rated sufficient. Insufficient was also the students' awareness about the natural resource base of the country.

The aforementioned study recommended, among other things, provision of a basic and aim-oriented education on issues pertaining to the use and management of natural resources in schools and TTIs so as to protect the environment and improve the livelihood of its inhabitants. In an attempt to put at least part of this recommendation into practice, it was decided to run a pilot project aimed at enabling schools to address key environmental problems. The project had been initiated and implemented by the Capacity Building and Community Empowerment Program of the Ethiopian Kale Heywet Church, in collaboration with Education Bureaus of Kembata-Tembaro Zone (KTZ) and Alaba Leyu Wereda (ALW). The project was started in 2004 and came to an end in 2007. The target group and direct beneficiaries of the pilot project were primary school teachers and students.

The following were the major activities planned to achieve the aims of the pilot project.

- Undertaking a baseline survey on the status of the environment and the place of environmental issues in the schools' curriculum;
- Preparing a handbook on natural resource management and environmental protection;
- Conducting a training program for primary school teachers on use and management of natural resources and environmental protection;
- Integrating issues related to use and management of natural resources and environmental protection into the daily lessons;
- Establishing environmental clubs in all the project schools;
- Establishing centers for development and dissemination of alternative energy technologies; and
- Establishing school nurseries.

This paper reports achievements and constrains related to two of the major activities, i.e., 'undertaking a baseline survey on the place of environmental issues

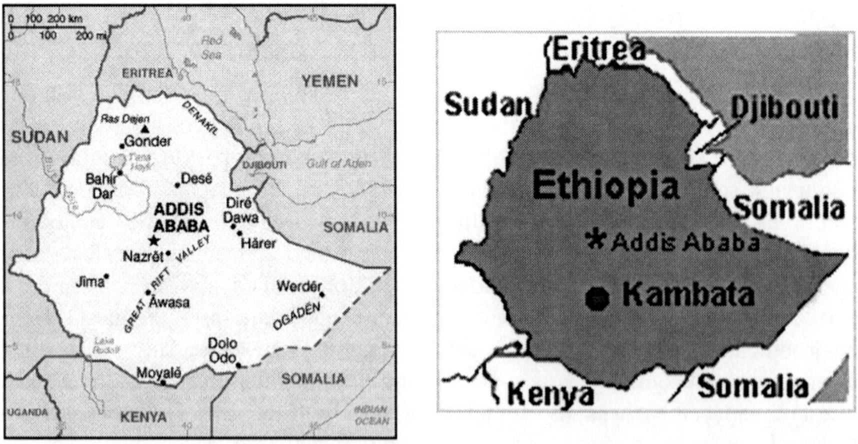

Fig. 1 Maps showing the location of Ethiopia in East Africa (*left*) and project area—Kambata (*right*)

in the schools' curriculum'; and 'establishing centers for development and dissemination of alternative energy technologies'.

Methodology

Selection of Participating Schools

The project schools had been identified on the basis of a survey on the state of the environment in the project area (Fig. 1). Accordingly, eleven out of the 104 (10.6%) higher primary schools had been selected by Wereda.[1]

Capacity Building, Education and Agriculture and Rural Development Offices. All the selected schools are located in villages that are seriously affected by environmental degradation.

Textbooks' Review

Student textbooks currently in use in the upper primary level (grade 5–8) have been reviewed by the staff of the project, based on a check-list prepared for this purpose. Attempt has been made to register topics related to energy and environmental protection.

---

[1] Kebele (peasant association) is the lowest level in the Ethiopian government structure. The other levels are **Wereda**, Zone and Region in ascending order.

Field Survey

Data relating to the dissemination of solar energy technologies has been gathered from all the eleven schools by the author with the help of one of the project officers. Two methods have been used to gather information: focus group discussion and observation of the alternative energy technologies. A total of 73 people participated in the focus group discussions which took three hours in one center, on average. The participants represented different segments of the community (school directors, teachers, training and education board members, parent-teacher association members and others like agriculture and health extension workers). The solar energy technologies (solar home systems, solar lanterns, solar cookers and driers) had been visited to check whether they had been operating as expected. Furthermore, all the eleven schools have been contacted by the project office once again in 2010 (three years after the official end of the project). The schools were asked to report on the status of the solar energy technologies. The aim here was to know whether the technologies were still operating after the project had phased-out.

Results

Integration of Issues Related to Energy and Environment

As part of the baseline survey, the exiting textbooks for grades 5–8 have been thoroughly examined by the project team to see the way in and extent to which issues related to energy and environmental protection have been addressed. The analysis shows that most of the school subjects (particularly social studies) have topics related to environmental issues in general (Table 1). In view of this paper, it is important to underline that topics related to energy use and associated environmental problems (including climate change) have been taken up in most of the school subjects. Some of the topics in this category include 'message from the tree' in Amharic grade five; 'generating biogas' in Amharic grade six; 'atmosphere' in Amharic grade eight; '*the tree*' in English grade five; '*water, deserts, greenhouse effect*' in English grade eight; '*energy*' in integrated science grades five and six; '*environmental pollution*' in Biology grade eight; '*air pollution,* fuels' in Chemistry grade seven; '*electric energy*' in Physics grade eight; '*the atmosphere, air pollution*' in Social Studies grade six; '*climate, destruction of natural vegetation, conservation of natural vegetation, the impacts of human beings*' in Social Studies grade seven; and '*the atmosphere, atmospheric pollution, causes and consequences of air pollution, methods used to control atmospheric pollution*' in Social Studies grade eight.

Table 1 thus indicates that the existing textbooks offer adequate opportunity to enable schools to address key environmental issues posing challenges to life in the

Table 1 Environmental issues included in students' textbooks

Subject	Grade	Issues
Amharic	Five	*Message from the tree*[a]
		Let our rivers be used for irrigation
	Six	Stream capping (development)
		Generating biogas
	Seven	Fish resources: Use and production
	Eight	*The atmosphere*
English	Five	*The tree*
	Six	Soils of Ethiopia
	Seven	Wild animals
		Health
	Eight	World problems (population, *water*, soil erosion, *deserts, greenhouse effect*)
Integrated science	Five	Air, water, plants, animals, *energy*
	Six	Air, water, animals, *energy*
Biology	Seven	Health, population and community
	Eight	The ecosystem *(environmental problems, pollution*, population explosion)
Chemistry	Seven	Environmental chemistry (air composition, *air pollution*)
		Natural resources (water, soil, *fuels*)
Physics	Eight	*Electric energy*
Social studies	Five	Unit One: The physical environment of Africa
		Unit Two: Social environment of Africa
		Unit Three: Economic activities of Africa
	Six	Unit One: The Earth and its environment
		Use of water bodies
		Atmosphere
		Plants and animals
		Air pollution
		Mining and industry
	Seven	Unit three: Man and his environment
		Climate
		Natural vegetation
		Destruction of natural vegetation
		Conservation of natural vegetation
		The impacts of human beings
	Eight	Unit Two: The natural and social environment
		The atmosphere
		Atmospheric pollution
		Causes and consequences of air pollution
		Methods used to control atmospspheric pollution
		The earth
		Water erosion and its outcomes
		Wind erosion and its outcomes

[a] Topics in bold and italicized are those related more directly to energy and climate change education

study area and the nation at large. A closer look at content of the textbooks and the way the issues have been presented reveals, however, a different fact. Most of the topics related to energy and environmental management have been presented neither comprehensively nor systematically. What is more, no effort has been exerted to critically analyze and discuss environmental issues in general and those related to energy in particular (Dalelo 2006). A handbook on natural resource management and environmental protection was therefore produced, as part of the pilot project, so as to supplement (not replace) the existing school textbooks. The handbook contained five modules: natural resource base of Ethiopia, natural resource base of the project area, population and environment, energy use and environmental pollution, and environmental health and sanitation. The writers of the modules were drawn from a pool of highly qualified persons mainly from Addis Ababa University, Debub University (now Hawassa University) and the Ethiopian Energy Authority (EEA).

Impacts of the Handbook

Gagliardi and Alfthan (1994) argue that environmentally sound attitudes must be supported by an understanding of environmental issues and a command of the methods and tools needed to manage the environment. Without such knowledge and mastery, the causes of environmental degradation cannot be understood nor can positive action be put into effect. In view of this, the preparation of the handbook, meant to complement the students' textbooks, seems to be highly justified. Both teachers and directors of the respective schools substantiated that the handbook had been used with a high degree of enthusiasm (Dalelo 2009). Teachers felt that the handbook was comprehensive enough to accommodate the diverse interests of the different subject areas. The level of difficulty of the handbook was also considered to be appropriate and the language was rated as understandable. The pictures included in the handbook, which depicted diverse aspects of the region, were particularly appreciated by the users. The teachers said they felt more confident to teach issues related to energy and environmental management as the handbook provided them not only with detailed explanations but also examples which were not available in the textbooks prepared by the Bureau of Education.

A teacher at Besheno School commented that the handbook helped them to have a more holistic view about the environment, which they were not able to develop during their college education (Dalelo 2009). The fifth module, which treated the natural resource base of the districts where these schools are located, was found to be particularly interesting to teach and easy to apply. Teachers were surprised to read stories about their own localities which was not the case in the centrally prepared textbooks. The SIM Canada final evaluation report also considers the handbook as "an excellent teacher's reference guide" (Paterson 2008, p. 29).

There were, however, some limitations observed in relation to the content and use of the handbook. The evaluation report referred to above also indicated that the handbook "provided relevant information at the local community level but little practical information as to what teachers and students could do to address these issues" (Paterson 2008, p. 29). The other challenge has to do with the high turnover of teachers who got the training (partly because of the unfavorable physical environment). In Sheshera School, for instance, only one out of the eleven trained teachers was still working in the school! In Hobicheka as well most of the teachers had been transferred. The positive aspect of the challenge was that teachers who got transferred after the training continued using the information they gained and some even established environmental protection clubs in the new schools which were not part of the pilot project (Dalelo 2009).

Dissemination of Solar Home Systems

Earlier studies indicate that SHSs have both economic and non-economic benefits (Ahm 1998). The direct economic benefits include income gneration[2] hrough development of cottage industries, saving expenditure on primary batteries (dry cells) as SHSs power radios, mitigation of urban migration (Kayastha 2000) and even attracting people back from urban areas, and a modest job-creation as the trained 'technicians' earn money through installing and maintenance of SHSs. Studies also indicate that customers attach a number of non-monetary values to SHSs (Cabraal et al. 1996; Ahm 1998), which contribute to the overall socio-cultural development. The major non-monetary values include convenient, instantly available light; higher-quality light, making such tasks as reading and study easier; cleaner indoor air due to reduced (or eliminated) soot and fumes from kerosene and candles; improved safety levels (elimination of dangers from accidental fires and burns from kerosene devices or candles); access to services such as TV and radio thus opening the window to the rest of the world and thereby reinforcing civil society; and an elevated social status associated with electrification.

All the eleven schools covered by the pilot project were provided with a solar home system and solar lanterns. The SHS package included a panel, battery, charge regulator, inverter, lamp holder, energy saving bulbs, cables, transmission wire and switches (Dalelo 2008). The price per system was 5,520 Ethiopian Birr (equivalent to 613 USD at the time of purchase). The initial purpose of providing the SHS was to supply schools with energy for lighting and operating radios/ cassette players. The ultimate goal was, however, to create access to information

[2] A study carried out in Nepal has indicated that SHSs are not helping to promote new kinds of income generating activities. They are, however, assisting the already existing businesses to generate more income (Kayastha 2000, p. 107).

about these alternative technologies, in communities where there is not such information, by using the schools as centers for information dissemination and skill development.

At the time of survey, all schools, except Durgie and Kerekicho, had been properly using the SHSs for lighting and operating educational radios/cassette players. Teachers also started using the light to make their lesson plans in the evening. Going to the school compound in the evening to make such preparations was hitherto unknown (Dalelo 2008). Education and training board members also used the opportunity to conduct meetings at night. This was particularly important as the members of the Board were all busy during the day doing their farm chores and/or engaged in other social issues. Access to electricity thus made it possible to use the evening hours more productively and this, in turn, improved the effectiveness of school administration. The energy was also used to power educational media (mini-media) used in schools to pass important information on current issues like environmental protection, HIV/AIDS, family planning, etc.

Before the installation of the solar panels, most of the schools had their school mini-media powered by dry cell batteries (Dalelo 2008). In Besheno School, for instance, they spent 21 Birr/week (84 Birr/month) on average for purchase of dry cell batteries. This was about 840 Birr/academic year. One can thus see that money spent only for dry cell batteries in 6 and half years is enough for buying SHS set with 40Wp. This is another evidence for the financial sustainability of the technology once its initial cost is covered by designing appropriate and region specific financing mechanisms.

What was not planned for but attracted the attention of the community to a great extent was cell phone (mobile) charging (Dalelo 2008). Six of the eleven schools had been using the solar energy for this purpose (Table 2). There were exciting stories with regard to mobile charging. In centers like Besheno the rate of purchase of mobile phones grew dramatically as a result of access to the new energy. There was a network coverage in the region but no access to energy. Under such a situation people used to send their phones to Kulito town (38 kms!) for recharging. Almost all the school committee members owned a mobile phone following the installation of the SHS; and this enhanced communication and planning. The total number of mobile phones grew from less than 20 to more than 300 in a matter of a few months. The mobile phones had also been used by some farmers and business people to gather market information and to contact lorry drivers when they wanted to take their products to the market place (Dalelo 2008).

Solar Cookers and Driers

Solar cookers and driers have been supplied to all the eleven schools by the pilot project so as to test their appropriateness and potential for wider dissemination. The schools have then made practical attempts to test the apparatus in their compounds and checked that the technologies were capable to give the expected

Table 2 Pattern of use of solar energy technologies

Name of the Kebele/village	Types of technology					
	SHSs (actual use)				Solar cookers (potential use)	
	Lighting	Powering educational media	Mobile changing	Others[a]	Cooking food	Boiling drinking water
Durgie	×	×	×	×	×	×
Mandoye	√	√	×	√	×	×
Zogoba	√	√	×	×	×	×
Adancho	√	√	×	√	√	√
Hobicheka	×	×	×	×	√	√
Kerekicho	√	√	√	×	×	×
Sheshera	√	√	√	√	√	√
Holegeba Zato	√	√	√	√	√	√
Gerema	√	√	√	√	√	√
Mekalla	√	√	√	√	√	√
Besheno	√	√	√	√	√	√
Total	9 of 11	9 of 11	6 of 11	7 of 11	7 of 11	7 of 11

[a] Lesson planning in the evening, conducting meetings, lighting the guard booth, etc

functions. Some of the schools also demonstrated the technology to the surrounding communities (Dalelo 2008). Seven out of the eleven villages saw a potential for use of solar technology for cooking and/or drying (Table 2). When it comes to the immediate application of solar cookers, the participants of the focus group discussion appeared a bit less enthusiastic. Most of the people in those villages use clay pots for cooking and these were felt unsuitable for solar cooking. The other perceived limitation of the technology is the relatively longer time it took to cook the same amount of food. The dependence of the technology on the strength of solar radiation also reduced its reliability. People felt that their plan to cook food could be affected by the weather condition, of which they have no prior information. In areas where there is no shortage of fuel wood, solar cookers seem to have no potential for dissemination. Solar cookers have been strongly recommended for boiling water in lowland villages like Adancho and Gerema. These are villages that depend on water from very dirty sources like ponds. The cost of the technology was, however, presented by many as a discouraging factor for poorer households.

The need for solar driers is evident and the experience is already there. Solar driers seem to have better chance to be disseminated in the villages if the cost is significantly reduced through local production and use of locally available materials (Dalelo 2008). The other possibility is use of solar driers for food processing and this could make major contribution to the improvement of household food security. The villages use different types of root crops like potato which could be dried and preserved for non-harvesting seasons. The teachers who participated in the focus group discussions felt that, with minimum training, solar cookers, driers and improved cooking stoves could be produced in their schools; and the technologies, particularly the improved cooking stoves, could then be used for income generation. The solar cooking and drying technologies are also believed to be excellent pedagogical aids especially for natural sciences.

Sings of Sustainability

All the elven schools have been contacted once again last year (end of 2010) to check whether the SHSs and solar lanterns are still operating. It was interesting to learn that the technologies have been operating and serving the intended purpose (in nine of the eleven schools) three years after the project phase-out, thereby signaling that schools can really play their role as centers for development and dissemination of environment-friendly technologies like the ones discussed in this paper.

Discussion

Jensen (2002, p. 332) empathically suggests that schools, as institutions for general education, "have a responsibility to help equip the members of society in their charge, their students, with the knowledge and commitment to take personally meaningful decisions and actions to address the challenges posed by both lifestyle and societal conditions". With regard to schools' role in encouraging appropriate decisions and actions related to energy use and conservation, three interrelated approaches have been identified (Schröder 1998). These are the pedagogical approach, the less intensive approach and the incentive-oriented approach. Success in attaining the goals of energy education in schools largely depends on the ability to combine the approaches in such a way that they produce long lasting effects.

The Pedagogical Approach focuses on integration of energy issues into the school curricula, working out appropriate methods to teach the content in the curricula, and establishment of school energy clubs (Schröder 1998). Project-oriented education is another way highly emphasized by the proponents of the pedagogical approach. As part of the pilot project discussed in this paper, attempts were made to review the existing textbooks which led to preparation of a handbook to bridge the information and methodological gaps identified. An in-service training was also given to teachers on the content of the handbook and its use. The handbook had positive impacts notwithstanding its limitations with regard to development of practical skills.

Supplementing the pedagogical efforts to energy conservation by practical actions like minimizing unnecessary use of light in a classroom or fuel wood at home would lead to an even better result. The second approach, namely, Less Intensive Approach, is based on this premise (Schröder 1998). In this regard, the pilot project introduced an improved cooking strove with a capacity to cut wood consumption by half (Dalelo 2008). The second approach encourages, in general, pupils to take part in some visible and tangible activities that contribute to a decrease in energy consumption. However, the approach does not propagate a massive and cost-intensive program on energy saving actions. The third approach, Incentive-oriented Approach, is entirely different from the other approaches. It is based on the provision of material incentives (bonus) to the participants of the energy saving activities (Schröder 1998). Provision of SHSs and solar lanterns by the pilot project was thus meant to give incentives to participating schools so that they gain the degree of knowledge about the technology and enthusiasm required to disseminate it in their surrounding communities. The sustained interest on the part of the schools to use the energy technologies both for educational and non-educational purposes attests to the fact that the incentive approach to energy education was effective.

Conclusion and Implication

Schools as institutions of formal education can also act as centres where pupils acquire practical, tangible and, at the same time, comprehensive information about the causes, consequences and cures of environmental degradation (Dalelo 2008). Besides, innovative ideas relating to and strategies for environmental protection can be first demonstrated in schools and then disseminated to the surrounding communities. The latter role of schools is all the more important in Ethiopia where contacting the grassroots population via other means is practically difficult and at times and in certain cases even impossible. This paper, based on a review of studies aimed at investigating the role of primary schools in dissemination of solar energy technologies, indicates that there is indeed a huge potential for dissemination of SHSs and solar lanterns in rural villages where there is no readily available grid electricity. It has also been substantiated that schools could be effective agents not only for creation of know-how related to energy and environmental issues but also dissemination of alternative energy technologies including, among others, SHSs, solar lanterns and improved cooking stoves.

The paper shows that solar energy technologies have both social and economic significance in the project area. Once a mechanism for paying the initial cost is put in place, the long-term comparative advantage of SHSs over kerosene and dry cell batteries proved to be self-evident as shown using a simple calculation based on the actual expenses for purchasing dry cell batteries to power school mini-media. The community in the study area requested for evening classes immediately after the installation of the panels though the schools could not respond positively because the nature of the energy (its technical and financial feasibility) was yet to be tested. In any case, the interest on the part of the community could be taken as a strong indicator for possible dissemination of solar energy in the villages. The appropriateness of SHSs for institutions like schools, churches and mosques; and that of solar lanterns for individual households has been repeatedly emphasized during the focus group discussions. Finally, it is important to underline that the SHSs and solar lanterns have been fully operational in nine of the eleven schools three years after the project phase-out. This is another evidence that the technology can endure test of time, given adequate capacity is built in the schools both technically and managerially.

The findings of this study seem to have a direct implication for climate change education. The growing challenge of climate change is often linked with the type and amount of energy used. Despite all the activities conducted to curb the effects of climate change, there is "still not an effective capacity to bring the understanding of the climate change facts to the public in manner that influences their day-to-day actions and habits (Ogbuigwe 2009, p. 21). In view of this, the grassroots initiative described in this paper could be taken as a good example of an effort to build the capacity required for binging 'the understanding of the climate change facts to the public'. The pilot project enhanced the achievement of the goals of climate change education in, at least, two ways. First, by disseminating solar energy know-how and technologies, it showed alternatives to use of energy sources that lead to or exacerbate the problem of climate change. Second, it empowered teachers to design and apply educational approaches that balance theory and practice using the same technologies as pedagogical aids which are rarely available in the project areas.

Acknowledgments This paper was written during my stay, as Alexander von Humboldt Research Fellow, at the Institute of Environmental and Sustainability Communication (INFU), Leuphana University of Lunenburg. I would like to express my gratitude to INFU and its Director, Prof. Dr. Gerd Michelsen, for providing office space, resources and much more.

References

Anderson T et al. (1999) Rural energy services: a handbook for sustainable energy development. IT Publications, New Jersey

Ahm P (1998) Preparing a sustainable market for PV solar home systems. In: Shrestha JN et al (eds) Role of renewable energy technology for rural development, proceedings of international conference on role of renewable energy technology for rural development, October, 1998, Kathmandu, Nepal, pp 12–14

Cabraal A et al (1996) Best practices for photovoltaic household electrification programs: lessons from experiences in selected countries. World Bank Technical Paper Number 324. The World Bank, Washington, DC

Dalelo A (2001) Natural resource degradation in Ethiopia: assessment of students' awareness and views, vol 11. University of Flensburg, Flensburger Regionale Studien

Dalelo A (2006) Enabling schools to address key environmental issues: opportunities and challenges. J Educ Dev 1(1):7–56 September 2006

Dalelo A (2008) Schools serving as centres for dissemination of alternative energy know-how and technologies: evidence from southern Ethiopia. Int Res Geog Environ Educ 17(3):250–267 2008

Dalelo A (2009) Efforts to empower teachers in Ethiopia to address local environmental problems: achievements and limitations. Int Res Geog Environ Educ 18(3):211–226 August 2009

Economic Commission for Africa and African Union Commission (2010) Report on Climate Change and Development in Africa. Available at. http://www.uneca.org/cfm/2010/documents/English/Report-onClimateChange-andDevelopment-inAfrica.pdf. Accessed on 12 Jan 2011

EPA (Environmental Protection Authority) (1997) The conservation strategy of ethiopia volume ii: federal policy on the environment. Addis Ababa

Gagliardi R, Alfthan T (1994) Environmental training: policy and practice for sustainable development. Oxford and IBH Publishing CO. PVT. Ltd, New Delhi

Gibe III HEP Office (2010) Environment & Social Issues Related to Gibe III. Addis Ababa, August 2010

Ham SH, Castillo L (1990) Elementary schools in rural Honduras: problems in exporting environmental education models from the United States. J Environ Educ 21(4):27–32

IPCC (2007) Climate change 2007: the physical science basis—summary for policymakers. Contribution of Working Group I to the Fourth Assessment Report of the Intergovernmental Panel on Climate Change

Jadhav HV (1997) Energy and environment. Himalaya Publishing House, Beijing

Jensen BB (2002) Knowledge, action and pro-environmental behavior. Environ Educ Res 8(3):325–334 2002

Kayastha Y (2000) The role of solar home systems in the promotion of income generating activities in selected villages of kavre district in Nepal. Unpublished MSc Thesis Presented to ARTES, University of Flensburg

Kazoora C, Wandera A (2003) Willingness to Pay to access and utilize electricity in rural Uganda: In: Bekure W (ed) Environmental management and local development in the horn and east Africa. RLDS, Addis Ababa University, Addis Ababa

Malassis L (1976) The rural world: education and development. Paris

Ogbuigwe A (2009) Sigtunna think piece 1: climate change education in Africa. South Afr J Environ Educ 26:19–26 2009

Paterson I (2008) Final report: evaluation of SIM Canada's 2004–2007 Program (S62896), April 2008 (unpublished)

Schleicher K (1989) Beyond environmental education: the need for ecological awareness. Int Rev Educ 35(3):257–281

Schröder D (1998) Anreizsysteme für intelligente Energienutzung an Schulen, in Energie sparen an unseren Schulen, IPTS Arbeitspapiere zur Unterrichtsfachberatung, 1998

Shakya B (2000) Natural Resources. In: Pande BD et al (eds) Environmental education source book for bachelor of education programme. Kathmandu: IUCN Nepal

Sokona Y (2002) Training and human resource development for the energy sector. In: Karekezi S, Mackenzie GA (eds) Energy options for africa: environmentally sustainable alternatives. Zed Books, London

Author Biography

Dr. Aklilu Dalelo is an Assistant Professor of Geography and Environmental Education at Addis Ababa University, Ethiopia. He is a geographer by training with a particular interest in environmental and sustainability education. Dr Aklilu made an extensive research over the last fifteen years mainly in energy and environmental education; and published more than ten articles in reputable national and international journals. Currently, he is an Alexander von Humboldt Research Fellow at the Institute of Environmental and Sustainability Communication, Faculty of Sustainability Sciences; University of Luneburg, Germany.

Chapter 18
Integrating Indigenous and Scientific Knowledge Systems for Climate Change Adaptation in Zambia

George Kasali

Abstract The projected impacts of climate change are expected to be adverse and particularly severe for the majority of Africans who currently rely heavily on climate-sensitive natural resources for their livelihoods and do not have the means to withstand shocks and multi-dimensional changes in their societies and ecosystems. Africans are, therefore, justifiably looking to their developed counterparts to assist them with technological options for adapting to climate change. In this quest, it is usually forgotten that indigenous knowledge systems can significantly contribute to location-specific sustainable climate change adaptation. This study used structured questionnaires, focus group discussions, participatory workshop techniques and literature reviews to explore the potential role of traditional knowledge and practices in climate change adaptation in Zambia. It was found that agricultural practices which embedded both scientific know-how and indigenous knowledge practices were by far superior in providing resilience to droughts and floods. Imported technologies that neglected ecological compatibility only served to provide improved short-term productivity but plunged communities into long-term livelihood hardships as they stripped the environment of its ecological sustainability. It was also found that local people used the phenology of plants and insects to forecast rainfall in the next season. The reliability and accuracy of this traditional knowledge has yet to be scientifically proven. It was concluded that the viability and utility value of indigenous knowledge can only be enhanced when integrated with scientific approaches and know-how.

G. Kasali (✉)
Department of Biological Sciences, School of Mathematics and Natural Sciences, The Copperbelt University, P. O. Box 21692, Jambo Drive, Riverside, Kitwe, Zambia
e-mail: kasali_george@yahoo.com

W. Leal Filho (ed.), *Experiences of Climate Change Adaptation in Africa*, Climate Change Management, DOI: 10.1007/978-3-642-22315-0_18,

Keywords Climate change · Adaptation · Indigenous knowledge · Conservation farming · Climate forecasting · Plant phenology · Indicators of rainfall · Drought · Flood

Introduction

Indigenous knowledge is referred to as local knowledge that is acquired and accumulated by communities and societies over generations and centuries and encompasses every aspect of human existence. It is generated through interactions of people with their external environment, resulting in outcomes that include technologies, skills and beliefs which are deployed in various livelihood strategies such as agriculture, natural resource management, medicine, communications, architecture and management of climatic hazards.

In fact it can be argued that the development of scientific knowledge and know-how is a natural and logical response to the limitations of indigenous knowledge systems. Moreover, both indigenous and scientific knowledge systems are driven by the innate human quest for survival and comprehension of the intricacies of life and its relationship with the cosmos. Actually, at one time or the other, the entire global humanity had to depend on indigenous knowledge systems for its survival and sustenance. However, due to various political, social, economic, environmental, cultural and climatic perturbations, some societies have discarded or simply lost their location-specific indigenous knowledge systems. In Africa there are communities that are still clinging to the traditional practicalities of meeting their livelihood needs and wants.

In Zambia there are communities or regions that have not abandoned their traditional agricultural practices up to this day in time. Of the nine provinces of Zambia, three are known for their notable indigenous technical systems of food production. These regions include the Northern, Luapula and Western provinces of Zambia.

The Bemba people of Northern Province of Zambia are known for their present day practices of a form of shifting cultivation known as *chitemene*. In this indigenous farming system, men lop, prune and coppice trees and shrubs and the cut vegetation is pulled by women who pile it into a circle that is burnt before sowing crops into the intermixed ash. Usually, the tree biomass is lopped from a much larger area (outfield) while the cultivation of crops is done in a smaller burnt area (infield). The outfield is left to regenerate while the infield can be used for 5–6 years and then left to fallow for 10–30 years (Mwale 1997). The chitemene system employs mixed cropping and crop rotations based on finger millet, sorghum, groundnuts, cucurbits, peas and beans. This system provides food security for subsistence farmers and moreover, the system is ecologically benign in that it ensures minimal loss of biodiversity and land cover, which are important factors in prevention of land degradation. These traditional systems, though historically without written documentation, are indicative of a people who understood their environment very well and only practiced that which was in conformity and consistent with the prevailing climatic and agro-ecological conditions. In fact the traditional *chitemene* system is in its practice and character very similar to modern day practices of agro-forestry. However, the viability of the *chitemene* system is now failing due to increased population growth and density which are limiting the availability of land for the required long fallow periods of the system. This calls for a scientific approach to ensure *chitemene* becomes sustainable for the people of northern Zambia.

The importance of climatic events in these communities is also reflected through oral documentation in the Bemba language whereby the months are named after climatic and livelihood events. For example, February is called *Kabengele kakalamba*, which implies a period of heavy rains when water courses begin to overflow, while May is the period of mild coldness (*Kapepo kanono*). Actually, all the twelve months of the year are named in such a way that they describe the pertinent climatic event and/or the predominant natural resources which can either be consumed or utilised during that particular climatic event. This traditional calendar ensured that knowledge about the relationship between climate and livelihood resources was passed from generation to generation. Scientists currently require this knowledge to understand the impacts of climate change on biodiversity in Zambia for purposes of devising sustainable adaptation strategies. Unfortunately, these traditional/livelihood calendars have now been replaced with western ones.

The people of Luapula province are known for their prowess as fishermen whose livelihoods are closely tied to Lake Mweru which is situated in the northern tip of Zambia. Lake Mweru has very few islands and when fishing at night several kilometres from the shore, people easily get lost as to the geographical position of their fishing camp sites. However, seasoned fishermen depend on their knowledge of astronomy and use stars in the sky to navigate to locations on the lake where nets are dropped and back to their camps at night. These stars have been given local names and are also used as clocks since specific stars only appear at specific times of the night. Fishermen have also developed early warning systems for stormy weather when the lake turns rough and becomes dangerous for fishing activities. In this regard, wind direction in a particular season is used to warn fishermen either not to venture into the lake or rush away from the lake. Seasoned fishermen also know which parts of the lake are productive in each particular season of the year and period of the moon cycle. This knowledge is crucial to understanding the vulnerability of fisheries to climate change. However, this unwritten indigenous knowledge is slowly dying as the older generation passes away and the Lake Mweru area becomes infiltrated by economic migrants from other regions of Zambia. Moreover, this lake has not yet been subjected to any climate change studies.

The Lozi peoples of Western Province reside in an agro-ecological region of Zambia that is characterised by wetland (Zambezi flood plain) and upland environments. The remarkable feature about the indigenous farming systems of the

Lozi is that they developed about eight different cultivation systems which were locally referred to as garden-types. These farming practices were intended to exploit the properties of the surrounding environment such as sandy, humic and clay soils, anti-hills, forests, thickets, silt deposition and perennial supply of moisture by the floodplain. The various garden-types were cropped with maize, cassava, sweet potatoes, pulses, fruit trees (like pawpaw and pineapples), sugar-cane, tobacco and vegetables (Kajoba 2008). They were continuously and intensively cultivated, ensuring all year round production of food. Nowadays, however, these indigenous practices are slowly dying out as western agriculture encroaches into Lozi society and the population becomes dependent on the most climate-sensitive crop of maize as its staple.

There are many other regions of Zambia where the application of indigenous knowledge in various livelihood strategies is being practised up to this day. However, this paper focuses on the Tonga people of Southern Province of Zambia and in particular Mazabuka District where a community-based climate change adaptation study was conducted. The emphasis is on peoples' experiences within the historical context of indigenous farming practices and the implications of these practices for climate change adaptation. The paper also provides information on traditional early warning systems for droughts and floods.

Materials and Methods

Description of the Study Area

Mazabuka district lies between latitudes 15 and 17°south of the equator and longitudes 27 and 29° east, and is administratively located in Southern Province of Zambia. The Kafue River forms the boundary with Kafue district of Lusaka province to the north while Siavonga is in the south-east. Gwembe and Monze districts are in the east and south-west of Mazabuka, respectively (Kalonga 2004). Mazabuka is situated about 125 km south-west of Lusaka, the capital city of Zambia, along the line of rail. The settlement pattern within the 22 administrative wards of Mazabuka district is scattered except for Mazabuka town and the three peri-urban settlements of Nakambala, Nega–Nega and Namalundu (Fig. 1).

Mazabuka recorded a population of 240,116 (CSO 2001) in 2000 comprising of 119,413 males (49.73%) and 120,703 females (50.27%). The rural population of Mazabuka is nearly 76.8%. Overall, the majority of the people of Mazabuka face enormous challenges in their livelihood pursuits. In the urbanised locations of Mazabuka, where people are able to access formal employment either as government, commercial or sugar plantation/refinery workers, the poverty levels are below 50%. However, in the typical rural areas of the district, over 70% of the population lives in poverty.

Mazabuka district covers an area of 6,687 km^2 and has three main geographical features which include the Kafue River and its flood plain, the central plateau area

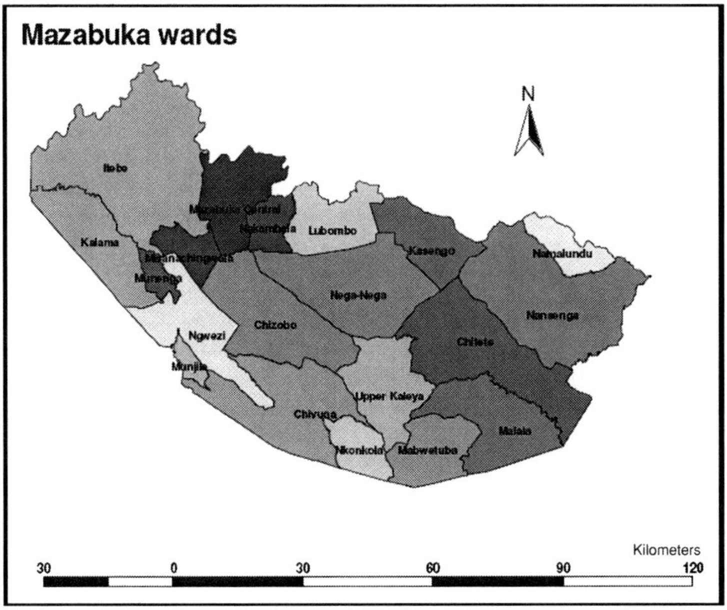

Fig. 1 Administrative wards of Mazabuka district. *Source*: Mazabuka municipal council, 2008

where Mazabuka town is located, and the hilly eastern part that forms part of the Zambezi escarpment. The plateau soils are comparatively more fertile and suitable for agriculture. In Mazabuka 80% of the land in the district is used for agricultural activities such as growing of maize (staple crop), sugar cane, coffee, cotton, groundnuts, dairy farming and fishing along Kafue River. Alongside private companies such as Zambia Sugar, Mazabuka has about 20,693 small-scale and 200 commercial farmers. Maize and sorghum are mainly grown for own consumption and it is only the excess that is sold for income. The rest of the crops are used for generating cash for the households. The livestock that are reared include cattle, sheep, goats, pigs and poultry.

Mazabuka District, like the rest of Zambia, experiences a year that can be divided into two distinct halves, a dry half from May to October and a wet half from November to April. There is, therefore, only one rainy season in Zambia. However, when temperature variations are included the year can be divided into three distinct seasons which include a dry-cold season (May to July) with daily temperatures that can drop to as low as 5.5°C, a hot-dry season (August to October) when daily maximum temperatures can rise up to 34°C and a hot-wet season lasting from November to April. The long-term mean annual rainfall stands at 754 mm with a range of 460–1080 mm. This high rainfall variability makes Mazabuka and the rest of the southern half of Zambia to be very prone to droughts and floods.

Research Approach

Structured questionnaires and focus group discussions were used to collect primary data from government officials and household members in the Munenga and Kalama wards of Mazabuka district. Secondary data were collected from various literature sources. A three-day workshop was also held for thirty Mazabuka farmers and villagers. Various participatory research techniques were used at this workshop to solicit information on traditional early-warning systems and other aspects of climate change adaptation.

Historical rainfall data for Magoye meteorological station in Mazabuka were collected from the Meteorological Department of Zambia. The occurrence of floods and droughts in Mazabuka was determined based upon the calculation of the drought and flood severity indices using the following equations:

$$\mathrm{RI} = (\mathrm{P} - \overline{\mathrm{P}})/\mathrm{S}$$

Where RI is the rainfall index, P is the annual rainfall and $\overline{\mathrm{P}}$ is the long term average rainfall of the time interval (1960–2008) and S is the standard deviation of the series of data. The RI was classified to define various categories of drought/flood intensities (severity indices).

Results and Discussion

Productivity of Indigenous Pre-Colonial Farming Systems

Written records of the history of indigenous farming practices of the Tonga people of Mazabuka and Southern province do not exist. What is known, therefore, about these systems comes from the writings of early Europeans who visited these territories and also from what has historically survived either orally or through livelihood practices.

The Tonga people originally practiced shifting cultivation which they combined with cattle rearing, before the coming of colonial rule. The major crops that they cultivated on fertile plateau soils were local maize, sorghum and pulses (Allan 1945). However, when they started to apply cattle manure in their gardens, they were able to develop a more elaborate and stable system of cultivation and fallowing that enabled families to work the same land for several decades. According to Trapnell and Clothier (1937), the Tonga evolved a system of crop rotation whereby maize, finger millet, sorghum and bulrush millet were rotated with groundnuts and other legumes that were nitrogen fixing.

The focus group discussions also revealed that traditionally the Tonga practised pot-holing or minimal tillage as a form of indigenous soil conservation tillage system. Pot-holing essentially involves slashing the vegetation or stover, leaving it

Table 1 Maize production/demand balance of southern province (1980–2000)

	1980	1990	2000
Demand (t)	107,520	151,360	208,426
Production (t)	470,439	287,456	200,574
Balance (t)	362,919	136,096	−7,852

Source: MAFF 2001

on the ground to dry and burning it to leave a clean seedbed. Sowing is then done without disturbing the soil, except for the planting holes that may be made by using a digging stick or hoe (Shetto 1999). The system is labour intensive and the Tonga had to create the required labour through polygamous marriages. These indigenous livelihood strategies ensured relative food security for Tonga society at a subsistence level, before the advent of colonial rule (Kajoba 2008).

In fact minimal till systems can improve soil quality, conserve water, reduce soil erosion and quadruple carbon sequestration in cropland soils. Globally, estimates indicate that technically about 6.0 Gt of CO2e in emissions could be reduced through less tillage of soils (World Bank 2010).

Impact of Western Practices on Local Food Security

Zambia became part of the British Crown in 1891 and by 1903 the first European farmers began to settle in Tonga territory of Southern province, especially around Kalomo, Choma and Mazabuka where land was fertile and free from tsetse fly. The settlers began to create native reserves for local people in order to set aside more land for further European settlement (Palmer 1973). In the case of Mazabuka this implied relocating local people from the fertile plateau soils to the sandy and less productive areas of the Kafue floodplain.

The settlers introduced what is referred to today as conventional tillage or farming techniques of using tractors, ploughs, cultivators and oxen to till the land. They produced hybrid maize and cattle on a commercial basis. The Tonga people adopted these new agricultural technologies such as hybrid maize, use of chemical fertilisers, and modern implements (especially ox- drawn ploughs), while indigenous farming techniques were on their way out (Kajoba 2008).

The adoption of European farming techniques by the Tonga communities greatly contributed to the generation of agricultural incomes and increased food security. The Tonga people were able to produce maize beyond basic subsistence requirements, and they even competed with European settler farmers for the maize market (Muntemba 1980). However, this trend continued until 1990 and by the year 2000 Southern province was unable to feed all its citizens as maize deficits began to occur (Table 1).

In fact Southern province (Tonga territory) became the bread basket of Zambia after independence in 1964, as it reigned as the highest producer of maize (staple

Table 2 Maize production by province in Zambia

Province	1981		1992 (Drought year)		2008	
	Production (t)	Yield (t/ha)	Production (t)	Yield (t/ha)	Production (t)	Yield (t/ha)
Central	465,300	2.99	172,794	1.70	313,694	5.00
Copperbelt	43,326	1.40	32,348	1.61	104,748	5.00
Eastern	238,590	1.60	82,317	0.32	267,596	4.00
Luapula	21,130	1.48	29,027	2.26	40,008	4.60
Lusaka	43,830	1.27	39,470	0.82	40,692	5.04
Northern	53,649	1.34	71,983	1.97	171,232	5.19
N/Western	11,655	1.17	16,602	1.39	60,561	4.84
Southern	553,230	4.03	25,214	0.19	106,891	2.17
Western	54,306	1.08	13,733	0.33	36,007	2.99
Zambia	1,485,016	2.39	483,492	0.70	1,141,429	1.95

Source: (MAFF 1997; CSO 2009)

crop) in the country. It is shown in Table 2 that by 1981, Southern province recorded the highest production of maize, accounting for 37% of the total production in Zambia. However, by 2008 this production had declined to only 9% of total production (Table 2).

While maize yields for other provinces have actually improved, it is evident from Table 2 that average yields for Southern province have significantly dropped from 1981 to 2008 by about 46%. In fact the maize yields shown in Table 2 are averages of both commercial and small-scale (peasant) production. The actual yields for peasant farmers, who are in the majority, are considerably lower than those presented in Table 2 above. The average production yield for small-scale (peasant) farmers is 1.2 t of maize grain per hectare as compared to that of commercial farmers of about 6 t/ha.

There are many reasons for the progressive decline in land productivity of the Southern province. However, one of the main causes is the severe land degradation resulting from the massive adoption of European farming techniques in the province. One critical aspect of these techniques is the requirement to uproot trees to enable mechanised tillage of land. Today, Southern province has the highest loss of land cover in the country resulting from deforestation through agricultural expansion. In 1986 the province had already recorded the highest annual deforestation rate in the country and was seconded by Lusaka (capital city) which was losing forests due to urban expansion (Fig. 2).

The incompatibility of European conventional tillage practices with local agro-ecological conditions in Zambia was also accentuated during the drought of the 1991/1992 rain season. Southern province, the birthplace of European agriculture in Zambia, suffered the worst crop losses amongst all the provinces of Zambia. It recorded the lowest maize yields of only 0.19 t/ha during the 1991/1992 drought as compared to the mean yield of 4.03 t/ha in 1981 (Table 2 above). This drought caused a 95% crop failure in Southern province. In fact Kajoba (2008) noted that the modernisation of agriculture that left the Tonga dependent on one single cereal

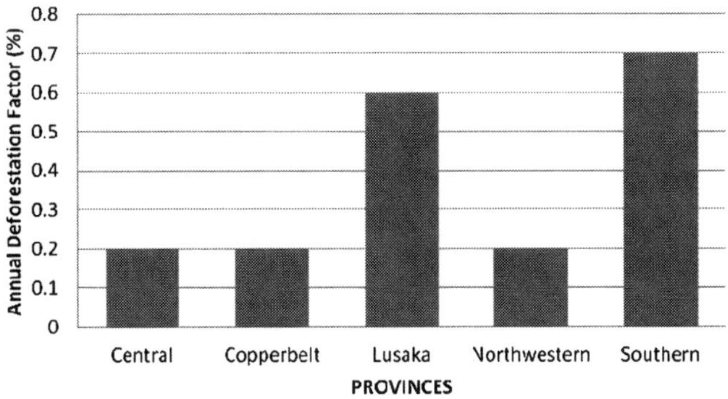

Fig. 2 Annual deforestation rates of provinces in 1986. *Source*: (MTENR, 1996)

of hybrid maize, which was less tolerant to drought, was a built-in factor of vulnerability to future environmental changes or shocks.

Recently, during the 2004/2005 agricultural season, the entire southern half of Zambia was hit by a drought causing a 56% drop in maize cereal production. In Mazabuka 82, 019 persons were at risk of starvation and required 7,874 t of cereal as relief food (ZVAC 2005). On the other hand, the floods of the 2007/2008 rainy season resulted in staple crop (maize) and cash crop (cotton and groundnuts) average losses of 53 and 45%, respectively for Mazabuka district. Estimates of income and food stock losses ranged between 40 and 60%. Additionally, cattle, goats, pigs and chicken died due to floods in Mazabuka (ZVAC 2008).

Overall, the experience of Southern province and Mazabuka district in particular has demonstrated that conventional tillage practices are not sustainable over the long term, especially when they are applied to rain-fed agriculture by resource-strapped communities in drought/flood-prone areas. The failure of conventional farming to provide long-term food security and protect peasant farmers against critical crop losses during droughts and floods triggered a search for novel climate-resilient approaches to farming in Zambia.

Combating Climate Change with Conservation Farming

It is now widely recognised in the southern and eastern African regions that conventional tillage, coupled with mono-cropping and bad husbandry practises leads to land degradation and to a situation where the soil can no longer support crops. There is no doubt also that the accumulative effects of inappropriate and unsustainable farming methods have exacerbated the effects of droughts in southern Zambia. Studies have shown that conventional tillage practices result in average annual losses of top soil amounting to about 35 t/ha, which is followed by a decline in organic matter, nitrogen and phosphate contents of the soil. In the case of maize, yields can decline by 4–11% annually (Siachinji-Musiwa 1999).

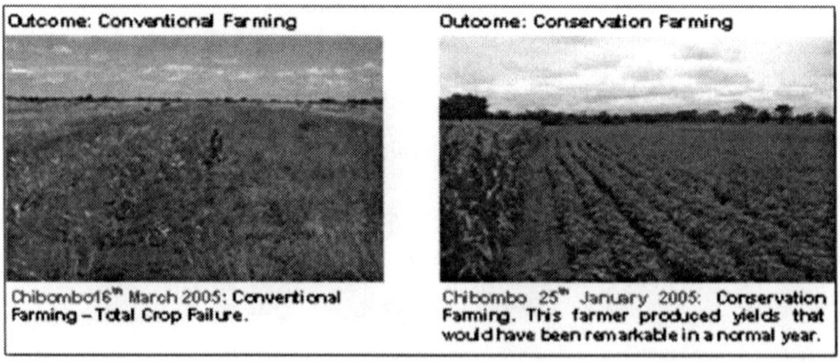

Fig. 3 The resilience of conservation farming to the drought of 2004/2005. *Source*: (ZNFU and GART 2007)

The most promising alternative to conventional farming is conservation farming which is based on a combination of indigenous/traditional and scientific crop and soil management practices of food production under fragile tropical conditions. The indigenous components of conservation farming include pot-holing (minimum tillage), mulching, crop rotations, intercropping, fallow cropping, compost manure practices and agroforestry. These practices have been with African communities in one form or the other for centuries. The scientific aspects pertain to knowledge about the biophysical properties of soils, soil organic matter formation, crop nutrient uptake and cycling and other agronomic aspects of agriculture. Actually, the advantages of conservation farming include reduced soil erosion due to minimum soil disturbance, increased soil biological activity due to adequate soil organic matter input, improved soil moisture retention due to in situ water harvesting and storage, increased soil organic matter content, as well as savings on labour input.

Conservation farming is reputed to be a more productive, efficient and environmentally sustainable way of farming. Its benefits occur in all years but are most dramatic in years of droughts or seasons of erratic rainfall distribution. The potential of conservation farming to improve the adaptive capacity of farmers to increased climate variability and change was demonstrated during the drought of the 2004/5 rainy season.

Farmers who practised conservation farming survived the drought while conventional farmers suffered almost total crop failure (Fig. 3). Conservational farming, when correctly and properly applied can actually increase maize yields for peasant farmers from the typical 1.2 t/ha to 6.5 t/ha and above (ZNFU and GART 2007). A study of conservation farming (CF) in Zambia found that increased maize yield after CF adoption was recorded among 65.7% of the respondents and that the gain in yield was three times more than the yield from conventional farming. It was, therefore, concluded that CF constitutes currently one of the major keys to increasing crop yield and productivity in Zambia (Kabamba and Muimba-Kankolongo 2009).

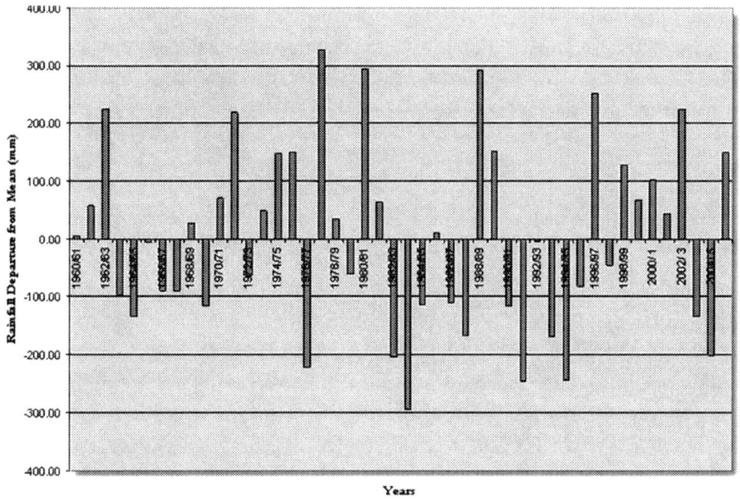

Fig. 4 Rainfall anomalies for Mazabuka. *Source*: Data obtained from department of meteorology, Zambia

The current target for Zambia is to increase the adoption of conservation farming from 120,000 to 240,000 by 2011. Indeed, it is evident by now in the country that it pays to integrate indigenous and scientific knowledge for purposes of enhancing the adaptive capacity of communities to climate change. It is also of critical importance to ensure that the adoption of any climate change adaptation technology is not only based on productivity but also on its ecological compatibility and sustainability.

Indigenous Indicators for Drought/Flood Prediction

Zambia has so far only 36 meteorological stations for a Zambia surface land area of 752,614 Km^2. This inadequacy in meteorological coverage of the country has compromised the accuracy and reliability of weather forecasts. During the focus group discussions and the participatory workshop, farmers complained that even though they listened to forecasts over the radio, they did not use them to plan their agricultural activities. They claimed that they found discrepancies between the official weather forecasts and actual rainfall events in their communities. They demanded to have weather forecasts that would be specific for their villages.

On the other hand, the people of Mazabuka have over decades suffered various episodes of droughts and floods as shown in Fig. 4 where negative and positive rainfall departures (anomalies) from the mean are indicative of the occurrence of droughts and floods, respectively. It was, therefore, likely that these people should have developed some form of traditional means of forecasting these climate extremes.

Focus group discussions with local communities in Mazabuka revealed that the people used some form of indicators as early warning systems to forecast the type of rainy season expected in a given year. The indicators that were mentioned by the communities are described below.

Temperature of Rock Formations

It was mentioned that in the olden days at a sacred site called *Ngombe Ilede* there was a rock formation which was used to predict whether the coming rainy season will be bad or good (normal). A group of experienced elders would go to the sacred site just before the onset of the rainy season and begin touching the rock formation at various points. The temperature of the stone would then be used to forecast the drought or normal rainfall for the coming season and this was communicated to the people. Currently, scientists are using sea-surface temperature levels to predict the El Niño phenomenon that causes droughts in Southern Africa. It can not thus be totally dismissed that one can use the heat content of geological formations for rainfall forecasting. Moreover, these formations, just like oceans, are a vital component of the energy budget of our solar system.

Excessive Flowering of Fruit Trees

The communities also observed that when they saw mango and wild fruit trees flowering excessively in a particular season then it was an indicator that the rainy season will be bad for food production. This observation was voiced by both young and elderly people who came from very different localities. In Zambia mangoes begin flowering in July which is about 4 months before the onset of the rainy season.

A study in neighbouring Tanzania (Chang'a et al. 2010) also found that local people used plant phenology to forecast rainfall, whereby a higher than normal flowering intensity of the *Erythrina abyssinica* trees during the months of July to November was indicative of good amount of well distributed rainfall in the upcoming season. However, a good fruits harvest from *Uapaca kirkiana* trees was a signal of impending drought in the upcoming season.

Direction of Winds

Specific wind directions were also identified as forecast indicators of droughts in the area. It was reported that the predominance of certain trade winds just before the rainy season was a sure indicator of poor rainfall in the coming season. Coincidentally, the national meteorological service also associates certain trade winds with either wet or dry weather.

Outbreaks of Flying Insects

Flying insects which were given a local name were also implicated as indicators of poor rainfall in the coming season. It was reported that the proliferation of these insects always signalled a poor harvest due to erratic rainfall or drought. This appeared to be a widely known phenomenon in the community. In fact the Tanzanian study (Chang'a et al. 2010) also found that the occurrence of more grasshoppers (*Hesperotettix* sp) in a particular year indicated less rainfall and hunger.

The indigenous indicators mentioned above appeared to be common knowledge in the community. However, their reliability and usefulness is questionable as nothing as yet been identified to show that communities respond to these signals and take steps to prepare for the impending climatic hazards. However, one underlying element in the indigenous rainfall forecast indicators reported by communities is the notion that the amount of rainfall in a given year can be forecast or projected based on observations of ecological processes occurring on Earth. It is scientifically well known that migratory birds and insects have genetic triggers that respond to photoperiod and weather patterns. It is, therefore, possible that plants and insects, whose reproduction cycles are closely linked to climatic episodes, may have evolved sensors for prior warning of rainfall conditions in the coming season. Additionally, it is a scientific fact that climatic and indeed rainfall patterns are largely influenced by global energy flows in the ground, air and waters of the Earth (Nordell and Gervet 2009). Surely, these indigenous indicators for drought/flood prediction deserve serious scientific investigations as to their credibility and validity.

Conclusion

The ultimate goal of all human knowledge is to ensure the survival and perpetuity of the human race on Earth. In this regard all knowledge is useful. In the African context, indigenous knowledge serves as a reservoir of undocumented technological and cultural experiences of communities arising from direct interaction with their external environment. This paper has shown that blind technology transfer without due consideration of the prevailing socio-ecological and climatic conditions can prove disastrous for local communities. Care should, therefore, be taken to ensure that international technology transfers for purposes of climate change adaptation do not turn out to be maladaptations for recipient countries. The case study of conservation farming has also demonstrated that the utility value of indigenous knowledge can be greatly enhanced through integration with scientific principles. Finally, current climatologically-derived drought/flood forecasting systems have proved very inefficient in averting starvation, poverty and misery amongst African communities whenever droughts or floods strike. The efficiency and efficacy of these systems could surely be improved by incorporating elements of indigenous forecasting systems.

Recommendations

Indigenous knowledge is generated and acquired through the co-existence and evolution of both human beings and the environment of a given locality. Disregarding or losing it is tantamount to creating a dent in the human development trajectory of any given society, because more often than not imported knowledge systems place demands on the environment that are not commensurate with its supply capacity. However, it is also acknowledged that the viability and utility value of indigenous knowledge can be greatly enhanced when integrated with scientific approaches and know-how. It is, therefore, justified to make the recommendations presented below.

- Multilateral conventions and protocols like the United Nations Framework Convention on Climate Change (UNFCCC) that has policy directives of technology transfer for climate change adaptation must include criteria for such transfers to enhance indigenous technical systems so as to avoid maladaptation in the long run.
- Indigenous knowledge is usually not written and is, therefore, disappearing in many parts of the world. It is thus imperative to develop an international network for documenting this indigenous knowledge in various socio-economic spheres around the world.
- Most national meteorological services in Africa have developed Radio and Internet technology (RANET) for disseminating climate-related and development information to rural communities. It is important that RANET includes indigenous knowledge components in the broadcast messages to improve accuracy and reliability of information, and;
- It is incumbent on all African countries to implement research into the scientific basis and validity of indigenous climate forecasting systems.

Acknowledgments The author is very grateful for the financial and material support received during the implementation of this study from the Climate Change Adaptation in Africa (CCAA) Research and Capacity Development Programme, funded by IDRC-CCAA, Canada/Nairobi, the International Institute for Environment and Development of London, The Copperbelt University and Energy and Environmental Concerns for Zambia.

References

Allan W (1945) Land holding and land usage among the plateau Tonga of Mazabuka district; a reconnaissance survey, Rhodes, Livingstone Institute No 14. Manchester Univ. Press, Manchester

Chang'a LB, Yanda PZ, Ngana J (2010) Indigenous knowledge in seasonal rainfall prediction in Tanzania: a case of the South-western Highland of Tanzania. J Geogr Reg Plann 3(4):66–72

Kabamba H, Muimba-Kankolongo A (2009) Adoption and impact of conservation farming on crop productivity among smallholder farmers in Kapiri Mposhi district of Zambia. J Anim Plant Sci 3(2):205–214

Kajoba GM (2008) Vulnerability and resilience of rural society in Zambia: From the view point of land tenure and food security, Working Paper on Social-Ecological Resilience Series, No 2008-003. Research Institute for Humanity and Nature (RIHN), Japan

Kalonga C (2004) Mazabuka district environmental policy situation analysis. Report for the Ministry of Tourism, Environment and Natural Resources, Kwacha House, Lusaka.

MAFF (1997) 1995/1996 Agricultural statistics bulletin. Ministry of Agriculture, Food and Fisheries, Lusaka

MTENR (1996) Zambia forestry action programme (ZFAP). Ministry of Tourism, Environment and Natural Resources, Lusaka

Muntemba MS (1980) Regional and social differentiation in Broken Hill rural district of Northern Rhodesia, 1930–1964. In: Klein MA (ed) Peasants in Africa: historical and contemporary perspectives. Sage Publishers, London

Mwale A (1997) The other route: an ITK based approach to sustainable land management. Department of Field Services, Kabwe, Zambia

Nordell B, Gervet B (2009) Global energy accumulation and net heat emission. Int J Global Warming 1(1/2/3):378–391

Palmer R (1973) Land in Zambia. In: Palmer R (eds) Zambia land and labour studies, vol 1. National Archives Occasional Paper No. 17

Shetto RM (1999) Indigenous soil conservation tillage systems and risks of animal traction on land degradation in Eastern and Southern Africa. In: Kaumbutho PG, Simalenga TE (eds) Conservation tillage with animal traction. A resource book of the Animal Traction Network for Eastern and Southern Africa (ATNESA), Harare

Siachinji-Musiwa JM (1999) Conservation tillage in Zambia: some technologies, indigenous methods and environmental issues. In: Kaumbutho PG, Simalenga TE (eds) Conservation tillage with animal traction. A resource book of the Animal Traction Network for Eastern and Southern Africa (ATNESA), Harare

SO C (2001) National census of population and housing. Central Statistical Office, Lusaka

SO C (2009) Zambia crop production for the 2007/2008 agricultural season. Central Statistical Office, Lusaka

Trapnell CG, Clothier JN (1937) The soils, vegetation and agricultural systems of North western rhodesia: report of the ecological survey. Government Printer, Lusaka

World Bank (2010) Development and climate change. World Development 2010 Report, Washington DC, p. 171, ISBN 978-0-8213-7988-5

ZNFU, GART (2007) Conservation farming and conservation agriculture, handbook for hoe farmer s in agro-ecological regions I & IIa—flat culture, 2007 edition, conservation farming unit, Zambia national farmers union and golden valley research trust, Lusaka/Chisamba

ZVAC (2005) Vulnerability and needs assessment, Final June 2005 VAC Assessment Report, Zambia Vulnerability Assessment Committee, Disaster Management and Mitigation Unit, Office of the Vice-President, Lusaka

ZVAC (2008) Southern province rapid flood impact assessment report, January, Zambia Vulnerability Assessment Committee, Disaster Management and Mitigation Unit, Office of the Vice-President, Lusaka

Author Biography

George Kasali is a Senior Lecturer in the Department of Biological Sciences, School of Mathematics and Natural Sciences, The Copperbelt University, Kitwe, Zambia, with a PhD in Environmental Biotechnology from Strathclyde University in Scotland. His areas of expertise include environmental microbiology, health and climate change, climate change mitigation and adaptation, energy and food security. He has been involved in a number of climate change and environmental research and consultancy projects for government and international agencies. He has already over 15 scientific publications in international journals with reviewers, announcements in international conferences and several articles published in bulletins and books.

Chapter 19
Building Nigeria's Response to Climate Change: Pilot Projects for Community-Based Adaptation in Nigeria

Ellen Woodley

Abstract Throughout history, human societies have had to effectively devise ways and means to adapt to climate variability by altering their lifestyles, agriculture, settlements and other critical aspects of their economies and livelihoods. The capacity to adapt enables societies to deal with a range of uncertainties. Coping and adaptation is a way of life in Nigeria, where climate variability is the norm and where planting cycles in a largely rain-fed agricultural system are affected by reoccurring droughts, floods and other extreme weather events. Climate change scenarios for Nigeria suggest a warmer climate and projected changes in precipitation suggest it will be wetter in the south along the coast and drier in the northeast. The climate models also suggest more extreme heat events will occur. Resource dependent people, such as farmers, hunters and fishers, who depend directly on the productivity of natural resources around them for their livelihoods, are the first to be impacted by these changes in local environmental conditions. The Government of Nigeria is a signatory to the UNFCCC and there is an initiative underway to develop a national strategy for community-based climate change adaptation. Since 2007, the Nigerian Environmental Study/Action Team (NEST) is an NGO has been implementing the project *Building Nigeria's Response to Climate Change* (BNRCC). Pilot Projects are one component of BNRCC, and are designed to test adaptation options on a small scale in order to strengthen the resilience of communities to climate change, increase their adaptive capacity and provide recommendations based on lessons learned from community-based adaptation projects to the national strategy. The projects involve seven partner organizations who are working directly with 15 vulnerable communities spanning Nigeria from the Sahel in the north east to the Coastal/Rainforest in the south east. The projects include but are not restricted to: increasing food security

E. Woodley (✉)
Nigerian Environmental Study/Action Team (NEST), Ibadan, Nigeria
e-mail: woodley.ellen@gmail.com

W. Leal Filho (ed.), *Experiences of Climate Change Adaptation in Africa*, Climate Change Management, DOI: 10.1007/978-3-642-22315-0_19,

by introducing improved crop varieties; testing alternative livelihood options such as aquaculture in order to provide a means of income and decrease reliance on dwindling forest resources; providing fuel efficient wood stoves; improving access to water sources to deal with water scarcity; and tree planting for ecosystem rehabilitation.

Keywords Community-based adaptation · Vulnerability · Livelihoods · Improved varieties · Water scarcity

Introduction

Building Nigeria's Response to Climate Change (BNRCC) is a five year project that began in 2007, finishing this year in mid 2011. It is funded by the Canadian International Development Association (CIDA), administered by Marbek Resources in Ottawa, Canada and implemented by the NGO, Nigerian Environmental Study/Action Team (NEST) in Ibadan, Nigeria. The BNRCC project has four components: Pilot Projects, Research Projects, Policy Development and Communications. The pilot projects were designed to provide community-based climate change adaptation information to a National Adaptation Strategy and Plan of Action (NASPA) for Nigeria. A Climate Change Adaptation Strategy Technical Report (CCASTR) was written as the policy component of BNRCC, which will be the resource document for the eventual NASPA. CCASTR is based on recommendations from the pilot projects, together with research projects in Nigeria and literature reviews. The report used climate change scenarios that were developed for Nigeria through a partnership between the Climate Research Analysis Group, University of Cape Town, South Africa and the Institute of Ecology and Environmental Studies at Obafemi Awolowo University, Ile-Ife. Scenarios of rainfall, temperature changes, heat waves, drought and extreme weather events were developed.

Nigeria's First National Communication (FNC) identified the nation's natural ecosystems, agricultural ecosystems and water resources including coastal and marine ecosystems, as highly susceptible to climate change effects. Specifically, the FNC and other relevant reports identified critical aspects of the nation's ecosystems that are vulnerable including:

1. The nation's over 850 km^2 coastal line which harbors highly diverse flora and fauna and important spawning grounds; and
2. Increasing drought leading to accelerated desertification and water scarcity in northern parts of Nigeria.

The report also noted that there is a lack of political will and financial resources to support adaptation measures as well as a lack of and access to requisite technology to reverse the current trend of degradation exacerbated by climate change, increasing population growth and an absence of policies to address climate change.

Climate Change Scenarios

A statistical downscaling approach was employed for simulations of nine global climate models (GCM) over 40 stations in Nigeria and two scenarios are used to assess the climate change impacts. Both models suggest a warmer climate in the future. The scenarios project a temperature increase ranging from 0.02 to 0.04°C per year from now till 2100, with one scenario predicting an increase of 0.08°C from 2010 to 2100. The coastal regions are predicted to warm less than the interior regions, due to the cooling effects from the Atlantic Ocean. The northerly regions are expected to be warmer than the south, and temperatures may be higher in February and March in the Guinea and Sudan savanna regions, with the highest increase in the northeast. The climate models also suggest more extreme heat events will occur: by 2046–2065, the number of days with temperatures reaching 38°C or more could increase by an estimated seven days per year in coastal zones and as much as 82 days per year in the Sahel.

The projected changes in precipitation vary across the country. Both scenarios suggest a wetter climate in the south along the coast, but a drier climate in the northeast. This is consistent with the increase in temperature which would result in more ocean evaporation to produce more rainfall over the coastal region. The warmer climate in the semi-arid region (i.e. northeast) would decrease the atmospheric humidity, and thereby reduce the chance of cloud formation and rainfall. The scenarios suggests a peak increase of about 2 mm/day in August over coastal and rainforest zones and about 1 mm/day in the same month over Guinea and Sudan savanna zones.

Over the coastal, rainforest and Guinea savanna zones, both scenarios project earlier rainfall onset dates and later cessation dates, hence the increase in the rainfall season could be up to two weeks. On the other hand the models predict a shorter rainfall season over the Sudan savanna, suggesting a decrease of more than one week.

Water is one of the most critical resources that determine the severity of climate change impacts in the Sahel region. The erratic short duration rainfall comes in heavy downpours and then drains off into the natural depressions infiltrating into the sandy soil while the remaining water quickly evaporates in the dry air, leaving little for human and livestock consumption. In the south, increased rainfall intensity leads to flooding. Variability in the rainy season has made it more difficult to predict the best planting dates for crops and this has also contributed to country-wide crop failure. The effects of heat stress and extreme weather events are also widely reported.

Fig. 1 Map of Nigeria with approximate locations of 15 pilot project communities (shown by red circles)

This chapter provides an overview of the pilot projects to date, presents lessons learned from the tested adaptation options and the recommendations made to the CCASTR.

Methods

Description of the Study Area

Pilot projects were implemented in fifteen communities in Nigeria (see Fig. 1), ranging from the Sahel in Yobe state in northeastern Nigeria to the coastal zone in Cross River State in the southeast.

Research Approach

At the inception of the BNRCC project, seven partner organizations were selected, based on a country-wide bid for proposals. Each partner organization chose two communities to work with (with the exception of one partner that works with three communities), for a total of 15 pilot project communities, that have a broad geographical representation in Nigeria. Communities were chosen on the basis of need: generally those that were facing the most critical water shortages, the greatest amounts of deforestation, or greatest loss of livelihoods, including crop failure. Partner organizations were the liaison between BNRCC staff and the communities, and partners made frequent and regular visits with the communities in order to facilitate the selected adaptation projects, in terms of providing resources such as knowledge, encouragement, finances and access channels to other needed resources. Social Analysis Systems (SAS) tools were applied in each community. These tools (developed by IDRC 2008) were used to ensure full community participation in selecting the adaptation options that addressed their most immediate needs due to perceived climate change impacts. The tools provided the means to elicit community natural resource use maps, an assessment of key stakeholders and their level of influence in the community and the impact of the proposed project on them. The exercises raised awareness of who in the community is not heard and the need for equality in project management and in who benefits. BNRCC staff monitored the projects through quarterly reports submitted by the partners and by two site visits to each community where focus group discussions (FGDs) were conducted. The projects were generally of 18 months duration and close to the end of the intervention by BNRCC, the partners were required to submit relevant information on the projects so that this information could then inform the technical report (CCASTR). Partners filled out tables that highlighted information on climate change hazards, impacts and adaptation options, as well as lessons learned.

As a means to increase the adaptive capacity of communities, climate change awareness workshops were organized by the partners and held in each of the communities at the onset of the project. The inclusion of local and traditional knowledge into climate change policies has also been encouraged as it can lead to the development of effective adaptation strategies that are cost-effective, participatory, sustainable and empowering for local communities (Robinson and Herbert 2001, as cited in Boko et al. 2007). A study in Nigeria, for example, shows that farmers use knowledge of weather systems such as rainfall, thunderstorms, windstorms, harmattan and sunshine to forecast weather patterns and to help plan farming activities (Ajibade and Shokemi, 2003 as cited by Boko et al. 2007). African women in particular are known to possess local knowledge which helps to maintain household food security, particularly in times of drought and famine. Similarly, the UNFCCC (2006) recognizes that people in many African countries have developed traditional adaptation strategies to face the great climate variability and extreme events. Monitoring the pilot projects has facilitated direct learning from on-the-ground activities undertaken by some of Nigeria's most vulnerable communities. This has enabled the policy recommendations made in the CCASTR to be directly informed by what is happening at the local level, thus supporting a bottom up approach and developing policy based on local realities.

The role of institutions has also been considered important in the pilot projects. The capacity to adapt is dependent on the role of traditional and new institutions. Research suggests that their role has been diminished by socioeconomic changes and government policies. As community members become further alienated from decisions about their local resources, dependence on government interventions tends to increase and poverty alleviation becomes more difficult (AIACC 2007). One of the key components of the pilot projects was to build in sustainability factor by ensuring the development of sustainable institutions. This began with the formation of Project Implementation Committees (PICs) which, made up of both men and women in management and decision making roles, was a precursor to Community-Based Organizations (CBOs). The CBOs, organized towards the end of the project cycle, are to oversee the existing project, to procure more funding arrangements and to liaise with local and state governments on expanding the project and developing policy on CBA.

In an attempt to begin to scale up project information, BNRCC employed the following three strategies to share the knowledge and learning so as to build lateral connections (with other communities) and to share experiences laterally with local, state and national decision makers:

1. *Peer education for lateral dissemination of information.* The lateral method of learning occurs at grass roots and involves using peer learning to share information and lessons learned using peer groups to pass information to the local populace. For example, focus groups among women's groups or youth groups or youth drama groups for community animation;
2. *Regional workshops.* Workshops held at the regional level link the state government and its relevant agencies and local government councils and other relevant stake holders. It is important to improve access to information and resources and ensure the continuity of projects after funding. Two regional workshops were held to share the findings from the pilot projects among the different stake holders and to create a link for the direct flow of information including small scale farmers.
3. *National workshop.* BNRCC intends to involve people from all the relevant sectors at the national level including politicians in key ministries, in an awareness raising workshop where relevant information and where lessons learned from the project and case studies will be shared.

Raising awareness on climate change and its local impacts among the people at that level is important since local government is often limited.

Definitions

Community-Based Adaptation (CBA) projects (adapted from CARE 2010) have, as their primary objective, to improve the capacity of local communities to adapt to climate change. This requires an integrated approach that combines local knowledge with innovative strategies that address current vulnerabilities. CBA projects also build the resilience of people to face these new challenges, and aims to protect and sustain the ecosystems that people depend on for their livelihoods. To build adaptive capacity, the process should incorporate four inter-related strategies: (1) promotion of climate-resilient livelihoods, including income diversification and capacity building for planning and improved risk management, (2) disaster risk reduction to reduce the impact of hazards, and (3) capacity development for local civil society and governmental institutions so they can provide better support to communities, households and individuals in their adaptation efforts; and advocacy, social mobilization and empowerment to address the underlying causes of vulnerability.

Adaptive capacity is the ability of a system (human or natural) to adjust to climate change, including climate variability and extremes, to moderate potential damages, to take advantage of opportunities, or to cope with the consequences (IPCC 2001).

Resilience is the ability of a system (human or natural) to resist, absorb and recover from the effects of hazards in a timely and efficient manner, preserving or restoring its essential basic structures, functions and identity (UNISDR 2009).

Results and Discussion

The accounts of the pilot projects illustrate the efforts of rural communities to determine the best adaptation options for them, and they also attest to their ability to fundraise, provide labour and the energy to ensure that the projects are put in place and that the knowledge derived from them is shared with others, both policy makers and other communities. The focus of the pilot projects is on empowering communities to take action on climate change impacts, based on their own decision making processes thus increasing their adaptive capacity. These bottom up community-based initiatives, discussed in this chapter are in contrast to the many 'top–down' energy-based interventions, which commonly dominate climate negotiations (Mitchell and Tanner 2006). The key adaption options implemented by the BNRCC pilot projects, from which recommendations were made to the CCASTR, are projects that promote meeting basic needs such as access to quality food and water, alternative livelihoods and ecosystem rehabilitation. Thus they are not unlike other "development" projects unless the connection is made specifically between climate change impacts and increased vulnerability. Some case studies, drawn from the 15 communities, are described below and grouped by ecozone. Appendix 1 lists the projects by the sector they address and the partners that are implementing them. Discussion is framed by the adaptation options the communities are testing and the lessons learned to inform policy on community-based adaptation.

Sahel

Improving Livelihoods and Stabilizing Sand Dunes in Two Communities in the Sahel of North Eastern Nigeria

Since December 2009, a team at the University of Maiduguri (UNIMAID) has been working with two communities, Tosha (pop. 5,000) and Sansan (pop. 400), in the Sahel in northeastern Nigeria on pilot projects which aim primarily to stabilize mobile sand dunes which are encroaching on farmland and houses; to improve the livelihoods of people most impacted by increasing aridity by reviving dry oases; and improve crop yield by introducing improved varieties.

The low annual rainfall in north eastern Nigeria, which varies from 250 mm annually in the extreme north to 750 mm in the south, makes this already arid region in Nigeria particularly vulnerable to climatic change and further ecological degradation. With the aridity, together with the exploitation of land through over cultivation, overgrazing, and over cutting of trees along with more variable rainfall due to climate change, the processes of desertification are escalating. The increasingly arid conditions in this region have caused many of the local oases around the community of Tosha to become completely dry. Of the 13 once thriving oases, now only three have water. Originally the oases had rich fertile soils and the seasonal fluctuations of rain and evaporation afforded the harvesting of commercially valuable potash. There were fertile uplands with forests and luxuriant grasslands, which supported abundant wildlife. These resources provided the early inhabitants in the region with enough resources for good livelihoods. Conditions started to decline after the 1972 Sahel drought and by 1980, environmental degradation was evident by the appearance of prominent sand dunes close to the village. These invading mobile sand dunes are now encroaching on farmland and grazing lands in both communities of Tosha and Sansan. In Tosha, the dunes have started to push into the community and destroy dwellings, with one death already reported from shifting sands crushing a home. Farmers in Sansan report that farm yields have decreased over the last 40 years as sand dunes have encroached on arable land used in the wet season. The dunes at Sansan are also partly the result of intensive farming activities near a natural water-holding depression. The presence of water in the depression attracts livestock to the area, thus triggering the process of dune formation, which is then exacerbated by aridity and severe weather (drought and sandstorms).

The Project

The project has started to re-establish oasis farming and livestock rearing through the use of shallow boreholes for irrigation and introduced improved crop varieties that withstand drought conditions. The project is reforesting sand dunes to both stabilize the dunes and provide sources of fuel, food and fodder. The use of biogas is being tested as an alternative to fuel wood but the biogas stoves are still in the experimental stage, with the units constructed and some initial training conducted. It remains to be seen if this alternative energy form will be helpful to these communities, especially for women who walk long distances to find and carry fuel wood, in a region where fuel wood scarcity has reached critical levels.

Early results show that the boreholes are reliable and dry season crops were successfully grown in the 2010 season, something which hasn't happened for years. The wells were also a water source for domestic purposes and for livestock. Over 800 head of cattle used the borehole every day at one oasis during the peak of the dry season. Improved varieties of cowpea, sorghum, millet groundnut, maize and rice were introduced. They are early maturing, drought resistant, high yielding, resistant to pests and diseases and, so far, acceptable to the communities. Yields were good this year, as reported by all farmers who tested them. Farmers are learning to record planting times, harvest times and yield, so that by next growing season, the yield of the introduced and the local varieties can be compared. Farmers are also being encouraged to grow resilient wild food tree species of high traditional value both on farm and in home compounds.

Sand dune stabilization is the other main thrust of the project and has been initiated with the planting of 15,000 seedlings, by community members, of the early colonizing, fast growing *Prosopis juliflora* on dunes that are threatening the communities and farmlands. This species was chosen by the communities as the most acceptable species to provide wood, food and livestock feed. Two months after planting, there was a 90% survival rate of the seedlings. Time will tell how the seedlings will survive the upcoming dry season in 2011 and if there is any danger of this non native species becoming invasive in this arid region.

Sudan and Guinea Savanna

Strengthening Community-Based Adaption to Climate Change by Increasing Food Security in the Sudan and Guinea Savanna Regions in Northern Nigeria

Since September 2009, a team at Abubakar Tafawa Balewa University (ATBU) in Bauchi State has been working with three rural communities, located in the Sudan savanna (Gorori, pop. 5,000) and the Guinea savanna (Bursali pop. 6,000 and Billeri pop. 3,000) to increase food security and reduce dependency on fuel wood. Rainfall data from the Bauchi meteorological station show that there is a sustained reduction in rainfall and year to year variability has been increasing since the 1970s. This adds uncertainty and risk to crop planting. Many farmers have to plant more than once, and when the first planting fails there is the risk of depleting the seed stock for that season, in addition to the increased labour of planting more than once. Overall, crop productivity is declining. For example, the women in Gorori observed that only ten years ago they could harvest ten bags of cowpea, but now they harvest very little or sometimes nothing at all. Women in Billeri say that they are getting only one quarter of the cowpea yield that they used to get.

The Project

The projects were designed to increase food security by introducing and testing improved varieties of the main crops of sorghum, millet, and cowpea; to assist in weather forecasting so planting times are more accurate and timed to consistent rainfall; and to reduce the demand for fuel wood by introducing and training women on the use and fabrication of fuel efficient stoves.

Food Security

Five women in each community were selected to receive improved seeds of cowpea and men were selected for improved seeds of millet and sorghum. These early maturing varieties have higher nutrient requirements, necessitating the use of fertilizers, inputs which are difficult to sustain year after year. Many farmers realize the benefits of organic fertilizers and their use is encouraged by the project. Yields of all three crops were above average in the 2010 growing season and all matured earlier, shortening the period of hunger that occurs before harvest each year. Some refinement in crop management is needed, however. The early maturing millet was subject to attack by birds, since it was the first available grain to mature in the area. The cowpea was subject to severe aphid infestation, more than the local, traditional variety, necessitating the regular application of pesticides. The women are in the position of having to decide whether to plant the local variety of cow pea the next growing season or the improved variety. If the higher yield of the improved variety is to be realized, then there is a need for an aphid-resistant variety of cow pea.

Risk Management: Weather Forecasting

Weather forecasters were given to four groups of five male farmers–making 20 direct beneficiaries–in each of the three communities. These small, easy to use hand held forecasters, help predict rainfall patterns, and were supplemented by "Drought Decision Support Tools". This is a software program used at ATBU to predict the best time for planting specific crop varieties, based on 30 years of local rainfall data and crop requirements. Used together, these tools ensured that farmers planted when rains were consistent enough to ensure germination and sustained growth. The farmers who used these tools did not have to replant this year as they have had to in the past when germination failed.

Energy Solutions: Fuel Efficient Wood Stoves

The reliance on the dwindling supply of trees for fuel wood has drastically reduced forest cover in Nigeria. To reduce the demand for fuel wood and to reduce women's labour, fuel efficient wood stoves, built by a local fabricator, were introduced to 20 women in each community. After two months of testing the stoves, the women were unanimous in stating that, while the stoves consumed less wood and the amount of smoke was reduced, they needed modification to meet their specific needs. In order to handle large pots of thick stew made of maize, called *tuwo*, the stove needs to be larger and more stable for them to want to use it over the traditional three stone stove.

Community-Based Climate Change Adaptation in Two Communities in the Guinea Savanna of Plateau State

Since September 2099, the Nigerian NGO, Catholic Archdiocesan Rural and Urban Development Program (CARUDEP), has been working with two communities, Dashe and Kwaikong, in the Guinea savanna to address issues of water scarcity and food insecurity. The higher elevation of the region around Jos in Plateau State has, in the past, ensured a climate cooler than most other regions in Nigeria. Nigerian Meteorological agency data (NIMET) from the state show that the annual average temperature has increased over the last 24 years, indicating that Jos and environs are warmer today than they were in the past. Farmers also report greater variability in amount and timing of rainfall, which affects the timing of planting and the length of the growing season. The water table is now lower and surface water such as dams, ponds and streams and rivers has been significantly diminished. This means that women and children have to travel up to 4 km to collect water. The people of Kwaikong have resorted to selling firewood for water, which has contributed to local deforestation. Schools have been forced to close because of lack of water. Higher temperatures have contributed heat stress leading to increased diseases in humans and livestock and lower crop yields. Farmers in the Kwaikong area have lost a high percentage of yam seedlings (the staple crop) due to intense heat and unpredictable rains. Women report a decline in their guinea corn harvest in the last five years, now only harvesting 20 "ties" compared to 100 they produced in years past.

The Project

The pilot projects targeted a total of 300 farmers in both communities, who were selected to receive five improved seed varieties, trees for planting initiatives and training on pest/disease control of crops. Eighty youth were also trained in water resources management, including construction of rainwater harvesting units and water filtration systems.

Water Resources Management

In Kwaikong, the dam reservoir was deepened in order to supplement dry season water supply. A new rainwater catchment system was installed so there are now two 46,000 l ferrous cement tanks in addition to a restored underground storage tank of 162,000 l, all of which are filled from eaves troughs installed on zinc roofing. Thirty household water purification systems were also distributed in Kwaikong: these are simple, easily constructed units, with filters made from sand and charcoal.

Food Security: Improved Crop Varieties

With low soil fertility and water scarcity threatening crop productivity in both communities, improved varieties of maize, cow pea, groundnut and soya bean were introduced. Results were very good in the 2010 growing season, with all four crops producing higher yields than the local variety. The use of fertilizers remains contentious, however, because of the dependence on purchased fertilizers and pesticides that are required for improved varieties. One community traditionally used cow dung and household waste to improve soil fertility but began to depend on purchased inorganic fertilizers after they were encouraged by the government. CARUDEP held workshops to encourage the use of organic fertilizers. They have also trained farmers on using soya bean to control the parasitic vine, *Striga asiatica*. In addition to improving overall fertility, soya bean reduces *Striga* infestation, since this parasite does not grow well on fertile soils.

Coastal Rainforest

Alternative Livelihood Options to Promote Adaption to Climate Change in the Rainforest/Coastal Ecozones of Cross River State, Nigeria

Since September 2009, the NGOs Coastal Life Initiative (COLIN) and Development in Nigeria (DIN), both based in Calabar, Cross River State have been working with four rural communities to reduce vulnerability by increasing awareness of resource over-use and climate change impacts, to increase food security and reduce dependency on wild fisheries, mangroves and the rainforest. The coast in this state is characterized by a large expanse of mangroves, which are important as buffers to the direct impact from storms and associated erosion from severe weather, as well providing habitat for diverse aquatic species. Mangroves are threatened by overuse, sea level rise and by the invasive species, Nipa palm, which has taken over much of the coastline in Cross River State. There is great concern in the coastal communities that not enough is being done to save the mangorves from the impacts of the invasive species, Nipa palm. The pilot project did not have enough funding to carry out a long term mangrove restoration project.

This state is also home to some of the last remaining stands of rainforest in Nigeria and the vast majority of the rainforest and derived savanna communities in Cross River State still rely on local forests for food products, including bush meat, and medicine (non timber forest products or NTFPs). Many forest animals, such as chimpanzees and gorillas, are now threatened or locally extinct.

In the two coastal communities, fishing supplements agriculture as the main livelihood activities. In the two rainforest communities, the collection of NTFPs and agriculture are the main livelihoods.

Fishing is traditionally men's activity and occurs in both the river estuaries and the open sea. The women then market the catch. Fishers in these communities are now reporting low fish stocks, the cause of which is not necessarily due to climate change, but other factors, including competition with fishers from outside the community and competition from those with bigger engines who are able to get to more areas quickly. There is also an increase in the number of people fishing, the destructive use of chemicals to catch fish and the use of small mesh nets that catch all sizes of fish, even young fish fry that are needed to replenish stocks. The loss of fishing as a key livelihood means that coastal communities are more vulnerable to additional stresses that may be caused by current or eventual climate change impacts–such as sea level rise and possible salt water intrusion onto farm lands.

In all four communities, agriculture is a key livelihood, the main crops being cassava, yam (*Dioscorea* sp) and cocoa yam (*Colocasia* sp). The communities have reported some reduction in crop yield, due to the unpredictability of rains during the growing season, but it is not as significant as in the crop decline reported in the savanna and sahel communities. The changes in livelihoods are causing many men and youth to leave the communities in search of other means of income, leaving the remaining women and children burdened with additional labour.

The Project

The projects focus mainly on increasing food security and reducing deforestation by testing alternative livelihoods, including aquaculture in the two coastal communities and testing cassava processing and snail production in the two rainforest communities. The projects have also introduced fuel efficient wood stoves to select households in order to reduce labour for women and to conserve local mangrove and rainforest ecosystems.

Food Security and Alternative Livelihoods

Aquaculture was the adaptation option chosen by the two fishing communities, in order to maintain the essence of their traditional livelihoods, to reduce pressure on the wild fishery and to provide income generating opportunities for both men and women. Two 10 m × 30 m ponds were completed in April 2010 and were each stocked with 1,000 tilapia and 2,000 catfish three months later in July. Training on fish farming was also conducted and COLIN produced a manual on aquaculture to address the most important concepts including sustainability, maintenance and harvesting. While still in the early stages, before the first fish harvest, there are some challenges that the community is has had to address. One is how to ensure that there is an equitable distribution of fish and income from marketing to all community members. The other issue is security for the pond, as the community wants to ensure there is no poaching of the fish so that stocks are sustainable and equitably used. Committees have been set up to deal with these two challenges.

In the two rainforest communities, alternative livelihood strategies include cassava (gari) processing and snail farming. Gari production provides income to both men and women. The snail farming, which is still in its early stages, is meant to be a readily available source of protein to replace bush meat, as well as a source of income. Both are intended to reduce the pressure on local forest resources. Income from gari can potentially replace the income that some men make from selling bush meat. One night of hunting can bring 10,000–12,000 naira (about 220 naira = 1 Euro) and one day of processing gari can bring N8,000–9,000. In addition to replacing bush meat, the snail farming is intended to reduce the number of destructive bush fires which are regularly set during the hunt for bush meat.

Energy Solutions: Fuel Efficient Wood Stoves

Reliance on the dwindling supply of trees for fuel wood has greatly impacted the mangroves and rainforests in Cross River State. Fuel efficient wood stoves were introduced and beneficiaries were trained on construction and use. A manual on how to construct, maintain and repair the stoves was also provided. The stoves have not been adopted on a wide scale as the materials to fabricate them are considered too expensive by the women and so they continue with the traditional stoves, despite the fact that they spend more money on fuel wood.

Appendix 1

Adaptation needs and options in the pilot project communities and the pilot project partners involved in implementation

Adaptation needs	Options	Local implementing partners
Water availability and use	Rainwater harvesting	Catholic Rural/Irban Development Programme (CARUDEP) in Jos, Plateau State
	Tapping of shallow aquifers (wash wells) for small scale irrigation	Greenwatch Initiative, Makurdi, Benue State
	Rehabilitation of boreholes	University of Maiduguri (UNIMAID), Borno State
	Building of dykes to retain and store water	

(continued)

(continued)

Adaptation needs	Options	Local implementing partners
Food security	Drought resistant and/or early maturing crops	CARUDEP
	Adoption of conservation farming methods, e.g. use of organic manure, mulching and cover crops	UNIMAID
	Natural pest control methods	Greenwatch Initiative
	Food preservation and drying to provide increased security and varieties of food	Abubakar Tafawa Balewa University (ATBU), Bauchi, Bauchi State
	Poultry production	Coastal live in Nigeria (COLIN), Calabar, Cross River State
	Snail production	Development in Nigeria (DIN)
	Fish production	Obudu, Cross River State
Ecosystem rehabilitation	Tree planting (using local varieties of fruit and forest trees)	CARUDEP
	Sand dune rehabilitation	UNIMAID
	Domestication of non-timber forest products to reduce over exploitation of forest	Greenwatch Initiative
	Switching from hunting to animal husbandry	ATBU COLIN DIN
Health	Simple sand filtration system to improve the quality of drinking water and reduce incidences of water borne diseases	CARUDEP
	Cloth filters on well water to reduce water borne disease	UNIMAID
Energy	Use of alternative types of energy such as livestock dung (biogas) and crop residue	UNIMAID, Greenwatch Initiative
	Increasing energy efficiency by using fuel efficient wood stoves and reducing negative health impacts	ATBU, COLIN, DIN
Income	Bee keeping	Greenwatch Initiative
	Snail farming	DIN
	Poultry production	CERCOPAN
	Aquaculture	COLIN
	Gari processing	DIN
	Spaghetti production	Greenwatch Initiative
	Millet grinding	UNIMAID
	Dry season farming	UNIMAID

(continued)

(continued)

Adaptation needs	Options	Local implementing partners
Early warning systems	Drought decision support tool	ATBU
	Simple weather forecasting tools	ATBU
	Formation of disaster risk management committees	DIN
Institutional development/ strengthening	Formation of project implementation committees in all project sites	All partners
	Inclusion of women in decision making at management levels of committees	
	Establishment of linkages between committees and local councils, relevant ministries and parastatals	
	Employing the peer learning tools to share information and build capacities of local population	

Conclusions, Lessons Learned and Recommendations

In all 15 pilot project communities, there is reported variability in rainfall, decline in crop yields and loss of biodiversity in local ecosystems. Livelihoods are impacted in all communities and in some, more than others, basic needs such as ample clean drinking water are simply not being met. Some projects aim to increase adaptive capacity by establishing alternative livelihood options, while some projects tackle the climate change impacts directly, by introducing drought resistant crop varieties and setting up water supply options.

There are lessons learned from the pilot projects after 18 months of implementation. One of the overarching lessons learned is the importance of time—essential in any community-based project. While pilot projects are, by their nature, small in scale, they ultimately require time in order to measure impacts. Only the most immediate outcomes can be assessed after a year and a half.

It is fair to say that, at this juncture, the projects were largely successful: the severe hunger that occurs in the months of August and September for the central and northern communities was averted for beneficiaries this year by the early maturing varieties of millet, bean and maize; there were higher yields of improved crops varieties this year in all communities; access to water was greatly improved for 3 of the 4 communities where water supply projects were undertaken; fuel efficient stoves, introduced in all but two communities, were well received, but results are mixed. Also, dry season farming was established in two communities and beneficiaries sold excess produce; efforts to stabilize dunes, restore vegetation along watercourses and conserve valuable trees on farms are well underway. Even ideology is changing: many women, who were initially reluctant to participate, have become more involved in some projects; the youth in one community and farmers in general are more committed to protecting trees, and there is less indiscriminate burning in the two rainforest communities because of increased awareness. Institutions are on the rise: it is anticipated that the newly formed CBOs will oversee the sustainability of the projects, that peer education programs will expand the knowledge of CBA and the workshops will sow the seeds for knowledge transfer between communities and policy makers. It is also at this juncture that communities must sustain the project themselves without the continued support of the BNRCC partners and it is only after some time that the project impacts and sustainability will be measured. At this point in the project cycle some of the key lessons learned are:

1. Factors that contribute to vulnerability of communities are complex. The projects illustrate that multiple interacting stressors underpin vulnerability to climate change, but there is a tendency among communities to attribute all negative impacts to climate change. It is essential to understand the range of factors that underlie vulnerability in order to identify solutions and it is important to understand the range of human impact on the environment and biodiversity that sustain us. For example, declining fishing stocks in coastal communities are due, in part, to overharvesting and unsustainable fishing methods; poor crop yields are due in part to declining soil fertility from overuse, and increasing deforestation and overhunting forest species all point to the need to break the "poverty cycle".

Recommendation: Efforts to increase adaptive capacity must include complimentary efforts to increase awareness within communities of the importance of local ecological functions for sustaining human life and livelihoods. In order to break the poverty cycle, increased awareness of human impacts together with community ingenuity can work in tandem with projects that build adaptive capacity.

2. Communities in the Sahel and Savanna ecozones are more vulnerable with regard to natural capital than communities in the south due to greater water scarcity, high aridity and fewer natural resources, such as forests, which act as a buffer to crop failure.

Recommendation: Due to the direct impacts of climate change on more vulnerable ecosystems in the north, there should be immediate efforts to increase adaptive capacity of these communities where basic needs are not being met.

3. The use of improved crop varieties to avert hunger and improve yields is not without a price. The emphasis on the use of inorganic fertilizers and pesticides may not be sustainable, as these inputs need to be purchased year after year to ensure optimal performance of the improved crop varieties and their continued use compromises soil structure and fertility.

Recommendation: More extension activities are needed to build capacity so that: (1) farmers are encouraged to use organic fertilizers to reduce the dependency on purchased ones, and to ensure good soil structure and fertility in the long term, and

(2) farmers develop a savings system so that they can afford to purchase inputs for as long as they are needed.

4. The adoption of fuel efficient wood stoves in the communities has met with challenges, pointing to the importance of careful monitoring so the project can quickly respond to results from initial trials. The stoves have been shown to reduce the amount of fuel wood used in all cases and, as a result, women spend less time and money procuring wood, and there is less pressure on local forests. Women beneficiaries also report that food cooks faster and there is less smoke produced, causing fewer respiratory problems. Despite the advantages, the adoption of the stoves is not immediate and widespread. The reasons are that the stoves are too costly and difficult to fabricate, in some cases they were given to men, and that they do not meet the specific cooking needs of the women who use them.

Recommendation: Women in rural communities, in their multiple roles as care-givers, educators and often farmers need to be included on all technical aspects of project interventions and provisions made so that they can give feedback on the products that they are the beneficiaries of.

5. The weather forecasting tools tested in the three communities in the arid central and northern regions were successful, in that the farmers who used them planted only once. Other farmers who did not use them had to plant twice, the first time being unsuccessful to due to lack of rain and germination failure. In this first season of testing, the handheld forecasters used in the field by the farmers supported by the drought decision support software used by the partner organization, helped farmers to have successful yields this year, while minimizing wasted seed and lost labour.

Recommendation: Based on the early success of the hand held weather forecasters and drought decision support software, there should be greater efforts to forecast weather for and by farmers by any and all means possible, including extension activities, radio programs and awareness, especially for those in the Sudan savanna and the Sahel, where planting times are most critical in a short and increasingly erratic rainy season.

6. Livelihood activities that are the emphasis in most of the pilot projects need to have an assured market for products: an increased supply of fish from aquaculture projects, gari from the processing of cassava, snail meat from household snail farms, for example, needs to have market outlets in order for the livelihoods to be successful.

Recommendation: Training and research in market chain analysis should be provided to back up the introduction of new and alternative livelihood projects to ensure that there is enough demand for the increased supply of these products on the market.

7. The development of community institutions and peer to peer education have so far shown to be empowering for community members and effective for increasing awareness within and between communities and to policy makers at local and state governments.

Recommendation: Community institutions should be encouraged and strengthened in order to ensure project sustainability, to liaise with policy makers to bring in new project support and to build community awareness of climate change impacts and adaptation options.

References

AIACC (2007) Assessment of impacts and adaptation to climate change. Summary of the final report of the AIACC project nov 2007. A global environment facility enabling activity in the climate change focal area project no. GFL-2328-2724-4330 Submitted by: The International START Secretariat, Washington DC, USA and The Academy of Sciences for the Developing World, Trieste, Italy Report Prepared by: Neil Leary, Jyoti Kulkarni and Clark Seipt

Boko M, Niang I, Nyong A, Vogel C, Githeko A, Medany M, Osman-Elasha B, Tabo R, Yanda P (2007) Africa: climate change 2007 impacts, adaptation and vulnerability. In: Parry ML, Canziani OF, Palutikof JP, van der Linden PJ, Hanson CE (eds) Contribution of working group II to the 4th assessment report of the intergovernmental panel on climate change. Cambridge University Press, Cambridge

CARE (2010) Community-based adaptation toolkit. Digital toolkit version 1.0 July 2010. In collaboration with IISD

IDRC (2008) SAS2 A guide to collaborative inquiry and social engagement. In: Chevalier JM, Buckles DJ. Sage, IDRC

Intergovernmental Panel on Climate Change (IPCC) (2001) Working group 2. Third assessment report, Annex B: glossary of terms

Mitchell T, Tanner T (2006) Adapting to climate change challenges and opportunities for the development community. Institute of Development Studies, IDS

UNFCCC (2006) (United Nations Framework Convention on Climate Change) Background paper on impacts, vulnerability and adaptation to climate change in Africa for the African Workshop on adaptation implementation of decision 1/CP.10 of the UNFCCC Convention. Accra, Ghana, Sept 21–23

UNISDR (2009) Terminology: basic terms of disaster risk reduction and IISD et al 2007. Community-based risk screening—Adaptation and livelihoods (CRiSTAL) User's Manual, Version 3.0

Author Biography

Ellen Woodley has a PhD in Interdisciplinary Rural Studies from the University of Guelph in Canada. She just finished working on a climate change adaptation project in Nigeria. As a consultant, most of her work has involved human-ecosystem interactions in an international context, which includes work with the UN (UNEP and UN FAO), the Canadian government and with local governments in the South Pacific.